Wearable Monitoring Systems

T0135168

Annalisa Bonfiglio • Danilo De Rossi

Editors

Wearable Monitoring Systems

 Springer

Editors
Annalisa Bonfiglio
Department of Electrical and
Electronic Engineering
University of Cagliari
Cagliari, Italy
annalisa@diee.unica.it

Danilo De Rossi
Interdepartmental Research Centre
"E. Piaggio"
University of Pisa
Pisa, Italy
d.derossi@ing.unipi.it

ISBN 978-1-4899-8156-1   ISBN 978-1-4419-7384-9 (eBook)
DOI 10.1007/978-1-4419-7384-9
Springer New York Dordrecht Heidelberg London

Springer is part of Springer Science+Business Media (www.springer.com)

Introduction

Progress in wearable technologies for monitoring is driven by the same factors that were behind the transition from desktop computing and communication tools to portable devices providing processing and ubiquitous connectivity, namely changes in social and economical factors. This transition is fuelled by the enormous technical advances in microelectronics and communication technologies as well as by the apparently never-ending process of miniaturization. This, in turn, is driven by dramatic changes in demography, lifestyle and the emergence of huge mass-markets that exert a great pulling force for development. As a result, applications of wearable technology will spread far and wide as dictated by these technological development trends, and, more specifically, these wearables will change our lives in becoming exoprostheses able to augment our perceptions of reality with physical, social and emotional contents.

The areas of application fostered by intense research or development activities that have been identified as being the most promising by market forecasts span from fashion and leisure, fitness and wellness, healthcare and medical, emergency and work to the space and military domains.

Despite the great thrust in evolution, there still remain several technical obstacles that hamper the development of a technology that fully satisfies the needs and expectations of end users. Confronting these obstacles necessitates major performance improvements and real breakthroughs at all levels of the essential sub-components of wearable systems including: sensors, actuators, low power on-board processing and communication, energy harvesting and storage. Most importantly, these components should all be integrated seamlessly into comfortable, easy-to-use and low cost clothing and garments, which also requires considerable work.

This book is a collection of contributions by renowned worldwide experts in research and development or applications of wearable systems that renders a broad overview and critical analysis of the field of wearable technologies.

The book is divided into 3 parts.

The first part is devoted to a review of the main components of wearables, including sensors, energy generation, signal processing and communications systems. Integration of previous components is also underlined. Recent work on wearable systems based on a e-textile technology is also reviewed and critically discussed.

Chapter 1 provides an overview on sensors technologies for wearable applications, focussing in particular on technologies for monitoring biomechanical and physiological signals. The chapter is centred on the illustration of strategies for an effective integration of such systems in wearable devices.

Chapter 2 is dedicated to energy harvesting systems for wearable applications. In particular, the principles and the applications of thermoelectric generation, that is presently the most suitable technique for integration in wearable applications, are reviewed in details. In Chap. 3 wireless communication systems for wearable applications are examined and different proposed solutions are compared.

The important issue of integration in wearable applications of all system components is the focus of Chap. 4.

On board signal processing in wearable systems, as discussed in Chap. 5, is also a topic of paramount important since highly efficient and fast algorithm are usually required for energy saving and synthetic output in transmitted information.

Dealing with large amount of data as might occur in some applications strictly requires proper development and use of data mining techniques; this topic is treated in Chap. 6.

The first part of this book ends with Chap. 7 in which textile based wearable systems are described in terms of technologies and application platforms. This subject area is expected to show considerable advances in the years to come.

In the second part of this book some key applications of wearable systems in a variety of fields are reported. These applications are the main driving force for the development of the components reviewed in the first part.

Applications in sports, wellness and fitness are reported in Chap. 8. Products in these areas are thought by many to be the closest to the market and they might guide development in other fields where market entry resistance is higher.

The important area of health monitoring and diagnostics is the focus of Chap. 9. Much research and development effort is nowadays devoted to the development of medical devices and products although strong technical financial and regulatory issues still hamper widespread use.

Protective garments for emergency and work in noxious and dangerous environments, as reported in Chap. 10, are also a very important area of application in which wearable systems offer incomparable advantages in terms of better operability and less risk to operators without interfering with comfort and ease of use of garments.

In Chap. 11 applications in space and planetary explorations are described. Although being definitely a niche area, developments related to use in such extreme conditions have proved in the past to be seminal to technology transfer to consumers applications.

Commercial, social and environmental factors affecting development and use of wearables will definitely condition the success or failure of these pervasive technologies as treated in the third part of this book.

In Chap. 12 an analysis of current developments and needs in Ambient Assistance Living is given. These scenarios should guide and enable technology development since user needs have to be seriously taken into account.

At the end, Chap. 13 provides arguments useful to analyse opportunities and barriers to commercial development. Present market penetration of wearable is still marginal and market forecast have not provided till now a reliable and widely shared analysis. The authors provide their view in this respect.

The broad and far-reaching range of technologies and applications covered by this book is intended not only to provide an authoritative coverage of the field of wearable monitoring systems, but also to serve as a source of inspiration for possible new technology developments and for new applications enabling the vision of a future of connected people anywhere anytime.

The editors would like to thank all the colleagues who contributed to this book and the publisher for the continuous support. The editors would also like to express their gratitude to Dr. Andreas Lymberis, Officer of the Information Society & Media Directorate-General of the European Commission for his passionate promotion and continuous support to the field of Wearable Microsystems.

Contents

Part II   Applications

Part III   Environmental and Commercial Scenarios

Contributors

Alessandro Tognetti
University of Pisa, Pisa, Italy

Alessandro Tognetti
Interdepartmental Centre "E. Piaggio", University of Pisa, Pisa, Italy

Antonio Lanatá
University of Pisa, Pisa, Italy

André Dittmar
Nanotechnologies Institute of Lyon, Villeurbanne Cedex, France

Annalisa Bonfiglio
University of Cagliari, Cagliari, Italy

Asta Krupaviciute
Université de Lyon, Bron, France

Balasubramaniam Natarajan
Kansas State University, Manhattan, KS, USA

Claudine Gehin
Nanotechnologies Institute of Lyon, Villeurbanne Cedex, France

Danilo De Rossi
Interdepartmental Research Centre "E. Piaggio", University of Pisa, Pisa, Italy

Dava Newman
Massachusetts Institute of Technology, Cambridge, MA, USA

Davide Curone
Eucentre, Pavia, Italy

Diana Young
Massachusetts Institute of Technology, Cambridge, MA, USA

Elena Castellano
ITACA, R&D Department, Valencia, Spain

Emanuele Lindo Secco
Eucentre, Pavia, Italy

Enzo Pasquale Scilingo
University of Pisa, Pisa, Italy

Eric McAdams
Nanotechnologies Institute of Lyon, Villeurbanne Cedex, France

Eric Guenterberg
University of Texas at Dallas, USA

Francois G. Meyer
University of Colorado, Boulder, CO, USA

G. Fico
Polytechnic University of Madrid, Madrid, Spain

George Kotrotsios
CSEM SA, Neuchâtel, Switzerland

Georges Delhomme
Nanotechnologies Institute of Lyon, Villeurbanne Cedex, France

Giovanni Magenes
Eucentre, Pavia, Italy

Hassan Ghasemzadeh
University of Texas at Dallas, USA

I. Peinado
Polytechnic University of Madrid, Madrid, Spain

JAD McLaughlin
University of Ulster, Belfast, Northern Ireland, UK

Jean Luprano
CSEM SA, Neuchâtel, Switzerland

Jocelyne Fayn
Université de Lyon, Bron, France

Maria Teresa Arredondo
Polytechnic University of Madrid, Spain

Paul Rubel
Université de Lyon, Bron, France

Roozbeh Jafari
University of Texas at Dallas, USA

Rita Paradiso
Smartex, Prato, Italy

Sergio Guillén
ITACA, Valencia, Spain

Steve Warren
Kansas State University, Manhattan, KS, USA

Vladimir Leonov
Imec, Leuven, Belgium

William J. Kaiser
University of California, Los Angeles, CA, USA

Part I
Components and Systems

Chapter 1
Sensors for Wearable Systems

Enzo Pasquale Scilingo, Antonio Lanatà, and Alessandro Tognetti

1.1 Introduction

When designing wearable systems to be used for physiological and biomechanical parameters monitoring, it is important to integrate sensors easy to use, comfortable to wear, and minimally obtrusive. Wearable systems include sensors for detecting physiological signs placed on-body without discomfort, and possibly with capability of real-time and continuous recording. The system should also be equipped with wireless communication to transmit signals, although sometimes it is opportune to extract locally relevant variables, which are transmitted when needed. Most sensors embedded into wearable systems need to be placed at specific body locations, e.g. motion sensors used to track the movements of body segments, often in direct contact with the skin, e.g. physiological sensors such as pulse meters or oximeters. However, it is reasonable to embed sensors within pieces of clothing to make the wearable system as less obtrusive as possible. In general, such systems should also contain some elementary processing capabilities to perform signal pre-processing and reduce the amount of data to be transmitted. A key technology for wearable systems is the possibility of implementing robust, cheap microsystems enabling the combination of all the above functionalities in a single device. This technology combines so-called micro-electro-mechanical systems (MEMS) with advanced electronic packaging technologies. The former allows complex electronic systems and mechanical structures (including sensors and even simple motors) to be jointly manufactured in a single semiconductor chip. A generic wearable system can be structured as a stack of different layers. The lowest layer is represented by the body, where the skin is the first interface with the sensor layer. This latter is comprised of three sub-layers: garment and sensors, conditioning and filtering of the signals and local processing. The processing layer collects the different sensor signals, extracts specific features and classifies the signals to provide high-level outcomes for the

A. Lanatà (✉)
Interdepartmental Research Center "E. Piaggio", Faculty of Engineering,
University of Pisa, via Diotisalvi 2, 56126 Pisa, Italy
e-mail: a.lanata@centropiaggio.unipi.it

A. Bonfiglio and D. De Rossi (eds.), *Wearable Monitoring Systems*,
DOI 10.1007/978-1-4419-7384-9_1, © Springer Science+Business Media, LLC 2011

application layer. The application layer can provide the feedback to the user and/or to the professional, according to the specific applications and to the user needs. Recent developments embed signal processing in their systems, e.g. extraction of heart rate, respiration rate and activity level. Activity classification and more advanced processing on e.g. heart signals can be achievable exploiting miniaturization and low-power consumption of the systems. Examples of data classification are [1, 2, 3]: classification of movement patterns such as sitting, walking or resting by using accelerometer data [4] or ECG parameters such as ST distance extracted from raw ECG data [5, 3]; another example is the estimation of the energy consumption of the body [6, 7]; in [8] the combined use of a triaxial accelerometer and a wearable heart rate sensor was exploited to accurately classify human physical activity; estimation of upper limb posture by means of textile embedded flexible piezoresistive sensors [9]. Examples of integrated systems for health monitoring are in [10, 11].

In the following paragraphs, two classes of sensors which can be easily integrated into wearable systems are reported and described. More specifically, inertial sensors to monitor biomechanical parameters of human body and sensors to capture physiological signs are addressed, describing the operating principles and indicating the possible fields of application.

1.2   Biomechanical Sensors

Biomechanical sensors are thought to be used to record kinematic parameters of body segments. Knowledge of body movement and gesture can be a means to detect movement disturbances related to a specific pathology or helpful to contextualize physiological information within specific physical activities. An increasing of heart rate, for example, could be either due to an altered cardiac behavior or simply because the subject is running.

1.2.1   Inertial Movement Sensors

Monitoring of parameters related to human movement has a wide range of applications. In the medical field, motion analysis tools are widely used both in rehabilitation [12] and in diagnostics [13, 14]. In the multimedia field, motion tracking is used for the implementation of life-like videogame interfaces and for computer animation [15]. Standard techniques enabling motion analysis are based on stereophotogrammetric, magnetic and electromechanical systems. These devices are very accurate but they operate in a restricted area and/or they require the application of obtrusive parts on the subject body. On the other hand, the recent advances in technology have led to the design and development of new tools in the field of motion detection which are comfortable for the user, portable and easily usable in non-structured environments. Current prototypes realized by these emergent technologies utilize micro-transducers applied to the subject body (as described in the current paragraph) or textile-based strain sensors (as reported in [16]). These latter are not treated in this chapter. The first category, instead, includes devices based on inertial sensors (mainly accelerometers and gyroscopes) that are directly applied on the body segment to be monitored. These sensors can be realized on a single chip (MEMS technology) with low cost and outstanding miniaturization. Accelerometers are widely used for the automatic discrimination of physical activity [17, 18] and the estimation of body segment inclination with respect to the absolute vertical [19]. Accelerometers alone are not indicated for the estimation of the full orientation of body segments. The body segment orientation can be estimated by using the combination of different sensors through data fusion techniques (Inertial Measurement Units, IMU). Usually, tri-axial accelerometers (inclination), tri-axial gyroscopes (angular velocity), magnetometers (heading angle) and temperature sensors (thermal drift compensation) are used together [19, 20, 21]. Main advantages of using accelerometers in motion analysis are the very low encumbrance and the low cost. Disadvantages are related to the possibility of obtaining only the inclination information in quasi-static situations (the effect of the system acceleration is a *noise* and the double integration of acceleration to estimate the segment absolute position is unreliable). Accelerometers are widely used in the field of wearable monitoring systems, generally used in the monitoring of daily life activities (ADL) [22, 23, 24, 17, 25]. Physical activity detection can be exploited for several fields of application, e.g. energy expenditure estimation, tremor or functional use of a body segment, assessment of motor control, load estimation using inverse dynamics techniques [26, 27] or artificial sensory feedback for control of electrical neuromuscular stimulation [28, 29, 30]. Usually, three-axial accelerometers are used. They can be assembled by mounting three single-axis accelerometers in a box with their sensitive axes in orthogonal directions or using a sensor based on one mass [31]. An accelerometer measures the acceleration and the local gravity that it experiences. Considering a calibrated tri-axial accelerometer (i.e. offset and sensitivity are compensated and the output is expressed in unit of g), the accelerometer signal (y) contains two factors: one is due to the gravity vector (g) and the other depends on the system inertial acceleration (a), both of them expressed in the accelerometer reference frame [19]:

$$\mathbf{y} = \mathbf{a} - \mathbf{g} \begin{pmatrix} y_1 \\ y_2 \\ y_3 \end{pmatrix} = \begin{pmatrix} a_1 \\ a_2 \\ a_3 \end{pmatrix} - \begin{pmatrix} g_1 \\ g_2 \\ g_3 \end{pmatrix} \qquad (1.1)$$

The inclination vector (z) is defined as the vertical unit vector, expressed in the accelerometer coordinate frame [4]. In static conditions, only the factor due to gravity is present and the inclination of the accelerometer with respect to the vertical is known. In dynamic conditions, the raw accelerometer signal does not provide a reliable estimation of the inclination, since the inertial acceleration is added to the gravity factor. This estimation error grows as the subject movements become faster (e.g. running, jumping). Many algorithms have been developed and tested to perform a reliable estimation of the subject body inclination: most of them use low pass filters with very low cut-off frequency in order to extract z [4] (i.e. introducing a considerable time delay), others implement more complex techniques which use a model-based approach mainly based on Kalman filter techniques [19].

An example of integration of these sensors in a garment was developed in the frame of the Proetex project (FP6-2004-IST-4-026987), which aimed at using textile and fibre based integrated smart wearables for emergency disaster intervention personnel. The ProeTEX motion sensing platform is used to detect long periods of user immobility and user falls to the ground and it is realized by means of two tri-axial accelerometer modules. One accelerometer is placed in the higher part of the trunk (collar level) in order to detect inactivity and falls to the ground. The second sensor is placed in the wrist region and its aim is to achieve more accuracy in inactivity detection, since an operator can move his arms while his trunk is not moving. The core of the motion sensor is the processing algorithm described in [32], which allows to perform a reliable estimation of the body inclination even in the case of intense physical activity such as running or jumping. This algorithm allows a good estimation of subject activities and generated fall alarms with very high sensitivity and extremely low level of false positives.

1.3 Physiological Sign Sensors

Wearable systems are generally thought to be used for health care, therefore necessarily including sensors to monitor physiological signs. Occasionally, it is possible to adapt commercial devices to be integrated into a wearable system, but mostly dedicated and customized sensors should be designed and embedded. Here sensors for respiration activity, pulse monitoring, galvanic skin response, thermal and cardiopulmonary radiant sensors, gas sensors and sensors for detecting biochemical markers are envisaged and described.

1.3.1 Respiration Activity

The most challenging vital sign to accurately record during continuous monitoring is the respiratory activity due to the fact that the signals are affected by movement artifacts and filtering or feature recognition algorithms are not very effective. Monitoring of respiratory activity involves the collection of data on the amount and the rate at which air passes into and out of the lungs over a given period of time. In literature, there are several methods to do this, both directly, by measuring the amount of air exchanged during the respiration activity, and indirectly, by measuring parameters physically correlated to breathing, such as changes in thorax circumference and/or cross section, or trans-thoracic impedance. Direct methods are based on a spyrometer that measures directly the airflow in the lung exchanged during inspiration and expiration, but of course it cannot be integrated into a wearable system because it employs a mouthpiece, which could interfere with the freedom of movements, disrupting the normal breathing pattern during measurement, thus causing discomfort for the user. Indirect methods exploit displacements of the lung that are transmitted to the thorax wall and vice versa, and therefore measurements of chest-abdominal surface movements can be used to estimate lung volume variation. In literature, a number of devices have been used to measure rib cage and abdominal motion including mercury in rubber strain gauges [33], linear differential transducers [34], magnetometers [35], and optical techniques [36], but almost all cannot be comfortably integrated into a wearable system. For reference only, it is worthwhile citing a more sophisticated technique, called stereophotogrammetry, which makes it possible to estimate the three-dimensional coordinates of points of the thorax, estimating therefore volume variations. Nevertheless, this system presents a considerable drawback in that it is cumbersome, extremely expensive, and can only be used in research environments or in laboratory applications. Indirect techniques that can be implemented in wearable systems are respiratory inductive plethysmography (RIP) [37], impedance plethysmography [38], piezoresistive [39] and/or piezoelectric pneumography. These systems are minimally invasive and do not interfere with physical activity. In the following, these four technics are described.

1.3.1.1 Inductive Plethysmography

The inductive plethysmography method for breathing monitoring consists of two elastic conductive wires placed around the thorax and the abdomen to detect the cross sectional area changes of the rib cage and the abdomen region during the respiratory cycles. The conductive wires are insulated and generally sewn in a zig-zag fashion onto each separate cloth band (see Fig. 1.1). They can be considered as a coil and are used to modulate the output frequency of a sinewave current produced by an electric oscillator circuit. As a matter of fact, the sinewave current generates a magnetic field, and the cross-sectional area changes due to the respiratory movements of the rib cage and of the abdomen determine a variation of the magnetic field flow through the coils. This change in flow causes a variation of the self-inductance of each coil that modulates the output frequency of the sinusoidal oscillator. This relationship allows for monitoring the respiratory activity by detecting the frequency change in the oscillator output signal. For accurate volumetric measurements using RIP, it is assumed that the cross-sectional area within the rib cage and the abdomen coil, respectively, reflects all of the changes occurring within the respective lung compartment, and further that the lung volume change is the sum of the volume changes of the two compartments. Under optimal situations, lung volume can be approximated with an error less than 10%.

Fig. 1.1 The respiratory
inductive plethysmography
system including the rib cage
and abdominal sensor bands

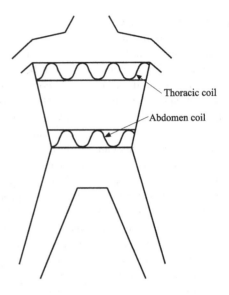

Thoracic coil

Abdomen coil

1.3.1.2   Impedance Plethysmography

This technique consists of injecting a high frequency and low amplitude current
through a pair of electrodes placed on the thorax and measuring the trans-thoracic
electrical impedance changes [40]. As a matter of fact, there is a relationship
between the flow of air through the lungs and the impedance change of the thorax.
The measurements can be carried out by using either two or four electrode configurations. Electrodes can be made of fabric and integrated into a garment or,
even, embedded into an undershirt. It is worthwhile noting that by measuring the
trans-thoracic electrical impedance it is possible to non-invasively monitor, in
addition to breathing rate also tidal volume, functional residual capacity, lung
water and cardiac output. In Fig. 1.2, the scheme of principle is depicted.

1.3.1.3   Pneumography Based on Piezoresistive Sensor

Piezoresistive pneumography is carried out by means of piezoresistive sensors that
monitor the cross-sectional variations of the rib cage. The piezoresistive sensor
changes its electrical resistance if stretched or shortened and is sensitive to the
thoracic circumference variations that occur during respiration. Piezoresistive sensors can be easily realized as simple elastic wires or by means of an innovative
sensorized textile technology. It consists of a conductive mixture directly spread
over the fabric. The lightness and the adherence of the fabric make the sensorized
garments truly unobtrusive and uncumbersome, and hence comfortable for the
subject wearing them. This mixture does not change the mechanical properties of
the fabric and maintains the wearability of the garment. Figure 1.3 shows where the
two conductive wires or bands could be applied.

Fig. 1.2 Principle scheme of impedance plethysmography system which can be integrated into a wearable system

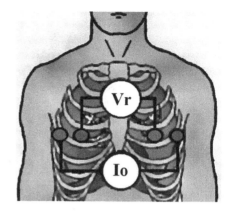

Fig. 1.3 Picture showing how two piezoresistive belts can be embedded into a garment to monitor abdominal and thoracic respiratory activity

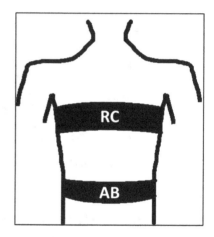

1.3.1.4 Plethysmography Based on Piezoelectric Sensor

This method is based on a piezoelectric cable or strip which can be simply fastened around the thorax, thus monitoring the thorax circumference variations during the respiratory activity. A possible implementation can be a coaxial cable whose dielectric is a piezoelectric polymer (p(VDF-TrFE)), which can be easily sewn in a textile belt and placed around the chest. In Fig. 1.4, a possible application is reported. The sensor is sensitive to the thorax movements and produces a signal directly proportional to the thorax expansion in terms of charge variation, which was converted in an output voltage proportional to the charge by means of a charge amplifier. A suitable local processor can enable implementation of the Fast Fourier Transform in real time and extraction of the breathing rate.

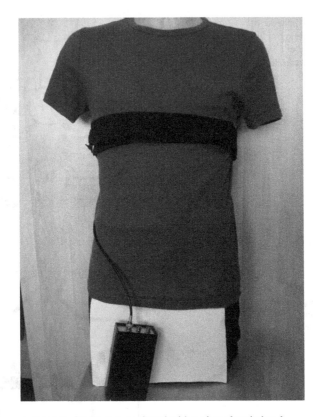

Fig. 1.4 Concept of a wearable system equipped with a piezoelectric band

1.3.2 Galvanic Skin Response

One of the most interesting measurements of the electrical body response is the Galvanic Skin Response (GSR), which was easily transformed from laboratory to wearable instrumentation, and has become one of the most used wearable devices especially for the high correlation that has shown with the most significant parameters in the field of neuroscience. It is a part of the whole ElectroDermal Response (EDR), which is also constituted of the measure of skin potentials. In deep, EDR is associated with sweat gland activity. Convincing evidence, indeed, was experimentally found in which a direct correlation is seen between EDR and stimulated sweat gland activity. Furthermore, when sweat gland activity is abolished, then there is an absence of EDR signals [41]. There are two major measures of the electrodermal response. The first, involving the measurement of resistance or conductance between two electrodes placed in the palmar region, was originally suggested by Féré [42]. It is possible also to detect voltages between these electrodes; these potential waveforms appear to be similar to the passive resistance changes, though its interpretation is less straightforward. This measurement was pioneered by Tarchanoff [43]. The first type of measurement is referred to as exosomatic, since the current on which the measurement is based is introduced from the outside. The second type, which is less commonly used, is called endosomatic, since the source of voltage is internal. Researchers also distinguish whether the measurement is of the (tonic) background level, or the time-varying (phasic) response type. An electrical equivalent model underlying EDR is represented in Fig. 1.5. This model provides only qualitative information. The active electrode is at the top (skin surface), whereas the reference electrode is considered to be at the bottom (hypodermis). R_1 and R_2 represent the resistance to current flow through the sweat ducts located in the epidermis and dermis, respectively. These are major current flow pathways when these ducts contain sweat, and their resistance decreases as the ducts fill. E_1 and R_4 represent access to the ducts through the duct wall in the dermis, whereas E_2 and R_3 describe the same pathway, but in the epidermis. Potentials E_1 and E_2 arise as a result of unequal ionic concentrations across the duct as well as selective ionic permeabilities. This potential is affected by the production of sweat, particularly if the buildup of hydrostatic pressure results in depolarization of the ductal membranes. Such a depolarization results in increased permeability to ion flow; this is manifested in the model by decreased values of R_3 and R_4. In particular, this is considered as an important mechanism to explain rapid-recovery signals. The potentials of E_1 and E_2 are normally lumen-negative. The resistance R_5 is that of the corneum, whereas E_3 is its potential. The phenomenon of hydration of the corneum, resulting from the diffusion of sweat from the sweat ducts into the normally dry and absorbant corneum, leads to a reduction in the value of R_5. The applications of the measure lie in the area of psychophysiology and relate to studies in which a quantitative measure of

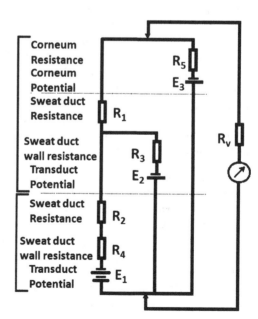

Fig. 1.5 A simplified equivalent circuit describing the electrodermal system. Components are identified in the text [44]

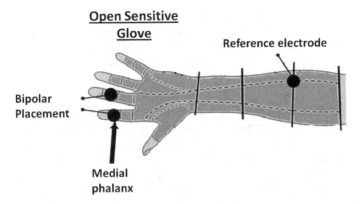

Fig. 1.6 Suggested electrodes site for the measurement of skin resistance and skin potentials

sympathetic activity is desired. The importance attached to such measurements includes the statement in one recent paper that palmar sweat is one of the most salient symptoms of an anxiety state and, for some, the single most noticeable bodily reaction [45]. Other suggested locations for electrode placement can be between two fingers. In this case, electrodes can be integrated into a glove (see Fig. 1.6).

1.3.3 Pulse Oximetry

Pulse oximetry was introduced in 1983 as a non-invasive method for monitoring the arterial blood oxygen saturation. Recognized worldwide as the standard of care in anaesthesiology, it is widely used in intensive care, operating rooms, emergency, patient transport, general wards, birth and delivery, neonatal care, sleep laboratories, home care and in veterinary medicine. Currently, several wearable pulse oximeters are being developed to transfer this standard technique to a most effective remote home-care monitoring. Being pulse oximeter non-invasive, easy to use, readily available, and accurate, the modern wearable system developed can supply information about blood oxygen saturation, heart rate and pulse amplitude. A pulse oximeter shines light of two wavelengths through a tissue bed such as the finger or earlobe and measures the transmitted light signal. The device operates according to the following principles [46]:

1. The light absorbance of oxygenated haemoglobin and deoxygenated haemoglobin at the two wavelengths is different. To be more precise, the set of associated extinction coefficients for the absorption of light for these wavelengths is linearly independent with great enough variation for adequate sensitivity but not so large that the blood appears opaque to either of the light sources. This model assumes that only oxygenated and deoxygenated haemoglobin are present in the blood.
2. The pulsatile nature of arterial blood results in a waveform in the transmitted signal that allows the absorbance effects of arterial blood to be identified from

those of non-pulsatile venous blood and other body tissue. By using a quotient of the two effects at different wavelengths, it is possible to obtain a measure requiring no absolute calibration with respect to overall tissue absorbance. This is a clear advantage of pulse oximeters over previous types of oximeters.

3. With adequate light, scattering in blood and tissue will illuminate sufficient arterial blood, allowing reliable detection of the pulsatile signal. The scattering effect necessitates empirical calibration of the pulse oximeter. On the other hand, this effect allows a transmittance path around bone in the finger.

Systems following the principles above shown provide an empirical measure of arterial blood saturation. However, with state-of-the-art instrumentation and proper initial calibration, the correlation between the pulse oximeter measurement, SpO_2, and arterial blood's actual oxygen saturation, SaO_2, is adequate-generally less than 3% discrepancy provided SaO_2 is above 70% for medical applications [47]. In general, when the calibration is difficult or impossible, these systems can be redirected at considering only a led and a photodiode so that the obtained measurement is a photopletismography. Really, most pulse oximeters on the market implement photoplethysmographic measurements. The signal for the photoplethys-mograph is derived from the same waveforms used to calculate SpO_2. The photo-plethysmograph may be used in a clinical setting in the same manner as a plethysmograph. However, the accuracy of the photoplethysmograph suffers from motion artifacts, and the patient must have adequate blood perfusion near place-ment of the pulse oximeter probe. Just as with the conventional plethysmogram, signal processing can derive heart rate from the photoplethysmogram waveform. Hence, most pulse oximeters also display heart rate. Similar to computing SpO_2, temporal low-pass filtering abates the effect of motion artifacts on heart rate estimation. Generally, pulse oximeters are applied to a fingertip (see Fig. 1.7), but as above-mentioned they are heavily affected from motion artifact (see Fig. 1.8) so that large part of the signal has to be strongly treated or completely removed, to avoid this signal lost latest research applications aim at positioning the sensor on the forehead (see Fig. 1.9), where it has been noted that the signal shows lower artifact noise and better characteristics.

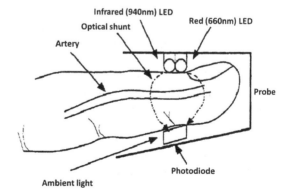

Fig. 1.7 Transmission pulse oximeter measuring the transmission of light by two LEDs through the finger of a patient

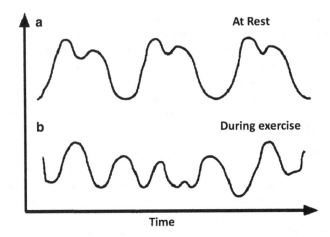

Fig. 1.8 The plethysmographic waveform of a subject at rest is periodic (**a**) and during exercise is not periodic (**b**)

Fig. 1.9 Reflectance pulse oximeter measuring the amount of light reflected back to the probe in forehead application

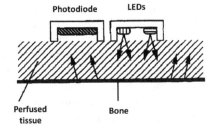

1.3.4 Radiant Thermal Sensors

The interest of the market devices for safety and security has rapidly grown over the last few years. In particular, the use of RadioFrequency (RF) technology for contact-less sensing has been promoted largely into several research projects [48, 49]. Body temperature is usually captured by means of thermal sensors placed in direct contact with skin. Skin temperature is strongly dependent on the body site and it is sensitive to local increasing of blood circulation. Reference body temperature, indeed, should be internal. Often skin contact with thermal sensors could be difficult and obtrusive, therefore radiant technology is preferred. The state of the art on radiant thermal sensors covers several high-potential commercial products. Meridian Medical Systems (http://mms-llc.com) is aiming at fabricating a radiometer as a Monolithic Microwave Integrated Circuit (MMIC) capable of detecting temperature of the heart. Although their research aims at implementing microwave radiometers for medical imaging, it seems they use a traditional approach based on MIC/MMIC. It is worthwhile mentioning that radiometer exists from a long time, and their approach using hybrid components is well known. Even though MMIC can reach good performance, their level of integration is limited traditionally to the analog-RF part only. Thermal stabilization and calibration circuits need to be implemented by means of external circuitry, resulting in bulky and expensive implementations inadequate for the mass-market. In fact, the system-on-chip implementation proposed in CMOS technology aims at implementing efficiently on the same die both the analog-RF and the digital calibration circuits. Only this result leads to consider that microwave radiometers can be implemented as a real system-on-a-chip device characterized by superior performance and highest level of integration. This is the real innovation expected for enabling microwave radiometry for the next-generation of mass-market wearable devices for medical imaging, and safety and security of emergency operators. Tyco Electronics (http://tycoelectronics.com) is developing a 24 GHz UWB radar sensor in MIC technology for short-range applications. Moreover, this device is targeted at general purpose applications (i.e. military, collision avoidance short-range automotive, etc.) and therefore only marginally related with our specific target. Anyway, as an additional consideration to all the limitations of the MIC approach cited above concerning the microwave radiometer and therefore still valid also in this case, it is worth mentioning that its bandwidth is limited to 500 MHz. A possible application of microwave radiometers could be designing a dedicated system to assist the fire fighters in their work, for instance by detecting a fire behind a door or a wall. This sensor can be mounted on two textile microstrip board. The former can contain the radiometric sensor and it is placed in the front side of the fireman jacket (this to detect the fire coming from the front). The latter can contain the low data-rate radio transceiver for sending out the information collected by the sensor, and this can be placed in the back side, for instance close to the neck. The system idea is shown in Fig. 1.10. The radiometer consists of a patch antenna array, a low noise 13 GHz radiometer module and a data acquisition and process unit. It is worthwhile noting that the sensor is mounted on the same microstrip board of the antenna. The ZigBee transceiver (*IEEE*802. 15. 4) transmits the data to the personal server (or a remote unit as well) of a wireless body area network (WBAN). The wireless platform allows collecting the data acquired by multiple sensors to realize an extended monitoring of the vital and environmental data. Moreover, such a wireless platform allows us to implement re-configurable systems, which can be managed by remote operators taking care of the safeguard of the rescue team. Hereinafter, how a fire in front of the subject with a separation wall between fire and subject can be detected. This inter-wall fire detection is tried in indoor environment to simulate a condition as close as possible to the operative scenario. In particular, the setup shown in Fig. 1.11 has been used for the proof-of-the concept. To model the scenario sensed by a microwave radiometer, the approach described in [50] could be adopted. This approach is based on the filling factor q, a quantity defined as the ratio between the area of the fire A_{FIRE} and the area of the antenna footprint A_{FOOT}:

$$q = A_{FIRE}/A_{FOOT}$$

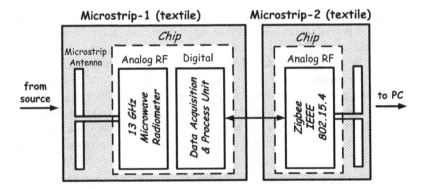

Fig. 1.10 Block diagram of the overall system

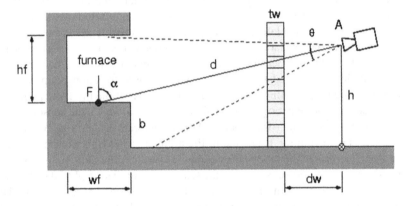

Fig. 1.11 Basic setup used for the inter-wall fire detection experiments

By considering Fig. 1.11, the footprint area can be evaluated approximating the main beam of the antenna with a cone of angular aperture θ (half power beam width) and by cutting this cone with the profile of the illuminated scene, the soil and the furnace in this case. The radiometric contrast ρ_T is defined as the increase of the antenna temperature due to the fire with respect to the condition without any fire. Applying the radiative transfer theory, the radiometric contrast can be derived as follows:

$$\rho_T = \tau_W[\epsilon_F T_F - \epsilon_S T_S - (\epsilon_F - \epsilon_S)\epsilon_W T_W]q, \qquad (1.2)$$

where τ_W is the transmissivity of the wall, ϵ_F, ϵ_S and ϵ_W are the emissivities of the fire, soil, and wall, respectively, and T_F, T_S and T_W their physical temperatures. In conclusion, the innovative low-cost, system-on-a-chip microwave radiometer could represent a very promising solution for the realization of a next-generation of wearable sensors. The SoC microwave radiometer will allow an extended detection capability in the cases where traditional devices, such as IR devices, fail.

1.3.5 Biochemical Markers

Last achievements in research have enabled the possibility of detecting biochemical markers through wearable instrumentation. CSEM[1] researchers have developed non-invasive biosensors for the detection of stress markers (such as lactate in sweat) and wound healing (focussing on pH and infection markers detection). The high level of miniaturization allows the integration in textile garments for the non-invasive monitoring of biological markers. In the frame of the BIOTEX[2] project, a miniaturized label-free system for application in wound dressing has been developed [51, 52]. Within the PROETEX project, CSEM researchers are currently realizing a wearable biosensor for real-time stress assessment of professional rescuers to improve their safety during the intervention. In both cases, the sensing principle is based on responsive hydrogels that shrink or swell in presence of the target marker to be detected. The hydrogels are sensitive to pH changes or they can be functionalized to the target molecule by incorporating specific enzymes. In the case of stress monitoring, the hydrogel is functionalized with lactate oxidase, the presence of the lactate changes the ionic concentration in the hydrogel and consequently the hydrogel volume by an osmotic effect. The volume modification of the hydrogel causes a modification of the refractive index of the structure. The sensitive hydrogel is then integrated on a waveguide grating chip. If the grating chip is interrogated with a light source it reflects the light at a specific wavelength. As the sensitive layer changes its refractive index, the wavelength of the reflected light is shifted in accordance. The detection principle is based on the measurements of the refractive index through an optical signal propagated along a wave guide. By exploiting this principle, a wearable optical bio-sensor has been designed and realized [51, 52]: it uses a sensitive layer on a waveguide grating chip, the biosensor is interrogated with a white light source (using a white led) and the reflected light is detected by a mini-spectrometer in order to measure the wavelength shift. Electrochemical sensors can be integrated into flexible (i.e. plastic, textile) substrates to develop wearable systems for the detection of biochemical markers. Some researchers are developing a portable electrochemical system based on Ion Sensitive Electrodes (ISE) integrated into a fabric substrate [53]. ISE can measure the sodium concentration in sweat and this measurement can be related to the operator dehydration that can lead to severe physiological consequences being able to go until the death. A portable electronic board connected to the sensing part has been developed. This board drives the electrochemical sensor, compensates the effect of temperature, performs analog acquisition and converts measurement data to digital value. Signal processing is implemented on board to correct raw data (gain, offset) and to convert them to ion concentrations. The system was evaluated in terms of sensitivity, selectivity and reproducibility initially in model solution and then in natural sweat, showing very promising

---

[1] Centre Suisse d'Electronique et Microtechnique SA, CH.
[2] Biosensing textile for health management, FP6-IST-NMP-2-016789.

performances. A more recent technique employed for biochemical marker detection in wearable systems is based on Organic Field Effect Transistors (OFET). Researcher at University of Cagliari developed a flexible OFET able to detect pH changes in chemical solutions thanks to a functionalized floating gate [54]. The sample solution is brought into contact with a portion of the floating gate, which is properly functionalized to achieve the sensitivity to a particular chemical species. The sensing mechanism is based on the detection of the electrical charge associated with the chemical species placed over the probe area. The charge immobilized on the floating gate generates an electrical field and thus induces a phenomenon of charge separation inside the electrode, affecting the channel formation in the transistor. This mechanism can be described in terms of a shift of the effective threshold voltage of the OFET. By properly functionalizing the floating gate surface, sensitivity to different species and the detection of different reactions can be achieved, with the same sensor.

1.3.6 Gas Sensors

Researchers from Dublin City University (DCU) are involved in the integration of sensing platforms into wearables for the detection of environmentally harmful gases surrounding emergency personnel. Special attention is being paid to carbon monoxide (CO) and carbon dioxide (CO_2). These gases are associated with fires and mining operations, and it is of the highest importance to warn and protect operators from potential harm caused by over-exposure to high concentrations of these gases. The goal is rapid detection of the status of an environment (low, medium or high hazard) and real-time communication of this information to the garment wearer. Critical in this identification of potential toxification is a reliable method of measuring CO/CO_2 exposure. Commercially available sensors have been carefully selected and are being integrated into the outer garments of firefighters. The sensors provide sufficient sensitivity to reliably alert users to the presence of these harmful gases. Another important aim is to achieve wireless transmission of sensor signals to a wearable wireless base station that gathers, processes and further transmits the data. When selecting the appropriate commercially available sensors for the gas sensing application, special attention was paid to sensor size, robustness, sensitivity and power requirement. Electrochemical sensors satisfy most of these requirements, especially in terms of size and power requirements. CO is detected using an amperometric sensor in which the current between the electrodes is proportional to the concentration of the gas. On the other hand, the CO_2 sensor is potentiometric. In this case, the reference and working electrodes are placed in an electrolyte that provides a reference CO_2 concentration. The measured potential is based on the difference in concentration between the reference electrode and the outside air. Both types of sensors are very sensitive and give an accurate reading (in parts per million). This means that both low concentrations of these gases (which can be hazardous over long periods of exposure) and high concentrations (which pose an immediate danger) can be accurately detected. The signal obtained from these sensors is transmitted wirelessly to the wearable base station using Zigbee. Power is supplied to the sensors using a nickel metal hydride rechargeable battery. The CO_2 sensor is placed in a specially designed pocket located on the firefighter's boot. The pocket is designed not to obstruct the firefighter's activities. The prototype currently used for testing is shown in Fig. 1; note the side pocket containing the CO_2 sensor along with the wireless sensing module and a battery. The pocket has a waterproof membrane that protects the sensor from humidity, but allows gas to pass through. The CO sensor will be integrated in the firefighter's outer garment (i.e. jacket). All sensed information will be fed to a wearable local base station that shares the data with a remote centralized base station. The ultimate goal is to achieve local communication between firefighters and civil workers in the operations area, as well as longer range communications between these personnel and the support team outside the operations area.

1.3.7 Cardiopulmonary Activity Systems

One of the most challenging points in the healthcare system is to use a single device to simultaneously gather cardiac and pulmonary information, which usually are both obtained from different systems and whose interdependences are left to the clinic experience only. An innovative cardiopulmonary wearable system that matches this dual request is based on Ultra WideBand (UWB) technology. The main advantage of this monitoring radar system is the absence of direct contact with the subject skin, dramatically reducing the typical disturbance due to motion artifact [55]. Before introducing the system concept of the system let us give a brief overview on the current state of the art. The most widespread system used to monitor the cardiac activity is the electrocardiograph (ECG), which provides information about the heart electrical activity. Another complementary technology for monitoring the cardiorespiratory activity is pulse oximetry, which measures the saturation level of the oxygen in the blood. Other systems for the monitoring of the cardiac activity are based on ultrasounds (echocardiograph or echo Doppler). Ultrasound-based systems are generally cumbersome and they can be used only by specialized operators. Anyway, all the presented measurement techniques require the direct contact with the body to carry out the measurement. Unlike the traditional techniques (electrocardiograph, echocardiograph and pulsed oximetry), radar systems allow the monitoring of the heart activity in a non-invasive and contactless way for the patient [56]. Microwave Doppler radars have been used to detect the respiration rate since 1975 [57]. These first devices were bulky and expensive, but the recent microelectronic advances led to develop CMOS fully integrated radars for non-contact cardiopulmonary monitoring [58]. Doppler radars typically transmit a continuous wave signal and receive the echo reflected by the target. The frequency of the reflected signal varies from that of the transmitted one by an amount proportional to the relative velocity of the target with respect to the radar. Another class of radar employed for the monitoring of vital parameters is based on pulse transmission. Pulse radars operate by sending short electromagnetic pulses and by receiving the echoes back-scattered by the target. The time delay between the transmission of the pulse and the reception of the echo is proportional to the distance of the radar from the target. Discrete prototypes of pulse radar for the detection of vital parameters are reported in literature [59, 60]. It is worthwhile mentioning that radar sensors monitor the mechanical movement of the heart wall instead of the electrical activity of the heart (such as the electrocardiograph), therefore when mechanical anomalies occur earlier than the electrical ones, this device can be used to prevent in advance possible cardiac failures. Moreover, the UWB pulses are not influenced by blankets or clothes [60]. From a circuit design point of view, UWB transceivers present a lower complexity with respect to traditional radiofrequency systems, leading to low power consumption for a long life of the battery. In fact, UWB systems do not require a stable frequency reference, which typically requires a large area on silicon die and high power consumption. Moreover, the extremely low level of transmitted power density (lower than -41.3 dBm/MHz) of the UWB radar should reduce the risk of molecular ionization [61, 44] (see Fig. 1.14). The main block of the novel wearable wireless interface for human health care described herein is the UWB radar sensor (see Fig. 1.12). The block diagram of the proposed radar sensor for the detection of the heart and breath rates is shown in Fig. 1.13. The radar exploits a correlation-based receiver topology followed by an integrator, which averages the received pulses to have an output signal containing the information on the heart and breath tones. The operating principle of a cross-correlator radar is explained hereinafter. An electromagnetic pulse is transmitted toward the target. The echo received from the target is multiplied by a delayed replica of the transmitted pulse; the output signal of the multiplier is then integrated. It is worthwhile noting that the output signal will reach its maximum in the case of perfect time alignment

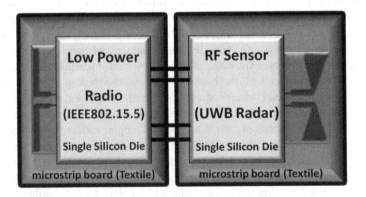

Fig. 1.12 Wearable Wireless UWB radar sensor interface for human health care

Fig. 1.13 Block diagram
of the UWB radar sensor

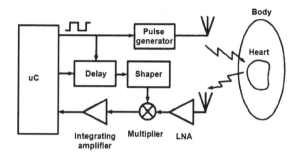

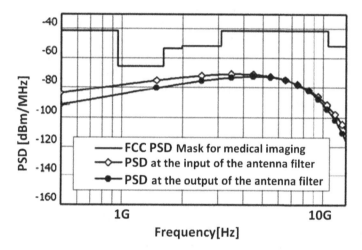

Fig. 1.14 FCC PSD mask for medical imaging and Power spectral density (PSD) of a pulse
sequence with PRF equal to 1 MHz vs. frequency

between the two signals at the input of the multiplier itself. In other terms, the cross-correlator has a frequency response equal to that of a matched filter. In particular, it can be demonstrated that the matched filter is the filter that allows obtaining the best signal-to-noise ratio at the output. Moreover, this has been confirmed by preliminary system simulations (by means of the Ptolemy simulator within Agilent ADS2005A). In detail, the CAD system analysis has shown that this topology allows us to achieve the best performance in terms of output signal-to-noise ratio (SNR) and sensitivity to small variations of the position of the heart wall with respect to other topologies, like that in which the receiver is simply turned on by the command given by the delayed replica of the transmitted pulse [61]. The principle of operation of the overall radar system shown in Fig. 1.13 is explained hereinafter [62]. A train of extremely short (about 200 picoseconds) Gaussian monocycle electromagnetic pulses is transmitted toward the heart. Since the heart muscle and the blood that flows inside have different characteristic impedance, a partial reflection of the energy associated with the radiated pulse occurs at the surface of separation of these two different media.

1.4   Conclusions

In this chapter, we gave an overview on sensors for physiological signals and biomechanical parameters, which can be easily integrated into wearable monitoring systems. The operating principle of each sensor was described as well as some applicative example was given. Generally, a wearable system has to comply with a series of requirements, e.g. minimallyinvasive, based on flexible technologies conformable to the human body, cost-effective, easy to use and customizable to the specific user. Several technologies can be easily adapted, but in several cases ad-hoc applications should be designed. Much work has to be done in this field, even if several effective sensing platforms are already available and promising for future improvements.

References

1. Paker O Low power digital signal processing. Citeseer
2. Piguet C (2003) Low-power systems on chips (SoCs). CMOS and BiCMOS VLSI De-sign'01. Advanced Digital Design, EPFL, Lausanne, Switzerland
3. Force T (1996) Heart rate variability: standards of measurement, physiological interpretation and clinical use. Task Force of the European Society of Cardiology and the North American Society of Pacing and Electrophysiology. Circulation 93(5):1043–1065
4. Karantonis DM, Narayanan MR, Mathie M, Lovell NH, Celler BG (2006) Implementation of a real-time human activity classifier using a triaxial accelerometer for ambulatory monitoring. IEEE Trans Inform Technol Biomed 10(1):156–167
5. Renevey P, Vetter R, Celka P, Krauss J (2002) Activity classification using HMm for improvement of wrist located pulse detection. In: Proceedings of the Biosignal, vol 2002, pp 192–196
6. Brage S, Brage N, Franks PW, Ekelund U, Wareham NJ (2005) Reliability and validity of the combined heart rate and movement sensor actiheart. Eur J Clin Nutr 59:561–570
7. Brage S, Ekelund U, Brage N, Hennings MA, Froberg K, Franks PW, Wareham NJ (2007) Hierarchy of individual calibration levels for heart rate and accelerometry to measure physical activity. J Appl Physiol 103:682–692
8. Curone D, Tognetti A, Secco EL, Anania G, Carbonaro N, De Rossi D, Magenes G (in press) Heart rate and accelerometer data fusion for activity assessment of rescuers during emergency interventions. IEEE Trans Inform Technol Biomed
9. Mattmann C, Clemens F, Troester G (2008) Sensor for measuring strain in textile. Sensors 8(6):3719

10. Inc V LifeShirt .[Online]. December 11, 2005
11. Jafari R, Encarnacao A, Zahoory A, Dabiri F, Noshadi H, Sarrafzadeh M (2005) Wireless sensor networks for health monitoring. In: Mobile and ubiquitous systems: Networking and services, 2005. The Second Annual International Conference on MobiQuitous 2005, pp 479–481
12. Sveistrup H (2004) Motor rehabilitation using virtual reality. J NeuroEng Rehabil 1(1):10
13. Baker R (2006) Gait analysis methods in rehabilitation. J NeuroEng Rehabil 3(1):4
14. Whittle M (2002) Gait analysis: an introduction. Butterworth, London
15. Menache A (2000) Understanding motion capture for computer animation and video games. Morgan Kaufmann, CA
16. Lorussi F, Scilingo EP, Tesconi M, Tognetti A, De Rossi D (2005) Strain sensing fabric for hand posture and gesture monitoring. IEEE Trans Inform Technol Biomed 9(3):372–381
17. Bouten CVC, Koekkoek KTM, Verduin M, Kodde R, Janssen JD (1997) A triaxial accelerometer and portable data processing unit for the assessment of daily physical activity. IEEE Trans Biomed Eng 44(3):136–147
18. Van Laerhoven K, Schmidt A, Gellersen HW (2002) Multi-sensor context aware clothing. In: Proceedings of the Sixth International Symposium on Wearable Computers, pp 49–56
19. Luinge HJ (2002) Inertial sensing of human movement. Unpublished PhD, University of Twente, Enschede, the Netherlands
20. Roetenberg D, Luinge HJ, Baten CTM, Veltink PH (2005) Compensation of magnetic disturbances improves inertial and magnetic sensing of human body segment orientation. IEEE Trans Neural Syst Rehabil Eng 13(3):395–405
21. Roetenberg D, Baten CTM, Veltink PH (2007) Estimating body segment orientation by applying inertial and magnetic sensing near ferromagnetic materials. IEEE Trans Neural Syst Rehabil Eng 15(3):469
22. Veltink PH, Bussmann HBJ, de Vries W, Martens WLJ, Van Lummel RC (1996) Detection of static and dynamic activities using uniaxial accelerometers. IEEE Trans Rehabil Eng 4(4):375–385
23. Busser HJ, De Korte WG, Glerum EBC, Van Lummel RC et al (1998) Method for objective assessment of physical work load at the workplace. Ergon Lond 41:1519–1526
24. Foerster F, Smeja M, Fahrenberg J (1999) Detection of posture and motion by accelerometry: a validation study in ambulatory monitoring. Comput Hum Behav 15(5):571–583
25. Uswatte G, Miltner WHR, Foo B, Varma M, Moran S, Taub E (2000) Objective measurement of functional upper-extremity movement using accelerometer recordings transformed with a threshold filter. Stroke 31(3):662
26. van den Bogert AJ, Read L, Nigg BM (1996) A method for inverse dynamic analysis using accelerometry. J Biomech 29(7):949–954
27. Baten CTM, Oosterhoff P, Kingma I, Veltink PH, Hermens HJ (1996) Inertial sensing in ambulatory load estimation. In: 18th annual International Conference of the IEEE-EMBS
28. Fisekovic N, Popovic DB (2001) New controller for functional electrical stimulation systems. Med Eng Phys 23(6):391–399
29. Willemsen ATM, Bloemhof F, Boom HBK (1990) Automatic stance-swing phase detection from accelerometer data for peroneal nerve stimulation. IEEE Trans Biomed Eng 37(12):1201–1208
30. Tong KY, Granat MH (1998) Virtual artificial sensor technique for functional electrical stimulation. Med Eng Phys 20(6):458–468
31. Lotters JC, Schipper J, Veltink PH, Olthuis W, Bergveld P (1998) Procedure for in-use calibration of triaxial accelerometers in medical applications. Sensor Actuator A Phys 68(1–3):221–228
32. Anania G, Tognetti A, Carbonaro N, Tesconi M, Cutolo F, Zupone G, De Rossi D (2008) Development of a novel algorithm for human fall detection using wearable sensors. 2008 IEEE Sensors, pp 1336–1339

33. Wade OL (1954) Movements of the thoracic cage and diaphragm in respiration. J Physiol 124(2):193
34. Mead J, Peterson N, Grimby G, Mead J (1967) Pulmonary ventilation measured from body surface movements. Science 156:1383–1384
35. Levine S, Silage D, Henson D, Wang JY, Krieg J, LaManca J, Levy S (1991) Use of a triaxial magnetometer for respiratory measurements. J Appl Physiol 70(5):2311
36. Peacock A, Gourlay A, Denison D (1985) Optical measurement of the change in trunk volume with breathing. Bulletin européen de physiopathologie respiratoire 21(2):125
37. Milledge JS, Stott FD (1977) Inductive plethysmography – a new respiratory transducer. J Physiol (Lond) 267:4P
38. de Geus EJC, Willemsen, GHM, Klaver CHAM, van Doornen LJP (1995) Ambulatory measurement of respiratory sinus arrhythmia and respiration rate. Biol Psychol 41(3):205–227
39. De Rossi D, Carpi F, Lorussi F, Mazzoldi A, Paradiso R, Scilingo EP, Tognetti A (2003) Electroactive fabrics and wearable biomonitoring devices. AUTEX Res J 3(4):180–185
40. Khalafalla AS, Stackhouse SP, Schmitt OH (1970) Thoracic impedance gradient with respect to breathing. IEEE Trans Biomed Eng 17(3):191
41. Fowles DC (1986) The eccrine system and electrodermal activity. Psychophysiol Syst Process Appl 1:51–96
42. Féré C (1888) Note sur les modifications de la résistance électrique sous l'influence des excitations sensorielles et des émotions. Comptes Rendus Société de Biologie 5:217–219
43. Tarchanoff J (1889) Décharges électriques dans la peau de lhomme sous linfluence de lexcitation des organes des sens et de différentes formes dactivité psychique. Comptes Rendus des séances de la Société de Biologie 41:441–451
44. 47 cfr part 15. Federal Communications Law Journal (2002)
45. Edelberg R (1968) Biopotentials from the skin surface: The hydration effect. New York Acad Sci Ann 148:252–262
46. Webster JG (1997) Design of pulse oximeters. Taylor and Francis, London
47. Severinghaus JW, Kelleher JF (1992) Recent developments in pulse oximetry. Anesthesiology 76(6):1018
48. Agranovskii AV, Evreinov GE, Berg OY (2004) Monitoring of vital functions using contactless sensors. Biomed Eng 38(1):13–16
49. Varshney U (2009) Wireless Health Monitoring: State of the Art. 119–146. Springer, NY
50. Luzi G, Coppo P, Ferrazzoli P, Gagliani S, Mazzoni T (1995) Microwave radiometry as a tool for forest fires detection: Model analysis and preliminary experiments. In: Solimini D (ed) Microwave radiometry and remote sensing of the environment. VSP Press, Utrecht, p 411
51. Pasche S et al Smart Wound Dressing with Integrated Biosensors
52. Pasche S, Angeloni S, Ischer R, Liley M, Luprano J, Voirin G Wearable Biosensors for Monitoring Wound Healing
53. Marchand G, Bourgerette A, Antonakios M, Rat V, David N, Vinet F, Guillemaud R (2009) Development of a dehydration sensor integrated on fabric. In: Proceedings of the 2009 Sixth International Workshop on Wearable and Implantable Body Sensor Networks. IEEE Computer Society, pp 230–233
54. Loi A, Manunza I, Bonfiglio A (2005) Flexible, organic, ion-sensitive field-effect transistor. Appl Phys Lett 86:103512
55. Klemm M, Troester G (2006) Textile UWB antennas for wireless body area networks. IEEE Transac Antenn Propag 54(11 Part 1):3192–3197
56. Staderini EM (2002) UWB radars in medicine. IEEE Aero Electron Syst Mag 17(1):13–18
57. Lin JC (1992) Microwave sensing of physiological movement and volume change: a review. Bioelectromagnetics 13:557–557
58. Droitcour AD, Boric-Lubecke O, Lubecke VM, Lin J, Kovacs GTA (2004) Range correlation and I/Q performance benefits in single-chip silicon Doppler radars for noncontact cardiopulmonary monitoring. IEEE Trans Microw Theor Tech 52(3):838–848
59. McEwan TE (1996) Body monitoring and imaging apparatus and method. US Patent 5573012

60. Immoreev IJ, Samkov SV (2002) Ultra-wideband (uwb) radar for remote measuring of main parameters of patients vital activity. Radio Phys Radio Astron 7(4):404–407
61. New public safety applications and broadband internet access among uses envisioned by fcc authorization of ultra-wideband technology. Federal Communications Law Journal (2002)
62. Zito D, Pepe D, Neri B, Zito F, De Rossi D, Lanatà A (2008) Feasibility study and de-sign of a wearable system-on-a-chip pulse radar for contact-less cardiopulmo-nary monitoring. Int J Telemed Appl 2008:10

Chapter 2
Energy Harvesting for Self-Powered Wearable Devices

Vladimir Leonov

2.1 Introduction to Energy Harvesting in Wearable Systems

Personalized sensor networks optionally should include wearable sensors or a body area network (BAN) wirelessly connected to a home computer or a remote computer through long-distance devices, such as a personal digital assistant or a mobile phone. While long-distance data transmission can typically be performed only by using the batteries as a power supply, the sensors with a short-distance wireless link can be powered autonomously. The idea of a self-powered device is not new and is actually known for centuries. The earliest example of self-powered wearable device is the self-winding watch invented in about 1770. However, typically not much energy is harvested in a small device, so that use of a battery, primary or rechargeable, is beneficial from practical point of view.

There are worldwide efforts ongoing on development of microgenerators that should eliminate the necessity of wiring and batteries in autonomous and stand-alone devices or in devices that are difficult to access. Energy harvesters are being developed for the same purpose. An energy harvester (also called an energy scavenger) is a relatively small power generator that does not require fossil fuel. Instead, it uses energy available in the ambient, such as an electromagnetic energy, vibrations, a wind, a water flow, and a thermal energy. These sources are the same as those used in power plants or power generators such as the ones for powering houses in remote locations, light towers, spacecrafts, and on transport (except those based on fossil fuels). An energy harvester is typically several-to-one centimeter-size power microplant that converts into electricity any primary energy that is available in the ambient. The reason to call them "harvesters" or "scavengers" is the new application area: they are used for powering small devices, such as sensors or sensor nodes. This way of powering them eliminates the need for cost-ineffective work, such as wiring or either

V. Leonov (✉)
Smart Systems and Energy Technology Imec, Kapeldreef 75, 3001 Leuven, Belgium
e-mail: leonov@imec.be

A. Bonfiglio and D. De Rossi (eds.), *Wearable Monitoring Systems*,
DOI 10.1007/978-1-4419-7384-9_2, © Springer Science+Business Media, LLC 2011

recharging or replacing batteries. An energy harvester could also be combined with a battery and serve a complementary source of power to improve energy autonomy of a device at limited size of the battery.

Three kinds of energy sources can be used for harvesting in wearable devices. These are the mechanical energy of people's own moving or accelerations on transport, an electromagnetic energy that is mainly light energy, and the heat flow caused by the difference in temperature between the human body and the ambient. There is a difference between truly unobtrusive energy harvesters such as photovoltaic (PV) cells and effort-driven micropower generators. The typical example for the latter is a flashlight that is to be shaken or pre-powered by using the embedded dynamo. A power of the order of Watts can be obtained in such effort-driven microgenerators. However, this way of powering BAN or wearable sensors should be rejected because of additional care required from the patient's side. The worst-case scenario for energy harvesting is a patient who stays in his/her own bed. Then, there is practically no mechanical energy to harvest. The light intensity at home is low. The heat flow minimizes because of a blanket and low metabolic rate, especially, in elderly people. Therefore, only a part of the head and, sometimes, wrists of the person is the only relatively small zone where the energy harvester of thermal or light energy can be located on such patients. The available power is low, too, because the illumination level indoors is low and the heat transfer from the person is determined by natural convection around the head. Nevertheless, even in such case, powering of, e.g., a health-monitoring sensor by using energy harvesters is feasible.

Preventive healthcare is considered as a way to potentially decrease the cost of healthcare, which is steadily on an upward trend. One of the strategies is to shift the health monitoring and management outside the expensive medical centers to family doctors and even to home. For example, the monitoring of chronic diseases while providing real-time data from and to the patient wherever he/she is and at any moment may offer significant potential for both cost reduction at the stage of monitoring and for making curative medicine cost-effective. Wireless healthcare systems, which could be an important component of so-called e-health or eHealth grids, are expected to focus on preventive care and effective provision of continuous treatment to patients, especially those living in remote locations and to elderly people. Real-time monitoring of patient's vital signs and patient-level health data requires use of wearable sensors and mobile devices. It would be good if such devices were small, unobtrusive, and maintenance-free for their entire service life.

2.2 Principles of Energy Harvesting by Using Human Body Heat

Warmblooded animals, or homeotherms, including humans constantly generate heat as a useful side effect of metabolism. However, only a part of this heat is dissipated into the ambient as a heat flow and infrared radiation, the rest of it is rejected in a form of water vapor. Furthermore, only a small fraction of the heat flow can be used in a compact, wearer's friendly and unobtrusive energy scavenger. For example, nobody would like to wear a device on his or her face. Therefore, the heat flow from the face cannot be used. The heat flow can be converted into electricity by using a thermoelectric generator (TEG), the heart of which is a thermopile. It is known from the thermodynamics that the heat flow observed on human skin cannot be effectively converted into electricity, although a human being generates more than 100 W of heat on average. Assuming that about 1–2% of this heat can be used, an electrical power of the order of milliwatts can be obtained using a person as a heat generator. If we recall that watches consume about 1,000 times less power, it is fairly good power.

The human body is not a perfect heat supply for a wearable TEG. The body has high thermal resistance; therefore, the heat flow is quite limited. This is explained by the fact that warmblooded animals have reached in the process of evolution a very effective thermal management. In particular, this includes a very high thermal resistance of the body at ambient temperatures below 20–25°C, especially, if the skin temperature decreases below the sensation of thermal comfort (Monteith and Mount 1974). At typical indoor conditions, the heat flow in a person depends on the location on the body and mainly stays within the 1–10 mW/cm^2. The forehead produces larger heat flow than the area covered by the clothes. Because of thermal insulation due to clothes, not much heat is dissipated from the skin and only about 3–6 mW/cm^2 is observed indoors, on average. Depending on the physical activity of a person, the heat dissipation in extremities "switched" either on or off. This is to preserve the temperature of the body core at low metabolic rate, and to dissipate the excess heat when body temperature rises due to increased physical activity.

The ambient air has a high thermal resistance, too. Indoors, it can be evaluated by using natural heat convection theory. The TEG placed at the interface between the objects with high thermal resistance, i.e., the body and air, must also have relatively high thermal resistance. This can be explained by using electro-thermal analogy, i.e., when voltage, current, and resistance are replaced with temperature difference, ΔT, heat flow, W, and thermal resistance, respectively. The corresponding thermal circuit is shown in Fig. 2.1 for the two cases: (1) a naked human being with no device, and (2) with a TEG on the skin. The human body as a heat generator and the

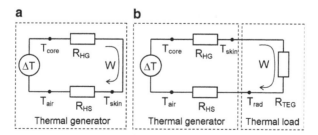

Fig. 2.1 Equivalent thermal circuits of: (**a**) a short-circuit natural thermal generator, and (**b**) the same generator with a thermal load. A relatively small surface of the skin, e.g., several square centimeters, is considered in both cases for the sake of simplicity

ambient air as a heat sink represent natural thermal generator that is shunted on the skin, i.e., at the interface between the body and air (Fig. 2.1a). If a TEG is placed on the skin (Fig. 2.1b), the device behaves as a thermal load of the thermal generator.

The thermal circuit of a wearable TEG placed in contact with the skin involves the thermal resistance of the body, R_{HG}, and of the ambient air, R_{HS}. These resistors are connected in series and represent the thermal resistance of the thermal generator. Despite the fact that the air is a heat sink, in terms of thermal circuit, its thermal resistance acts in the same way as the one of the body, i.e., of the heat generator, and must be included into the thermal generator. In other words, the thermal resistance of the body and air is the thermal resistance of the environment surrounding the TEG. The heat flow in the circuit, W, is the ratio of the temperature difference between the deep body temperature, or core temperature, T_{core} and the ambient air with the temperature T_{air} to the thermal resistance of the circuit. The normal core temperature in humans is about 37°C with a day-to-night variation of 0.5–1°C. Animals, in general, have similar core temperatures, but in cattle it is frequently a little higher, up to 39°C. In camels and baby animals, it can further raise up to about 41°C. The highest core temperatures, up to about 45°C, have been registered in small birds. Typically, the bird temperature ranges between 38°C and 42°C. At night, however, birds have the lowest temperature, which is called nocturnal hypothermia. In general, the smaller the animal, the smaller wearable TEG is needed to produce the same power. The smallest TEG is required on a bird because of a high heat transfer coefficient from it during flight (forced air convection), which is good for the bird.

It is obvious from Fig. 2.1b that the available temperature difference $\Delta T = T_{core} - T_{air}$ can never appear on the TEG because of high thermal resistance of the ambient air and, frequently, of the body. The ratio $R_{TEG}/(R_{HG} + R_{HS} + R_{TEG})$ determines the part of available temperature difference to be obtained on a TEG, i.e., $\Delta T_{TEG} = (T_{skin} - T_{rad})$, where T_{rad} is the temperature of the outer surface of the TEG, which is called radiator. The thermal resistors composing the thermal generator are variable and depend on each other, and on the thermal resistance of a TEG. Therefore, T_{skin} and T_{rad} in Fig. 2.1b are not the same as in Fig. 2.1a at the same ambient conditions. The increased thermal resistance of the circuit in Fig. 2.1b due to a thermal load causes also the heat flow W to decrease.

Because of specific conditions of a thermopile application discussed above, there are specific requirements to both the thermopile and the TEG in most of the energy harvesters including wearable devices. First, the optimal thermal resistance of a thermopile, R_{tp}, required for maximum power generation must be equal to:

$$R_{tp} = \frac{R_{pp} R_{TEGopt}}{R_{pp} - R_{TEGopt}}, \tag{2.1}$$

where R_{pp} is the parasitic thermal resistance of a TEG, and R_{TEGopt} is the optimal thermal resistance of a TEG, at which power generation reaches its maximum. The parasitic thermal resistance, R_{pp}, is always observed due to: (1) air inside the TEG, (2) holding mechanical components interconnecting the cold and hot sides of a TEG, i.e., the elements connected thermally in parallel to the thermopile, and (3) a heat exchange due to infrared radiation. The thermal resistor R_{pp} is connected thermally in parallel to the thermopile between its hot and cold junctions. Actually, it may include some thermal resistance associated with parasitic heat transfer from the heat source to the radiator or to the boundary layer through convection and radiation outside the TEG. The optimal thermal resistance of a TEG can be obtained from the equation of its thermal matching with the ambient:

$$R_{TEGopt} = \frac{(R_{HG} + R_{HS}) R_{em}}{2(R_{HG} + R_{HS}) + R_{em}},$$ (2.2)

where R_{HG} is the local thermal resistance of human body between the body core and the chosen location on the skin, R_{HS} is the thermal resistance of a heat sink, i.e., the thermal resistance due to convection and radiation on the outer side of TEG, and R_{em} is the thermal resistance of a TEG which could occur if the TEG would be "empty," namely, with no thermoelectric material in it. Equation (2.2) is a thermal equivalent of electrical matching of a generator with its load. The last requirement is that the thermal insulation factor N, defined as

$$N = R_{em}/(R_{HG} + R_{HS}),$$ (2.3)

must preferably be more than one. This ratio depends on the area of radiator, the contact area with the skin, and on the thickness of a TEG. The thinner the TEG, the less power it regrettably produces due to thermal shunting of a thermopile through the air and holding components. The maximum power takes place at the optimal temperature difference between the cold and hot thermopile junctions, ΔT_{tp}. The latter can be expressed as:

$$\Delta T_{tp} = \frac{\Delta T}{2(1 + 1/N)},$$ (2.4)

so that at $N = 1$, only 25% of ΔT can be obtained on the thermopile. If $N \gg 1$, ΔT_{tp} approaches a half of ΔT like in the other reversible heat engines. The thermal conductivity of air is significantly less than that of thermoelectric material and can therefore be neglected. In this case, one can obtain the expression for the power that can be reached in a wearable TEG, P_{max}, as:

$$P_{max} = \frac{Z}{8} \frac{\Delta T}{(R_{HG} + R_{HS})} \Delta T_{tp},$$ (2.5)

where Z is the thermoelectric figure-of-merit.

From (2.1) to (2.5), a compact wearable TEG should be semiempty, where the thermopile must occupy only a minor part of the device volume. The rest must be filled with air or with a material showing thermal conductivity less than the thermal conductivity of air. The radiation heat exchange between the hot and cold components of a TEG must preferably be minimized through the use of materials with low emission coefficient in long-wave infrared spectral region, i.e., metals. The requirement of a "semiempty" TEG offers a good chance to body-powered power converters to be embedded in pieces of clothing. Such low-weight devices could be user-friendly and comfortable while being worn.

2.3 Calculated Characteristics of Wearable TEGs

The factor N, as follows from (2.4) to (2.5), must exceed one for satisfactory power generation. This places a barrier for the minimal thickness of a TEG at a fixed area that it occupies on the human body. The thermal resistance of the thermoelectric material and air between the two plates of a TEG is proportional to the distance between the plates (Fig. 2.2). However, decreasing the thickness of a TEG does not essentially affect the thermal resistance of thermal generator (Fig. 2.1). As a result, e.g., a thermopile weaved in clothes cannot produce satisfactory power levels. This is because N becomes much less than one. Therefore, unacceptably low $\Delta T_{\rm tp}$ is developed on the thermopile. It could have a thermal resistance of a few cm^{2}K/W, while for reaching the power maximum it should be hundred times higher.

There are two basic ways to maximize the power. The first way is to make a thin TEG, say, 3-mm-thin, and provide a very good thermal isolation between the plates of the TEG (Fig. 2.2b). This increases numerator of (2.3), the factor N, and the power. The two plates larger than the area occupied by a thermopile are required to

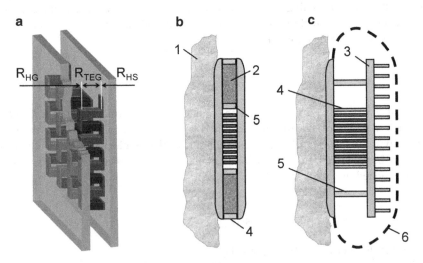

Fig. 2.2 A thermopile between the hot and cold plates of a thermoelectric generator (**a**), and the cross-section of wearable thermoelectric generators: (**b**) a thin TEG on the human skin (1) filled with the material (2) with a thermal conductivity much less than that of air, and (**c**) a thick air-filled TEG with a radiator (3). The other thermally isolating and shock-protecting components are: (4) encapsulation wall, (5) rigid supports such as pillars, and (6) a thermally isolating protection grid that allows air convection and being transparent for infrared radiation

decrease the thermal resistance of the thermal generator (i.e., of the environment) and to get optimal temperature difference on the thermopile. Because of fragility of thermoelectric materials, the device must be enforced by using stiff supports, such as pillars or an encapsulating wall, placed in between the plates. In principle, filling of such a TEG with the material having thermal conductivity less than that of air could be advantageous for further lowering parasitic heat exchange between the plates. The device could be integrated into a piece of clothing. However, such a TEG does not reach the best power that can be obtained on a person because of a low factor N. Furthermore, accounting for technological limitations in industrial fabrication process of thermopiles, only a low voltage, much less than 1 V can be obtained in a compact device. As a result, only several thermoelectric devices connected electrically in series could guarantee an output voltage of the order of 0.5–1 V, which can be effectively used for powering electronic devices. As an alternative, the TEG could be made thicker, e.g., 1–2-cm thick. Despite complications related to integration of such units into clothes, it could reach much higher N, and therefore better power per unit area of the skin. As a result, thicker units would produce higher voltage and it becomes possible to use only one unit for powering a wearable device, of course, if the TEG produces power enough for the particular application.

The second way to maximize power is to decrease the denominator in (2.3), i.e., the thermal resistance of the thermal generator. This can be done by using a fin radiator, or the one with pins. Of course, such radiator consumes some volume of the TEG. The device with a radiator cannot be therefore thin. However, in a TEG that has a thickness of 1–2 cm, the radiator helps to further increase the factor N and the power.

As a numerical example, let us analyze a wearable TEG resembling a big button of 3 cm in diameter. In the calculations, we will vary the thickness of such unit and determine the dependence of maximum power on its thickness. The device resembles the one shown in Fig. 2.2b; however, the empty space between the plates is filled with air, i.e., (2) is air. Two rigid metal plates with a thickness of 1 mm will provide stiffness to the device and good thermal conductance from the human skin to the thermopile and from the latter to the ambient air. The small temperature drop related to limited thermal conductivity of the plate material is neglected. It is assumed that the unit is integrated in a piece of clothing and is located on the chest or arm of the person. We assume that the heat transfer to the ambient is described by natural convection and radiation. The heat transfer correlations are used for a vertical plate with a characteristic length of 30 cm (Incropera and DeWitt 1996) while assuming that the heat transfer from the outer surface of the device is the same as from the clothed human being. The calculations are performed for the distance between the plates from 0.5 to 8 mm, so that the thickness of the TEG varies from 2.5 to 10 mm. The other parameters are: air temperature is 22°C, the deep body temperature of a subject is 37°C, the thermal resistance of the body is 250 cm^2K/W, $Z = 0.003 \ K^{-1}$, the supports and encapsulation together have a thermal resistance of 400 K/W per 1 mm distance between the plates, the emission coefficient of the outer surface of the TEG is 90%, no radiation heat transfer between the polished aluminum plates, and no convection inside the TEG, i.e., it is encapsulated.

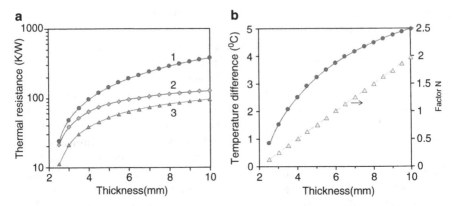

Fig. 2.3 Calculated dependence of thermal characteristics of optimal both the thermoelectric generator and the thermopile on the thickness of a TEG: (**a**) the optimal thermal resistance of an empty TEG (1), of the matched TEG (2) and of the thermopile (3), and (**b**) the temperature difference on a thermopile (1) and the factor N (2)

The results of modeling show that the thermal resistance of an empty TEG that scales linearly with its thickness results in a decreased thermal resistance of the thermally matched TEG if it is thin (Fig. 2.3a). The factor N becomes small and the temperature difference on the thermopile decreases to about $1\,^{\circ}C$ even in the optimized TEG (Fig. 2.3b). At the thickness less than 6 mm, even a half of the theoretical power, (2.5), cannot be reached because $N < 1$.

Ideally, a wearable device and its power supply should be small. Therefore, the power produced per unit volume of a TEG is of primary importance. Under the conditions specified above, it has a maximum in a 4–5-mm-thick device (Fig. 2.4a). The absolute power produced in a thicker device increases (Fig. 2.4b); however, the volume increases more rapidly than the power. Analysis shows that increasing the thickness from 2.5 to 6 mm causes an increase of the power because of increase in numerator of (2.3). In a thicker device, on the contrary, decreasing the denominator of (2.3) could effectively help to further increase the power. Therefore, a second device has been modeled, which resembles the TEG shown in Fig. 2.2c. In the modeled device, there is no protection grid. Then, the only difference with the first modeled device is that a part of its volume is occupied by a radiator. The results of such modeling are shown in Fig. 2.4, too. The radiator size increases up to 40% of the device volume in a 10-mm-thick TEG. It enables keeping the maximum power generation independent of the volume (Fig. 2.4a). Therefore, power generated by such TEG increases linearly at least up to 10-mm thickness.

One should not expect linear increase of the power in devices thicker than 1 cm. Actually, in such devices, the other effects that have been neglected in the above modeling start to be important. Application of the radiator results in local increase of the heat flow in humans. The larger the radiator, the larger is the heat flow and the lower is the skin temperature under the TEG. The radiator temperature decreases below the temperature of the outer surface of a clothed person. Therefore, the heat transfer becomes less effective than it was assumed in the model. We can

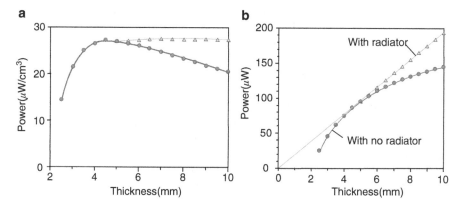

Fig. 2.4 Calculated dependence of the power on a thickness of an optimal TEG with no radiator (*circles*) (Fig. 2.2b), and with the radiator of an optimal size (*triangles*) (Fig. 2.2c): (**a**) power per unit volume, (**b**) power produced in a TEG of 3 cm in diameter. A *dashed line* in (**b**) is the guide for an eye

conclude from Fig. 2.4 that an optimized small wearable TEG can produce about 25 μW/cm^2 and about 25 μW/cm^3 indoors, i.e., with no wind, no sunlight, no pieces of clothing worn on top of the TEG, and in the location on the human body, where the thermal resistance of the latter is 250 cm^{2}K/W.

The measured performance characteristics of wearable TEGs are close to their theoretical analysis performed in this section. However, if the TEG is located on an open skin surface, the radiator temperature is significantly less than the skin temperature. Consequently, the power per unit volume decreases as compared with calculations performed in this section due to higher temperature of the convection layer formed around the human body. Based on both theoretical and practical results (still to be discussed below), we conclude that a correctly designed unobtrusive TEG in the right location on the human body can produce approximately 10–30 μW/cm^2 of electrical power in moderate climate, on 24-h average. The produced power depends on the thickness of a TEG and its size: the thicker the TEG, the better is power generation while the larger the TEG, the less power per unit area is produced. It also very much depends on the location on the human being therefore the latter requires particular attention.

2.4 Human Body as a Heat Source for a Wearable Thermoelectric Power Supply

Medical studies of the properties of a human being, in particular, of heat flows and its thermal conductance are typically performed on the whole human body or on its parts such as the head, arm, hand or trunk (Hardy et al. 1970; Itoh et al. 1972). Furthermore, they are mainly conducted on naked skin surface. Clothes change the overall heat flow from the human body and its pattern. Clothes have a tremendous effect on the heat transfer from the body at ambient temperatures less than 25–28°C. All three main channels of heat rejection, namely, convection, radiation, and evaporation from the skin surface are affected by clothes. The lower the ambient temperature, the larger is the percentage of heat dissipated from open skin, i.e., from the face. The trunk has much more stable temperature at different ambient conditions (temperature, wind, and sunlight) than the head and extremities. This is because people choose appropriate clothes depending on the weather conditions. However, even indoors, at typical temperatures of 20–25°C, certain variations of the skin temperature are observed on the scale of centimeters. An example of the temperature map of the wrist and hand is shown in Fig. 2.5a. The temperature profile around the wrist is shown in Fig. 2.5b as measured at two indoor ambient temperatures. The temperature reaches maximum close to the radial and ulnar arteries. Local heat flows also change from place to place. If a TEG is attached to the body, especially the one with a radiator, the heat flow depends not only on the skin temperature, but also on the local thermal resistance of the human body. The latter is defined as a thermal resistance between the body core and the chosen location on the skin.

As an example, the skin temperature has been measured in the middle of the forehead before attaching a TEG and under attached TEG. At 21.5°C, a heat flow of 9.5 mW/cm^2 and a thermal resistance of 380 cm^{2}K/W have been measured by using a thermopile with a thermal resistance of 50 cm^{2}K/W attached to the forehead. A skin temperature of 34.7°C has been measured, but a deep brain temperature of 37.5°C has been assumed to obtain the thermal resistance. Then, a TEG with a fin radiator of 1.6 cm × 1.6 cm × 3.8 cm size has been attached on the same place. The contact area between the TEG and the skin was 4 cm^2. The heat flow has increased to 22.5 mW/cm^2, the thermal resistance of the forehead has decreased to 227 cm^{2}K/W, and the skin temperature under the TEG dropped to 30.9°C.

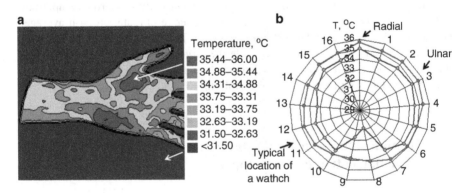

Fig. 2.5 (a) Temperature map of the hand (palmar view). The infrared image is taken with calibrated radiometric camera within the 8–12 μm spectral range. (b) Temperature profiles around the wrist with a circumference of 17 cm at two ambient temperatures, 27°C (*circles*) and 22.3°C (*diamonds*), measured indoors

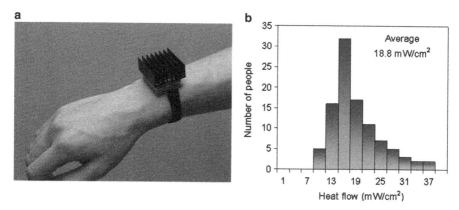

Fig. 2.6 (**a**) Thermoelectric generator on the wrist and (**b**) the heat flow through the TEG per square centimeter of the skin measured indoors on the wrist of 100 people sitting still at a mean room temperature of 22.3°C

The increase of heat flow due to radiator has caused a decrease in the thermal resistance of the forehead by a factor of 1.7.

The thermal resistance of the wrist under a large-size TEG with a fin radiator of 1.6 cm × 3.6 cm × 3.8 cm size (Fig. 2.6a) has been measured in the distal forearm of students at 22.7°C. In the typical location of a watch, its average value measured on 77 volunteers sitting still for several tens of minutes is 440 cm^{2}K/W. The volunteers have been asked to attach the TEG to the wrist with tightness according to their preferences. Therefore, the contact area of the hot plate of a TEG with the skin varied a little in uncontrollable way. Therefore, the statistical data presented below in Fig. 2.6b account for the user-related tightness, which is useful for designing TEGs. The heat flow through a TEG was 200 mW, on average; however, it depended on the skin temperature. The latter measured on 77 persons shows variations within the 27.5–32.5°C range with 30°C on average. The mean heat flow varied with the skin temperature from 15 to 24 mW/cm^2. However, the standard deviation, σ, due to difference between subjects was large, with a σ/mean of 17%. Therefore, the corresponding thermal resistance largely varied. In 90% of studied subjects, the thermal resistance of the body combined with the skin-to-TEG interface contact resistance was within the 200–650 cm^{2}K/W range.

To understand importance of the thermal resistance of the body for designing a TEG, we divide the thermal resistance into two components. The first one, R_{c-r}, denotes the thermal resistance between the body core and the arterial blood in the wrist. The second component, R_{r-TEG}, denotes the thermal resistance between the arterial blood and the hot plate of the TEG. At a blood temperature of 35.8°C, on estimate, the R_{c-r} and R_{r-TEG} can be evaluated, assuming a core temperature of 37°C (Fig. 2.7a). As one can see, only R_{r-TEG} strongly depends on the skin temperature. Therefore, within the measured range for skin temperatures, the thermal resistance in the wrist observed between arteries and the skin dominates over the vasomotor

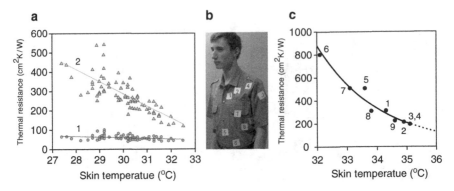

Fig. 2.7 (**a**) Estimated thermal resistance per square centimeter of the skin measured at 22.7 ± 0.5°C on the wrist of 77 people under the attached TEG: (1) is the thermal resistance between the body core and arterial blood in the wrist, (2) is the thermal resistance between the arterial blood and TEG. (**b**) Nine locations, where the thermal resistance of the human being has been measured, and (**c**) is the thermal resistance of the human body at 23°C depending on its location on the trunk

response, thermal resistance of the cardiovascular system, and interface contact resistance between the skin and the TEG.

Two experiments have been conducted to demonstrate the importance of accounting for the thermal resistance of the body. In the first one, its comparative measurements have been performed in three locations: on the forehead, on the wrist (on the radial artery), and on the chest, left side, on lowest ribs. A TEG with a size of 3 cm × 4 cm × 0.65 cm and with a thermal resistance of 580 cm^2K/W has been used in this experiment. The heat flow at 22.8°C was the same, however the skin temperature was different, from 33.8°C in the wrist to 35.8°C in the forehead. The corresponding thermal resistance of the body varied from one location to another by a factor of three. The same TEG has also been integrated sequentially in nine locations in a shirt (Fig. 2.7b). The corresponding thermal resistance shows large variations (Fig. 2.7c). Therefore, the power generation varies within a factor of three over the nine measured locations.

In the second experiment, the ability of the human being to provide large heat flow has been studied. A thermopile with a thermal resistance of 50 cm^2K/W has been attached to the wrist in two locations, namely, on the radial artery and in the typical location of a watch. A large piece of aluminum maintained at room temperature has been provided on the outer side of the thermopile and served as almost perfect heat sink. The experiment shows that, on the radial artery, a heat flow of 90 mW/cm^2 exceeds by a factor of three the heat flow that the human being can provide in the location of a watch. However, in indoor applications, the heat flow exceeding 15–30 mW/cm^2, depending on the location of a TEG, causes sensation of cold. Therefore, an acceptable heat flow of 10–20 mW/cm^2 through a wearable TEG supplied with a radiator seems the maximum indoors.

In cold environment, certain zones of the human body allow larger heat flow with no sensation of cold. As measured outdoors on the neck, near an artery, the maximum heat flow of 60 mW/cm^2 was acceptable at air temperature of 0°C. At a temperature of −4°C, a heat flow of 70 mW/cm^2 was still quite comfortable on the front side of the leg, about 25 cm above the knee of a person wearing jeans. The maximum comfortable heat flow of 100–130 mW/cm^2 has been registered on the radial artery in the wrist, at air temperatures of −4°C to +2.3°C. Therefore, the TEG with a thickness of 3 cm unobtrusively produced in this location a power of 1–1.4 mW/cm^2 on a walking person.

2.5   TEGs in Wearable Devices

The first wearable TEG serving as a power supply for a simple wireless sensor worn on a wrist has been fabricated in 2004 (Fig. 2.8a, b). At 22°C, it produces a power of 100 μW transferred into the electronic module of a sensor node. This is the only 40% of the generated power because of low efficiency of the voltage up-converter. The latter is a necessary circuit component because the output voltage from the TEG fluctuates indoors within the 0.7–1.5 V range. At 0.7 V output, the power is not enough for the sensor, while at 1.5 V, too much power is produced. Therefore, at the system level, a short-or long-term power reserve must be provided in the form of rechargeable battery or a supercapacitor to avoid power

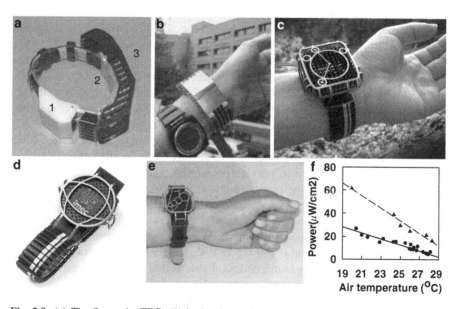

Fig. 2.8 (**a**) The first wrist TEG: (1) is the electronics module, (2) is a hot plate, and (3) is a radiator. (**b**) A similar TEG worn next to a watch. (**c**) A TEG with a pin-featured radiator. (**d**) A waterproof TEG for outdoor use. (**e**) A TEG in the wireless sensor for measuring the power generated by people in real life. (**f**) The power produced by the device shown in (**c**) in the office on a sitting (*circles*) and walking (*triangles*) person

shortages. By using such energy storage element, the power gained by a TEG on occasional basis can be uniformly redistributed and consumed at near-constant rate over a long period of time. In the first wireless sensor, the electronic board was powered by two NiMH cells (2.4 V). The power generated at daytime was enough for powering the electronics and a 2.4 GHz radio, and for transmitting several measured parameters to a nearby PC every 15 s.

In 2005–2006, watch-size wrist TEGs of three different designs have been fabricated (Fig. 2.8c–e). The power generated in the office on a person sitting still for a while is shown in Fig. 2.8f. At 20–22°C indoors, the TEG produces 200–300 μW at an open-circuit voltage of 2 V. This power decreases to about 100–150 μW at night or on a person resting for a long period of time, i.e., at low metabolic rate. However, it rises in a few minutes of walking indoors to 500–700 μW. This power increase is explained by the forced air convection on a walking person. On the same reason, i.e., because of wind and more physical activities, wearable TEGs work better outdoors. Taking into account adverse illumination conditions at home, on transport and at night, these TEGs are much more powerful, on 24-h average, than the best PV cells because the majority of people spend indoors most of the lifetime. Higher voltage allows direct charging of an NiMH cell. However, at temperatures above 26°C, the TEG is mismatched with the cell and the efficiency of power transfer decreases. Therefore, some wireless sensors have been made with a supercapacitor as the charge storage element instead of a battery because the former can start storing energy at lower voltage. This means that the battery-less device works more efficiently at higher ambient temperatures than the device with a battery.

An example of such sensor node is shown in Fig. 2.8e. The 4-stage thermopiles used in the TEG have an equivalent aspect ratio of thermocouple legs of 35. At a distance of 7 mm between the hot plate and the radiator, this aspect ratio is the optimal one. Therefore, at ambient temperatures within the 20–22°C, the TEG produces more than 25 μW/cm^2, i.e., almost the maximum possible power. In the best orientation of the TEG, namely, facing the radiator down, the power reaches 30 μW/cm^2. The battery-less wireless sensor node has been designed to track the power produced by the human being at high ambient temperatures in real life. To make it functional at such temperatures, the duty cycle for the radio transmission bursts is made variable to prevent power shortages. On the other hand, it allows consumption of all the produced power at typical ambient temperatures, where otherwise the supercapacitor would be saturated and the harvested energy would not be transferred into it. By varying the interval between transmissions from 0.1 to 100 s, the voltage on the charge storage supercapacitor is maintained always near the matching point. The sensor node has been tested up to an ambient temperature of 35°C. The measurement results show that a temperature difference of 2–3°C between the skin and air provides enough power for the sensor. An interesting observation is that due to fluctuations of air and skin temperatures, different activities of a person, and variable both sunlight and wind, a battery-free device is able to work at any ambient temperature, at least, a part of the time. Even at a mean ambient temperature equal to the skin temperature, the average power production is not zero.

It is interesting to compare the performance characteristics of the TEG shown in Fig. 2.8d with the modeling results obtained in Sect. 2.3. This TEG has a distance between the plates of 7 mm, the 0.5-mm-thick hot plate and the 1.5-mm-thick cold plate, so that the TEG has a thickness of 9 mm. According to Fig. 2.4, it can produce up to 20 μW/cm^2. There are some differences between the calculated case and the TEG shown in Fig. 2.8d. First, the TEG has a larger cold plate of 3.4 cm in diameter, but a contact area of about 6 cm^2 between the hot plate and the skin is less than in above calculations. It has also been tested on the wrist therefore the heat transfer coefficient is better than in the modeled device. However, the protection grid adversely affects the power and partially decreases the convective and radiation heat transfer from the TEG. The measurements of the power generation have been performed at ambient temperatures of 23–25°C. The device produces by about 17% less power per unit area than its thicker version shown in Fig. 2.8c. This corresponds to a power of 140 μW, or 15.8 μW/cm^2 at 22°C, a pretty close to the modeled 140 μW (Fig. 2.4). More exact modeling of this device on the wrist performed earlier has predicted 160 μW, or 17.8 μW/cm^2 at 22°C, but with no protection grid.

A TEG similar to the one shown in Fig. 2.8c has been used as an energy supply for the first body-powered medical sensor, namely, a pulse oximeter or SpO$_2$ sensor. The device noninvasively measures the oxygen content in arterial blood by using a commercially available finger sensor (Fig. 2.9). This battery-free device is fully self-powered at an output update rate every 15 s. Its power consumption in this case is 62 μW, while the TEG typically produces more than 100 μW. About 47% of power is used for the signal processing, 36% is consumed by two LEDs, 12% is used for a quiescent power, and 5% for the radio. The device switches automatically on if there is enough voltage on the supercapacitor. In case of fully discharged supercapacitor, it starts in about 15 min after putting the device on.

The signal processing in the pulse oximeter is performed onboard therefore a minimal power is required for the radio transmission. In case of monitoring biopotential signals, the waveform must be transmitted. In this case, the radio consumes most of the power. To demonstrate the possibility of creation of more complex battery-less wireless devices, a two-channel electroencephalography (EEG) system has been fabricated (Van Bavel et al. 2008). It consumes 0.8 mW

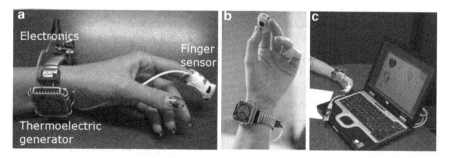

Fig. 2.9 Wireless pulse oximeter (**a, b**) and the application running on a laptop (**c**)

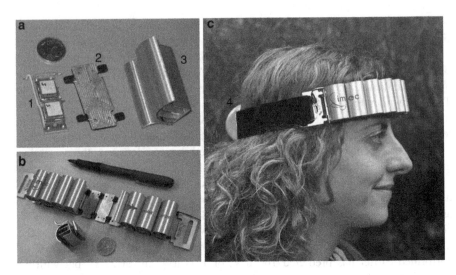

Fig. 2.10 Wireless electrocardiography system powered by the body heat: (**a**) and (**b**) show the TEG components in the assembling stage, and (**c**) is a completed device. (1) is a thermopile module, (2) is a hot plate, (3) is a radiator and (4) is the electronics module

therefore the TEG must provide more power at 22°C to make sure that there will be no power shortages at higher ambient temperatures. Taking into account that the limit of the power calculated and measured in TEGs of 1–1.3-cm thick is about 25 μW/cm^2, the device must occupy a relatively large area. Therefore, the TEG has been divided into 10 units. The units are connected to each other in a track resembling those of crawler-type tanks or big bulldozers (Fig. 2.10).

The thickness of radiators has been increased as compared to wrist TEGs to increase the power per unit area of the skin. The TEG has a thickness of 29 mm. The size of a hot plate track is 4 cm × 20 cm with a contact area to the skin of 64 cm^2. The measured power at 22°C is about 2.5 mW, or 30 μW/cm^2. The system has been designed for the indoor use at 21–26°C. At a temperature below 18–19°C, the heat flow through the TEG increases and the device is considered by users as too cold. (At a temperature of 19°C, the power increases to 3.7 mW.) Therefore, to make it acceptable for outdoor use at low ambient temperatures, the heat flow must be decreased, i.e., the radiators must be smaller. As a result, at high ambient temperatures the TEG would not produce enough power and its size would further grow. Therefore, in the TEG acceptable at low ambient temperatures, PV cells could be added. In a device with fixed dimensions, they compensate for a lack of power from the TEG at high temperatures. Furthermore, PV cells are more efficient outdoors and can gain a significant energy to be stored in a battery.

The EEG system, pulse oximeter, and the other sensors described in this section have power consumption less than the power generated by a TEG in the worst application scenario. However, at ambient temperatures of 35–38°C, thermoelectric power minimizes. To provide enough power in such situation, a secondary battery must be provided. As it has been shown, smart power management together with decreased duty cycle and power consumption in case of energy deficit enable body-powered devices in a wide temperature range. If high ambient temperatures are expected for long periods of time for a particular application, it is also beneficial to hybridize a TEG with PV cells.

2.6   Hybrid Thermoelectric-Photovoltaic Wearable Energy Harvesters

Hybrid energy scavengers have been fabricated for EEG systems in 2008 primarily to avoid sensation of cold induced by a TEG in cold weather. Figure 2.11a illustrates the principle of hybridization of a TEG and PV cells. The latter are mounted on the outer surface of radiators and serve as their external heat dissipating surface. The TEG and PV cells are connected in two parallel electrical circuits and charge one supercapacitor. Additional power gained by PV cells enables decreasing heat flow through the TEG (and the produced power, too) thereby making it comfortable in harsh weather conditions. One of the systems is shown in Fig. 2.11b.

The hybrid power supply provides more than 1 mW in most of the situations. This is more than enough for the two-channel EEG application consuming 0.8 mW. The absolute and relative input power gained from the thermoelectric and PV power supplies constantly varies, thereby reflecting variations in both the illumination level and the heat transfer from the head. A power of 45 mW was generated by PV cells in direct sunlight (March, Belgium), while a power of 0.2 mW has been measured in the office, far from the window in a cloudy day. The TEG provides much more uniform power output than PV cells because it depends mainly on air temperature and wind speed. At 22°C, indoors, the TEG generates 1.5 mW, while outdoors, at 9.5°C with no wind, the power increases to 5.5 mW. The EEG system is

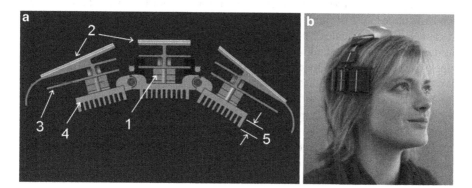

Fig. 2.11 (a) Cross-section of the hybrid thermoelectric-photovoltaic generator unit used in an EEG system: (1) is a thermopile, (2) are PV cells, (3) is a radiator, (4) is a hot plate with (5) thermal shunts. (b) Two-channel EEG system with a hybrid power supply (reproduced with permission from Van Bavel et al. 2008)

battery-free, so the power exceeding 1 mW is typically wasted. However, using a supercapacitor instead of secondary battery allows demonstration of a nice system feature: in less than 1 min (typically, in 10–30 s) after putting it on, the charge storage supercapacitor is charged from the fully discharged state and the system is self-started by the body heat.

As tested outdoors at a temperature of 7°C, the device is still very comfortable for the user. As a rule of thumb, at 10°C outdoors, PV cells generate eight times more power than the TEG while indoors the latter offers eight times more power than PV cells in the office. By using a two-way power supply that exploits both the heat dissipated from person's temples and ambient light as energy sources, the dimensions and weight of the TEG are reduced. The location on the hair is much more convenient, according to user's responses. In addition, the EEG system works much more reliably at high ambient temperature like 28°C (with available light).

Comparison of a TEG with PV cells of the same area shows that the latter generate much less power on average, because not much light is available indoors, where the authors and the reader of this book are resting at this moment. In addition, the quantum efficiency of high-efficiency PV cells at low illumination rapidly decreases. If high efficiency is obtained in PV cells indoors, they could become competitive to a thin TEG. The power in a TEG scales proportionally to its thickness, at least within the 4–10 mm range. However, as modeled in Sect. 2.3, even in a 4-mm-thin TEG, it can reach 10 $\mu W/cm^2$. This is still much better than the power generated by high-efficiency monocrystalline silicon cells, especially on a 24-h average.

2.7 TEGs in Clothing

A system integrated in a piece of clothing must be thin, lightweight, and should sustain repeated laundry and pressing. Therefore, it must be waterproof, either bendable under load or rigid, and sustain high temperatures. High accelerations in modern washing machines up to about 300 g together with mechanical shocks during use of devices set additional requirements for the mechanical strength and shock protection. Photovoltaic cells are thin and even if enforced with a rigid or a flexible metal plate, have a thickness of about 1 mm. TEGs can also be made flexible, i.e., with thin plates. However, as pointed out in Sect. 2.3, the TEGs must not be thinner than about 2 mm, otherwise, the area occupied by the TEG would dramatically enlarge. The system components must also provide the sweat path from the body to prevent wetting of the skin at high metabolic rate, e.g., during exercise, and in a summer season. At a system level, a part-time use of a piece of clothing suggests that the devices must hibernate during long periods of nonuse and perform auto-start when in use.

To demonstrate the feasibility of such devices, an electrocardiography (ECG) system has been integrated into an office-style shirt in 2009 (Fig. 2.12a). Unlike the EEG and SpO_2 sensors described in previous sections, it is powered by a secondary

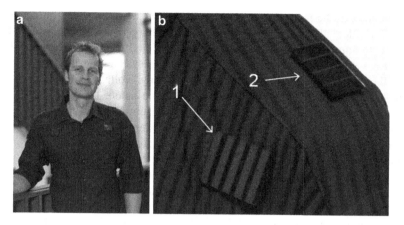

Fig. 2.12 (**a**) Electrocardiography system integrated in a shirt. (**b**) One of thermoelectric modules (1) and the left-side PV cell (2)

battery. The battery is constantly recharged using the wearer's body heat. The power consumption of the energy-efficient ECG system is 0.44–0.5 mW depending on the sample rate. Given the best demonstrated power efficiency of 75% of voltage up-converter, the only 0.6–0.7 mW are required from the TEG. The sample rate is set automatically depending on available power. To be comfortable for the user, the TEG is built on modular approach. Fourteen 6.5-mm-thin TEG modules with outer metal plates of 3 cm × 4 cm size acting as radiators have been integrated into the front side of the shirt (Fig. 2.12b). They occupy less than 1.5% of the total area of the shirt. According to the modeling results (Fig. 2.4a), the TEG modules must produce near-maximum power per unit volume. In the office, the TEG typically provides the power within the 0.8–1 mW range at about 1 V on the matched load during person's usual sedentary activity. On a person walking indoors, the power production increases to 2–3 mW due to forced convection. The radiators of TEG modules have been painted like chameleon into the shirt colors, except one module, which is done to show the module size. The wiring and the other modules of ECG system are located on the inner side of the shirt. Because of high thermal resistance of thermally matched TEG modules, they are never cold. As measured at about 10°C outdoors on a person wearing a thick jacket, the power typically increases a little at low ambient temperatures.

Two charging circuits, one with a TEG, and the other with PV cells are connected to the power management module. Two amorphous silicon solar cells of 2.5 cm × 4 cm size each has been integrated in the shirt on its shoulders. PV cells have been added to the system because if the shirt is not worn for months, the battery can be emptied due to its self-discharge. When the shirt is not used for a long period of time, more than a month, it must be stored in an environment where light is available periodically, e.g., in a wardrobe with windows. The power provided by solar cells is enough to compensate for the self-discharge of the battery and for the standby power. In this way, even after months of non-use, the electronics is maintained in the ready-to-start state, waiting for the moment the shirt is used again. If accidentally the battery is completely discharged, the shirt is still not lost. Its PV cells must be just placed in direct sunlight and charged for several days. By using a wake-up button, the operability of the system can then be verified. Once the battery reaches the minimum working voltage, the up-converter becomes functional and the ECG shirt can be worn again. During its daily use, the produced power typically exceeds the power consumption, so the battery will be fully charged in a course of several days. The system components, i.e., a TEG, PV cells and electronics in a flex circuit, have waterproof encapsulation and sustain machine washing with drying cycle at 1,000 rpm. If the TEG voltage drops to near-zero, which happens when the shirt is taken off, the system switches into a standby regime with 1 μW power consumption. The self-start of the system takes place within a few seconds after touching the skin while the shirt is being put on again.

At a conversion efficiency of 75%, the system functions up to 25–29°C, depending on the activity of the user. The harvesting still takes place up to 31°C during a walk. This does not mean that the system will stop at an ambient temperature of, e.g., 35°C. In such case, the battery will provide the power until the user enters an air-conditioned room. Furthermore, a temperature of 35°C outdoors with a high probability means that there is a plenty of sunlight, so that PV cells instead of a TEG will be the main power supply for a while.

2.8 Development of New Technologies for Wearable Thermopiles

The theory (see Sect. 2.2) does not require large-size thermopiles for the maximum power generation in a wearable TEG. The only requirement for a TEG is that it must have a thermally matched thermopile with high thermal resistance per square centimeter of the skin. However, the power per unit area, at least within the 4–10 mm thickness of a TEG, scales linearly with its thickness (Sect. 2.3). If a small-size thermopile is used in such TEG, the distance between the two plates of a TEG (Fig. 2.2) must be kept the same. The design of a TEG changes a little (Fig. 2.13a), i.e., one or two thermal shunts must interconnect a small-size thermopile with the plates. (A thermal shunt is a thermally conducting element such as a spacer, a fin or a pillar that thermally shunts a part of the environment.) Then, a miniaturized thermopile can produce about the same power in a TEG of a fixed thickness as obtained by using large-size thermopiles purchased on the market. This does not mean that any small-size thermopile is good for wearable devices. Still both the electrical contact resistance and the thermal conductance parallel to the thermopile in a TEG must be minimized because these are parasitic factors that adversely affect its performance characteristics (Sect. 2.2).

The modeling of a thermopile in a wearable TEG shows that due to scaling laws, the smaller the thermopile, the lower aspect ratio is required to provide its thermal matching (Fig. 2.13b). (An aspect ratio is the ratio of the length of thermocouple

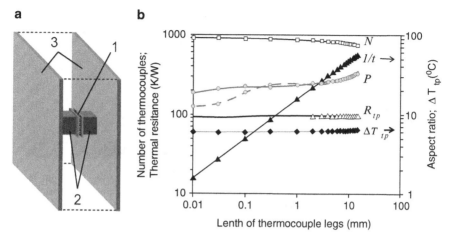

Fig. 2.13 (a) Design of a wearable TEG with a small-size thermopile (1) that is thermally connected to the plates (3) by using thermal shunts (2). (b) Modeled dependence of optimal parameters of a thermally matched thermopile in a wearable TEG of 3 cm × 3 cm × 1.7 cm size on the length of thermocouple legs: N is the optimal quantity of thermocouples, l/t is an aspect ratio of thermocouple legs, R_{tp} is the thermal resistance of a thermopile, ΔT_{tp} is the temperature difference between hot and cold junctions, and P is the power calculated at a contact resistance of 10 Ω μm^2 (*solid line*) and 100 Ω μm^2 (*dashed line*) between semiconducting legs and metal interconnects

legs, l, to their lateral dimension, t.) Commercially available thermopiles require high aspect ratio. For the device modeled in this section (Fig. 2.13b), it must exceed 20, at $l = 2$ mm, and 50, at $l = 15$ mm. Both values essentially exceed capabilities of industrial technologies. Therefore, the only practical solution has been found to build the TEGs described in this chapter, namely, the use of multistage thermopiles. Decreasing the thermopile size causes proportional decrease of an aspect ratio that is required for the same thermal resistance. As a result, at a length of thermocouple legs of about 10 μm, the optimal aspect ratio decreases to values acceptable in microelectronic and microelectro mechanical systems (MEMS) technologies. At larger dimensions, thick-film and inkjet printing technologies could be used instead in thermopiles fabricated on a polymer tape (Stark 2006) as well as in membrane-based and membrane-less thermopiles (Van Andel et al. 2010).

With microelectronic technologies, the aspect ratio required for 6–15-μm-long thermocouples can be reached using projection lithography because a critical dimension of 1–3 μm is sufficient. One of the possible designs is shown in Fig. 2.14a. A height of 6 μm with inclined thermocouple legs has been already reached in the technological process developed for the polycrystalline SiGe (Fig. 2.14b) (Su et al. 2010). This height corresponds to about 12 μm length of thermocouples. A research is ongoing toward practical demonstration of poly-SiGe thermopiles with high aspect ratio. The required low contact resistance between semiconducting legs and metal interconnects, i.e., less than 100 Ω μm^2 (Fig. 2.13b) seems feasible (Wijngaards and Wolffenbuttel 2005). Alternatively, an on-chip vacuum packaging can enable required performance characteristics even at larger contact resistance

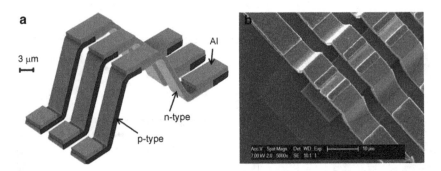

Fig. 2.14 Surface micromachined arcade thermopile. (**a**) The conceptual design of a thermocouple. Three thermocouples are shown. (**b**) An SEM picture of a poly-SiGe thermopile test structure with a 6 μm topography and a critical dimension of 3 μm (reproduced from Su et al. 2010 with permission from Elsevier)

(Xie et al. 2010). To obtain performance characteristics shown in Fig. 2.13b, a film technology for BiTe materials still must be developed. Thick-film BiTe processes have already been demonstrated (Böttner et al. 2004; Snyder et al. 2003). In the near future, thermoelectric properties of film-based BiTe are expected to approach those of bulk materials, but this is not an easy technological task.

Miniaturizing of thermopiles offers a potential for essential reduction of the fabrication cost. The expected production cost of micromachined thermopiles is by a factor of 100–1,000 less than the cost of today's thermopiles on the market because only 1–2 mm^2 of the wafer is required for a compact wearable TEG. A film-based thermopile on a polymer tape could require several square centimeters of the tape, a hundred times larger area. However, the low cost of thick-film and inkjet technologies, and the tape itself could result in a low cost, too. Therefore, wearable thermopiles can be very competitive on cost with the batteries in mass production.

2.9 Conclusions

The theory of a wearable TEG shows that a power of 10–30 μW/cm^2 can be produced for a typical person indoors. These values have been also practically obtained in different prototypes of wearable self-powered wireless sensor nodes powered either thermoelectrically or by using hybrid thermoelectric-PV generators. The evolution of body-powered devices during 6 years of their development indicates that only low-power applications, i.e., those consuming below 1 mW, can be unobtrusively powered indoors by using human body heat. This means that practically none of medical devices existing on the market can be turned into self-powered ones. On the other hand, it has been shown that most of the wireless health monitoring and medical devices can work at a power of less than 1 mW with no loss in the signal quality. Further miniaturizing energy scavengers can be done in case of electronics with less power consumption and with lower power radio. The related research is ongoing worldwide. A simple wireless sensor consuming 10 μW has already been demonstrated (Pop et al. 2008). Such sensor can be powered by a very small TEG, because only 1–3 cm^2 of the human body area is needed to get the required power. However, to obtain a voltage of at least 1–2 V in such a small TEG, film-based miniaturized thermopiles must be developed. In the near future, an optimized wearable TEG is expected to outperform any existing battery of the same weight in less than 1 year of its use. A possibility of low-cost fabrication technology and green energy are also very attractive features of the discussed devices. Therefore, a TEG can become a good candidate for serving as a lifetime power supply for low-power wearable electronics in the near future.

References

Böttner H, Nurnus J, Gavrikov A et al (2004) New thermoelectric components using microsystem technologies. IEEE J Microelectromech Syst 13:414–420

Hardy JD, Gagge AP, Stolwijk JAJ (eds) (1970) Physiological and behavioral temperature regulation. Charles C Thomas Publisher, Springfield

Incropera FP, DeWitt DP (1996) Fundamentals of heat and mass transfer. Wiley, New York

Itoh S, Ogata K, Yoshimura H (eds) (1972) Advances in climatic physiology. Igaku Shoin Ltd., Tokyo; Springer-Verlag, Berlin

Monteith J, Mount L (eds) (1974) Heat loss from animals and man. Butterworths, London

Pop V, van de Molengraft J, Schnitzler F et al (2008) Power optimization for wireless autonomous transducer solutions. In: Proceedings of the PowerMEMS and MicroEMS Workshop, Sendai, Japan, 9–12 November 2008, pp 141–144

Snyder GJ, Lim JR, Huang C-K, Fleurial JP (2003) Thermoelectric microdevice fabricated by a MEMS-like electrochemical process. Nat Mater 2:528–531. doi:10.1038/nmat943

Stark I (2006) Thermal energy harvesting with thermo life. In: Proceedings of the International Workshop on Wearable and Implantable Body Sensor Networks (BSN), Boston, 3–5 April 2006. doi:10.1109/BSN.2006.37

Su J, Vullers RJM, Goedbloed M et al (2010) Thermoelectric energy harvester fabricated by stepper. Microelectron Eng 87:1242–1244

Van Andel Y, Jambunathan M, Vullers RJM, Leonov V (2010) Membrane-less in-plane bulk-micromachined thermopiles for energy harvesting. Microelectron Eng 87:1294–1296 (Proceedings of the 35th International Conference on Micro & Nano Engineering, Ghent, Belgium, 28 September to 1 October 2009). doi:10.1016/j.mee.2009.10.003

Van Bavel M, Leonov V, Yazicioglu RF et al (2008) Wearable battery-free wireless 2-channel EEG systems powered by energy scavengers. Sens Transducers J 94:103–115. http://www.sensorsportal.com/HTML/DIGEST/P_300.htm

Wijngaards DDL, Wolffenbuttel RF (2005) Thermo-electric characterization of APCVD Poly-Si$_{0.7}$Ge$_{0.3}$ for IC-compatible fabrication of integrated lateral Peltier elements. IEEE Trans El Dev 52:1014–1025

Xie J, Lee C, Feng H (2010) Design, fabrication, and characterization of CMOS MEMS-based thermoelectric power generators. J Microelectromech Syst 19:317–324

Chapter 3
Wireless Communication Technologies for Wearable Systems

Steve Warren and Balasubramaniam Natarajan

Wireless communication technologies and wearable health-monitoring systems are inexorably linked, as wireless capabilities allow devices worn by ambulatory patients to share data (and often power) in real time with other wireless nodes. Such systems incur little-to-no cost in terms of patient/device interaction and are a dramatic improvement over traditional store-and-forward wearable monitoring devices, such as Holter monitors. In addition, the ad hoc networking functionality supported by emerging plug-and-play wireless standards points to the inevitable reality of patient environments that host pervasive networks of wireless health care devices whose primary role is to increase their quality of life. Clearly, the realization of these wireless tools does not come without cost. Wires that may have once provided high data throughput and access to power, data storage, and processing resources are no longer available. The engineering challenge is therefore to incorporate adequate power, data storage, and processing capabilities on the wireless devices to balance the throughput and operational lifetime needs of the monitoring application. Concurrently, one must consider security issues associated with broadcasting previously local information.

This chapter provides an overview of wireless technologies relevant to wearable health monitoring systems, including categories of wireless devices, networking topologies, and standards (both existing and emerging) that will help to ease the transition into health monitoring environments pervaded by wireless wearable, nearby, and desktop medical sensors. These thoughts are followed by a synopsis of safety and security issues that inevitably arise from using wearable, wireless health monitoring systems as well as the design considerations that come into play when working with wireless links. Next, the authors briefly address technologies available to a developer to implement wireless solutions. This is followed by a synopsis of a few recent efforts that utilize wireless technology for human/animal health monitoring.

S. Warren (✉)
Department of Electrical and Computer Engineering,
Kansas State University,
Manhattan, KS, USA
e-mail: swarren@ksu.edu

A. Bonfiglio and D. De Rossi (eds.), *Wearable Monitoring Systems*,
DOI 10.1007/978-1-4419-7384-9_3, © Springer Science+Business Media, LLC 2011

3.1 System-Level Considerations

Wireless communication links have been a mainstay in commercial and military domains for many years. Wearable health monitoring technologies also have a notable history of several decades in the form of devices, such as electrocardiogram, Holter monitors (Raciti et al. 1994; Thakor 1984), and accelerometer-based pedometers (Meijer et al. 1991). The merger of the two technology classes, however, is relatively recent. It was not until the mid-to-late 1990s that medical-sensor-laden body area networks (BANs), a.k.a. personal area networks (PANs) or wireless personal area networks (WPANs) (IEEE 2004; Istepanian et al. 2004; Istepanian and Laxminaryan 2000; Yao et al. 2005), became a focus of discussion because of their potential to support continuous monitoring of high-risk patients without imposing severe mobility limitations affiliated with desktop clinical, home care, and telemedicine systems. Since that time, advances have occurred in the areas of wireless networking technologies and wearable devices that in aggregate make mobile health [a.k.a. "m-Health" (Istepanian et al. 2004), "unwired e-med" (Istepanian and Laxminaryan 2000)], or wireless personal health ["p-Health" (Teng et al. 2008)] a feasible health monitoring alternative as opposed to an esoteric possibility. Excellent surveys of wearable, wireless health monitoring efforts, and the clinical potential for these technologies can be found in the aforementioned references in addition to (Bonato 2009; Hung et al. 2004; Lymberis 2003a, 2003b; Nakajima 2009; Ooi et al. 2005; Pantelopoulos and Bourbakis 2010).

At the system level, wireless wearable systems by nature imply wireless transmission ranges of 10 m or less, ad hoc network creation, plug-and-play interoperability, the ability to stream data from multiple sensors when needed, and the capacity to operate at very low power levels for long periods of time. These requirements were not well met by traditional wireless technologies, including those that supported consumer applications, such as garage door openers and television remote controls. The embedded biomedical systems community therefore started to look toward emerging standards, such as Bluetooth (BluetoothSIG 2010; Chen et al. 2003; IEEE 2004; Kroc and Delic 2003; Park and Kang 2004; Yao et al. 2003) and, more recently, ZigBee (Huang and Park 2009; IEEE 2010a, 2010b; ZigBeeAlliance 2010; Zito et al. 2007a, 2007b) to meet those needs. During the same time frame, radiofrequency identification (RFID) (Lehpamer 2007; Pereira et al. 2008; Tuttle 1997; Ze et al. 2008) technologies were developed for numerous security and inventory tracking applications, yet their promise for biomedical monitoring applications was clear. Even more recently, cell phone technology has become a focal point of discussion (Lam et al. 2009), as cell phones have the potential to offer both the functionality and the economy of scale desired for wearable data loggers that serve as BAN hubs.

This section presents a comparative analysis of enabling wireless technologies for BAN implementation and the networking topologies they support. It then concludes with notes about information security and patient safety that are unique to wearable health monitoring systems.

3.1.1 Body Area Networks

Many high-level architectures have been proposed for health monitoring systems that employ sets of wearable sensors and their supporting communication and information storage networks (Cheng et al. 2008a, 2008b; Crowe et al. 2004; Durresi et al. 2008; Galeottei et al. 2008; Gialelis et al. 2008; Hung et al. 2004, 2009; Jovanov et al. 2005; Jung et al. 2009; Kim and Lee 2008; Kroc and Delic 2003; Lam et al. 2009; Lee et al. 2007; Massot et al. 2009; Mendoza and Tran 2002; Montgomery et al. 2004; Sagahyroon et al. 2008; Thiruvengada et al. 2008; van de Ven et al. 2009; Wang et al. 2009a, 2009b; Warren et al. 2004a, 2004b; Wu and Xiaoming 2007; Yao et al. 2005). The most common system-level, functional architecture [see Fig. 3.1 (Jovanov et al. 2005)] includes (1) a network of sensors on the body that stores its data to a wearable or handheld data logger/hub that then communicates wirelessly with a local base station or Internet gateway, (2) a central command center that receives data from these gateways, and (3) a network backbone infrastructure that facilitates the exchange of information between the command center and the appropriate medical service centers. While issues such as security and reliability as addressed in this chapter apply to all three levels of network communication, this chapter focuses on wireless technologies for the BANs themselves [item (1) above]. This is because architectural topologies at

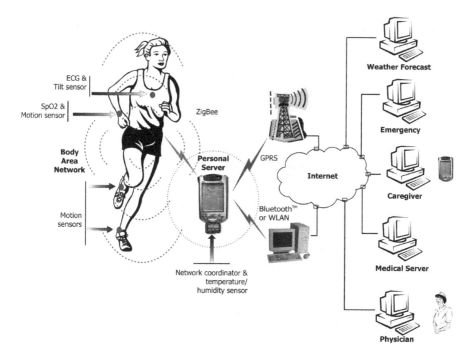

Fig. 3.1 Depiction of a body area network with its supporting information infrastructure. Courtesy Dr. Emil Jovanov, University of Alabama at Huntsville, USA. Originally published in (Jovanov et al. 2005)

the body level are quite limited, promoting a focused discussion in a single chapter. At the body level, sensors communicate bi-directionally with wearable/handheld data loggers, which then interact with the resources external to the wearer. Mesh sensor topologies, wireless routing schemes, and other more complex topological approaches are typically unnecessary at the body level due to the close proximity of the wireless nodes. In addition, the peripheral technologies that support the local gateways and the extended backbone network infrastructure (which may include cellular systems, wireless local area networks, or HomeRF implementations) are mature, and treatments of such subsystems are addressed in numerous wireless texts.

3.1.2 Wireless Standards Comparison

WPANs are a class of networks that subsumes BANs. Specifically, WPANs are short-range networks usually confined to the personal operating spaces of users. The first WPAN effort was initiated in 1998 with the emergence of Bluetooth technology (BluetoothSIG 2010). In 1999, the IEEE 802.15 standards working group was created, and this group has since been actively involved in developing wireless standards for the personal area network arena. The IEEE 802.15.1 standard based on Bluetooth technology was the first standard to address WPANs. This was followed by efforts of the IEEE 802.15.3 task group which led to the ZigBee standard (ZigBeeAlliance 2010). Finally, the IEEE 802.15.4 task group is developing a standard for low-data-rate sensor network applications (IEEE 2010a, 2010b). In the following subsections, we provide a brief discussion of these enabling wireless technologies, which are expected to provide most BAN implementations in the near future. Table 3.1 provides a comparison between features of candidate WPAN technologies (Ashok and Agarwal 2003).

3.1.2.1 Bluetooth – IEEE 802.15.1

In 1998, a group of companies sought to develop technology to enable wireless interconnectivity between personal devices (e.g., laptops, personal digital assistants, digital cameras, etc.), respectively, separated by 10 m or less. Bluetooth technology was developed to meet this need. In 1999, the IEEE 802.15.1 working group adopted the Bluetooth specifications as the basis of a 1 Mbps WPAN standard (IEEE 2004). (The name Bluetooth comes from the Danish king Harald Blatand (Bluetooth), who united the Scandinavian people in the tenth century.)

Bluetooth links operate in the 2.4 GHz unlicensed band. More specifically, the 2,402–2,480 GHz industrial, scientific, and medical (ISM) band is divided into 79 equally spaced 1 MHz channels, which each Bluetooth device employs using random hops. During transmission, these hops occur at a rate of 1,600 hops/s. The modulation scheme uses binary Gaussian-shaped frequency shift keying (GFSK) at a 1 Mbps rate. Typical transmitter output power is about 1 mW,

Table 3.1 WPAN comparison (Ashok and Agarwal 2003)

	Bluetooth (802.15.1)	WiMedia (802.15.3)	ZigBee (802.15.4)
Operational spectrum	2.4 GHz	3.1–10.6 GHz	2.4 GHz and 868/915 MHz
PHY layer details	Frequency hopping spread spectrum (FH-SS), 1,600 hops/s	Multiband OFDM	DSSS with BPSK or MSK
Channel access	Master slave polling	CSMA/CA	CSMA/CA
Max data rate	1 Mbps	480 Mbps	868 MHz to 20 kbps 915 MHz to 40 kbps 2.4 GHz to 250 kbps
Coverage radius	<10 m	<10 m	<20 m
Typical current drain	1–60 mA	Up to 400 mA	Very low (20–50 μA)
Relative cost per node	Low	Medium	Very low

From AGRAWAL/ZENG. *Introduction to Wireless and Mobile Systems*, 2E. ©2006 Cengage Learning, a part of Cengage Learning, Inc. Reproduced by permission. http://www.cengage.com/permissions

with an option to scale up to 100 mW for larger transmission ranges. The low transmission power makes Bluetooth devices suitable for battery operation. Additionally, when operating within the 10 m communication range, Bluetooth devices can organize themselves into piconets. Each piconet has a master node which controls and communicates with at most seven slaves; this master–slave mode of operation avoids contention within the system. Piconets can operate independently, and their configurations can change spatially and over time. Every new device that wishes to enter a piconet must register with the master node. If the master node is already full with seven slave nodes, the new device registers in "parked" mode. Devices may also exist in standby mode when they do not associate with a piconet. Communication among members of a piconet is defined in the IEEE 802.15.1 standard, and details regarding all of the protocol layers in a Bluetooth system can be found in (IEEE 2004). Wearable, wireless systems based on the Bluetooth specification can be found in (Chen et al. 2003; Crk et al. 2009; Galeottei et al. 2008; Hung et al. 2004; Kroc and Delic 2003; Park and Kang 2004; Sagahyroon et al. 2008; Shen et al. 2008; Warren et al. 2006; Wu and Xiaoming 2007; Yao et al. 2005; Yao and Warren 2005).

3.1.2.2 IEEE 802.15.3 and WiMedia

Since the Bluetooth standard is not intended for high-rate multimedia applications, the IEEE 802.15.3 working group pursued a high rate (11–55 Mbps) WPAN standard in 2003 with a focus on medium access control (MAC) and physical layer (PHY) specifications (IEEE 2009). This standard specifies an ad hoc PAN topology similar to the master–slave Bluetooth architecture: devices can assume either master or slave functionality and exit or enter the ad hoc network without complicated setup procedures. The MAC is based on CSMA/CA (Carrier Sense Multiple Access with Collision Avoidance) (Rappaport 2002). In CSMA, each device first listens to the channel for a predetermined amount of time to verify if it is idle. If the channel is idle, then the device transmits; if it is not, the device defers its transmission. The PHY layer of the 802.15.3 standard specifies operation in the unlicensed 2.4 GHz frequency band with four 15 MHz channels. Depending on the data rate supported (11, 33, 44, or 55 Mbps), one of the five modulation techniques is used. The base modulation is differentially encoded QPSK (quadrature phase shift keying) with options to switch to uncoded QPSK, trellis coded QPSK, and 16/32/64-QAM (quadrature amplitude modulation). Transmit power complies with the FCC specification of 0 dBm, and range is limited to 10 m.

The 802.15.3a group was formed to create a high-speed enhancement of the 802.15.3 PHY layer. The focus of this group was to develop a standard that would support bit rates up to 480 Mbps using principles of ultrawideband (UWB) communication (Reed 2005). However, the task group was unable to unify the two candidate proposals for the standard (multiband orthogonal frequency division multiplexing (MB-OFDM)-based UWB and direct sequence UWB) and was therefore dissolved in 2006. The WiMedia alliance, a consortium of more than 350 organizations, has adopted MB-OFDM-based UWB as the enabling technology for high-rate PANs (WiMedia 2010). Wireless USB is based on the WiMedia Alliance's MB-OFDM-based UWB PHY layer and is capable of sending up to 480 and 110 Mbps at distances up to 3 and 10 m, respectively. It operates in the 3.1–10.6 GHz frequency range. Recently, the Wimedia Alliance released an enhanced UWB PHY specification that supports up to 1,024 Mbps. A wearable, wireless systems based on the WiMedia specification is described in (Shaban et al. 2009).

3.1.2.3   IEEE 802.15.4/ZigBee

The IEEE 802.15.4 specification defines a standard for low-rate, low-power WPANs that are well suited for body area and home networking applications (e.g., home automation, security, etc.). In 2000, the ZigBee Alliance and the IEEE 802 working group came together to build the specifications for low-rate PANs (ZigBeeAlliance 2010). The IEEE working group addressed the PHY and MAC layers, while the ZigBee Alliance defined the higher layer protocol specifications. IEEE 802.15.4 supports both star and peer-to-peer network topologies. The MAC layer is once again based on CSMA/CA. The PHY layer is based on direct sequence spread spectrum techniques operating in the unlicensed 2.4 GHz ISM band worldwide. There is also a provision for use of the 868/915 MHz bands in Europe and in the United States, respectively. The 2.4 GHz band supports data rates up to 250 kbps, while the 868/915 MHz band supports rates of 20 and 40 kbps. BPSK modulation is specified, and spectrum spreading is accomplished using a 15-chip m-sequence. A choice of aggregating four bits and using minimum shift keying (MSK) modulation is also provided. A typical coverage radius for each of these devices is envisioned to be around 10–20 m. Wearable, wireless systems based on the ZigBee specification can be found in (Huang and Park 2009; Sukor et al. 2008; Thepvilojanapong et al. 2008; Zito et al. 2007a, 2007b).

3.1.3 Device and Information Surety

"Surety" is a broad term that aggregates the concepts of security, safety, reliability, verifiability, redundancy, usability, and other system properties that speak to the ability of the system to always work as anticipated and in the best interest of the user. In the case of wireless, wearable health monitoring systems, surety is essential. For example, the transition from wired to wireless connections makes it more difficult to control who can access patient data. These wireless data should therefore be encrypted – a straightforward design problem given encryption facilities provided in cellular phone products and base wireless standards, such as Bluetooth and ZigBee. A system designer must also make reasonable decisions about on-body connections, balancing the power and computational costs of encryption and on-body key distribution against the likelihood that, e.g., sensor-to-data-logger transmissions will be intercepted (Dağtaş et al. 2008). Means must be in place to authenticate the identity of patients that use these devices, since data corruption (e.g., storing health data in the wrong patient record) is arguably in some contexts a greater concern than a breach of patient confidentiality.

Patient-authentication approaches germane to wireless, wearable systems include device pre-activation for a unique user, smart cards (Kim et al. 2008), implantable RFID pills (Perakslis and Wolk 2005), fingerprint/palm recognition (Corcoran et al. 2007), i-Button-based device activation protocols (Daradimos et al. 2007), shared cryptographic keys based upon acquired physiologic values (Venkatasubramanian et al. 2008) and biometric approaches based upon the waveforms acquired from the patients (e.g., electrocardiograms (Sriram et al. 2009) and photo-plethysmograms (Gu et al. 2003; Shu-Di et al. 2005; Wan et al. 2007)). The concept of "owner-aware" sensors is tightly coupled to the authentication of the patient in wireless, wearable health monitoring systems (Warren and Jovanov 2006; Warren et al. 2005a, 2005b).

Wireless transmissions are also more susceptible to data corruption and packet loss when compared to their wired counterparts. Data integrity measures, such as cyclic redundancy checks, are well-known approaches for addressing data loss and are treated by numerous wireless communication texts and articles (Durresi et al. 2005; Sklavos and Zhang 2007). Like encryption, a designer must balance the computational, power, and throughput costs of these approaches against the available system resources, which can be minimal at the BAN sensor level. A large body of work exists that addresses information surety as it relates to patient safety within the context of health monitoring systems (Venkatasubramanian and Gupta 2006), and guidelines published by the U.S. Food and Drug

Administration are changing to address the use of wireless, wearable systems in
home environments (Lewis 2001). In a more general way, the Health Insurance
Portability and Accountability Act (HIPAA) of 1996 (DHHS 2010; Meraki 2009;
O'Dorisio 2003) includes a broad mandate to protect data in health networks from
unauthorized access and tampering. Excellent treatments of security issues in low-
power, ambulatory wireless devices can be found in texts, such as (Stavroulakis
and Stamp 2010).

3.2 Lower-Level Tradeoffs

3.2.1 Wireless Technology Categories

BAN developers have an array of short-range, low-power wireless technologies at
their disposal. Technologies for on-body communications fall into three general
categories:

1. Radiofrequency (RF) links
2. Inductively coupled (magnetic) links, and
3. Intrabody communication approaches

Tools for data-logger-to-base-station interactions are generally RF links, where
Bluetooth, ZigBee, and WiMedia are all viable alternatives depending on
throughput and power consumption requirements (see Sect. 3.1.2). Cell phones
will also fill this role. The paragraphs below speak primarily to the technologies that
will be used on-body.

Category 1 – Radiofrequency links are by far the most commonly used on-body
technology set; many commercial off-the-shelf tools exist to implement RF links.
As noted in Sect. 3.1.2, operational ranges for BAN carrier frequencies have been
specified as part of the broader collection of ISM radio bands, and multiple "form
factors" exist with this broader category of links. First, RF links can be designed
with custom transmit/receive electronics and antennas, where these specialized
designs may be targeted toward specific power, size, and frequency specifications
(Zito et al. 2007a, 2007b). Second, chip-level solutions driven by industry standards
can be purchased that offer a development kit which can be used to test driver
firmware and the viability of the wireless link prior to incorporating the chip in a
board-level design. Examples of such tools include the Nordic nRF24L01 single-
chip 2.4 GHz transceiver (Chen et al. 2008; Nordic 2010; Wong et al. 2009) and the
Jennic JN5139 kit (ZigBee and 6LoWPAN at 2.4 GHz) (Li and Warren 2010).
Recent activities have also targeted ultra-low-power MEMS at the microwatt level
as chip-level candidates for wearable wireless sensors (Enz et al. 2007; Otis et al.
2004). Higher-level designs such as the Linx 916 MHz modules allow one to easily
interface a wireless link to a host microcontroller through a serial port (Jovanov
et al. 2001). For even greater ease of use with point-to-point connections, some

wireless tools serve as simple RS-232 or USB dongles, allowing one to disconnect a serial line and insert a set of two wireless modules, where the modules effectively function as a cable segment, setting up their wireless connection in the background without the knowledge of the host system. Early applications for this technology were driven by replacements for printer cables.

A more sophisticated RF toolset that includes fully functional nodes (with radios, storage, preprogrammed networking capabilities, and often on-board sensors) would fall within a grouping of wireless tools commonly referred to as "motes." These tools allow programming access to higher-level network functionality, freeing the developer from the details of the low-level network interactions and allowing them to concentrate on BAN sensor interfacing, development of custom sensor designs, and data analysis tasks. In other words, motes allow one to create a simple, stable wireless network with relatively little effort compared to a custom design, assuming that the physical form factors, data rates, and power requirements for these mote-based nodes are suitable for the intended application environment. Many mote platforms already incorporate sensors for ambient temperature/humidity, light levels, three-axis acceleration, and global position. Additional sensors can be integrated with a parent mote through the use of a small daughter card. This class of technology is available through companies such as CrossBow (e.g., MICA2 motes at 868/916 MHz and TelosB motes at 2.4 GHz; Crossbow 2010), Moteiv (e.g., Telos motes at 2.4 GHz; Jovanov et al. 2005; Moteiv 2010), and Intel [e.g., next generation motes (Intel 2010) and Shimmer Motes (Patel et al. 2009)]. The primary advantages of this approach include the ability to quickly establish a robust wireless network, software-only interaction with the wireless units themselves, the availability of a host operating system (e.g., TinyOS) on the wireless board, and the potential for high-volume use via a purchase order. Disadvantages include a limited set of form factors (size, computational capabilities, and power draw that may be an awkward match for a given application), the expense of a network at current prices, and resource inflexibility.

While the aforementioned resources are general purpose wireless networking tools, new RF products are emerging that are intended to fulfill BAN roles. A good example is the Texas Instruments eZ430 Chronos unit: a wireless sports watch (based on the TI CC430 915 MHz radio) that is accompanied by a development kit at an overall cost of $50 U.S. (TexasInstruments 2010). It is specifically geared toward use as a hub for personal area networks (e.g., to store pedometer or heart rate data) or a node for remote data collection. The watch contains an integrated pressure sensor and a three-axis accelerometer.

Category 2 – Inductive links rely on magnetic coupling to convey information between a transmitter and a receiver (Bunszel 2001; Zierhofer and Hochmair 1990). On the transmission side of an inductive link, the electronics drive an electric current (containing the carrier signal and patient data) through a loop antenna, which creates a magnetic field that surrounds the entire transmitter. This directional magnetic field decreases in intensity as $1/r^3$ in its near field, where r is the distance between the source and receiving antennas (Microchip 1998; Simons and Miranda 2006).

When a receiver loop antenna is placed within the magnetic field, a current is generated that is fed to the receiver electronics. Ferromagnetic cores and multi-turn windings are used to optimize the transmit/receive coupling, and the transmit/receive LC tanks must be tuned to their resonant frequencies for optimal transmission. Most inductive links for through-tissue transmission operate in the frequency range of 125 kHz to 300 MHz.

Inductive links hold an advantage over traditional RF transmissions in their ability to send data through tissue, where the associated sensors may be ingested, implanted, or simply on opposite sides of the wearer's body. RF transmissions in the 900 MHz to 2.4 GHz range are essentially useless in a tissue medium due to water-based attenuation (Sharma and Guha 1975). In a magnetically coupled link, the transmission range is more affected by the transmit/receive winding radii and their numbers of turns, core permeabilities, and relative orientations than by the presence of tissue/water within the transmission region (Microchip 1998).

Because of their ability to transmit power wirelessly within their near field, inductive links are attractive means to recharge pacemakers (Papastergiou and Macpherson 2008), implanted neural prostheses (Troyk and DeMichele 2003), and implanted insulin pumps (Furse 2009). They can also provide power to (and receive data from) passive RFID chips (Microchip 1998), some of which can be injected into a host, and other implanted devices (Simons et al. 2006). These links can be used to provide commands and firmware upgrades to implanted devices (Liang et al. 2005) and to receive signal/image data from ingestible gastrointestinal pills (Chirwa et al. 2003; Swain 2003) and implantable sensors (Mackay 1961; Strömmer et al. 2006). Animal applications (e.g., swallowable health monitoring pills for cattle) have pushed the start of the art for such designs, since transmit/ receive distances can be on the order of 3–4 ft between, e.g., a swallowable pill in the reticulum and a receiver coil on a halter (Hoskins et al. 2009; Martinez 2007; Martinez et al. 2006; Warren et al. 2008a, 2008b).

Category 3 – Intrabody communication links are relatively new and use the body tissue as a transmission medium (Cho et al. 2007; Fujii et al. 2006; Gao et al. 2009a, 2009b, 2009c; Goldstein 2006; Hachisuka et al. 2003; Ruiz and Shimamoto 2005; Sasaki et al. 2004; Sasamori et al. 2009; Shinagawa et al. 2004; Sun et al. 2007; Xu et al. 2009). In this approach, signals are transmitted between on-body or intra-body transceivers using either electromagnetic waves (Cho et al. 2007; Hachisuka et al. 2003, 2005; Ruiz and Shimamoto 2005; Sasamori et al. 2009) or electric fields (Fujii et al. 2006, 2007; Gao et al. 2009a, 2009b, 2009c; Shinagawa et al. 2004; Wegmueller et al. 2007, 2009, 2010). The near-field, electromagnetic method treats the human body as a waveguide and employs both two- and four-electrode schemes [e.g., over a frequency range of 200–600 MHz (Ruiz and Shimamoto 2005) and at 10.7 MHz (Hachisuka et al. 2003, 2005)]. Different modulation schemes (e.g., FSK, ASK, BPSK, and MSK) for the electromagnetic solution operate within this context to achieve high transmission rates and low error rates. Near-field electric methods treat the human body as a conductor wrapped in an insulator. A current loop is established by the transmitter electrode, the body channel, the receiver electrode, and the capacitive return path through ground. This allows the use of a distributed resistance-capacitance model to analyze the channel characteristics as a function of frequency and channel length (Cho et al. 2007). A transceiver for this approach (based on an electro-optic sensor design) that is suitable for detection of small and unstable electric fields produced by the human body was reported in (Sasaki et al. 2004, 2009; Shinagawa et al. 2004). Shinagawa's module enables IEEE 802.3 (IEEE 2010a, 2010b) communication at a 10 Mbps data rate over a 150 cm distance through tissue with a 0.04% packet error rate.

Intrabody communication shows promise in that (1) data can potentially be transferred to handheld computers or embedded terminals simply through a single touch ("Touch-And-Play"; Hyoung et al. 2006) and (2) communications are confined to the body area, providing more secure transmissions and less interference with other wireless signals. Applications that utilize this technology have started to emerge. A touch-and-play JPEG image printer application was developed, where an image file can be digitally transferred from one hand to the opposite hand with a 1 Mbps data rate (Hyoung et al. 2006). A guidance system for blind people was also conceived to send a signal from a foot sensor to an earphone (Ruiz and Shimamoto 2005). Finally, a distributed implanted sensor network using galvanic coupling was built to enable wireless communication between implanted devices (Wegmueller et al. 2009, 2010). Drawbacks of intrabody communication are that the method has no compelling applications that are not already served by other RF technologies, it does not yet work reliably, and it may face perception problems among the general public because of the ability to send and extract data with a single touch (Goldstein 2006).

3.2.2 Signal Throughput

One of the challenges when designing a reliable wearable wireless system is to understand the throughput variability that occurs in practice. A common approach for designing wireless solutions is to first calculate throughput requirements based on (1) the number and types of sensors used, (2) the sampling rate, and (3) the amount of information that is to be exchanged. This is followed by identifying wireless modules that claim to provide a certain data rate and then integrating them with the sensors. Unfortunately, more often than not, this exercise does not offer the expected performance. Practical data rates tend to be lower than the maximum rates indicated in the specifications. This is due to a number of factors. First, a system's usage environment plays a major role. For example, Bluetooth and wireless LAN devices operate in the unlicensed 2.4 GHz ISM band. Many household devices such as cordless phones, microwave ovens, and car alarms also operate in this band and serve as interferers that can lower transmission rates. Second, network specifications usually denote the aggregate data rate supported and state that the rate per device may not be guaranteed as the number of deployed devices increases. Finally, the rated throughput values typically indicate "raw data rates", whereas actual data rates must accommodate information transmission overhead such as (1) channel coding, (2) training data for channel estimation, (3) frame synchronization packets, (4) addressing/authentication information, and (5) higher MAC/Network/-Application-layer-related information. In the off-the-shelf, standards-based wireless chipsets and solutions discussed earlier, the end-user cannot control the amount of transmission overhead. As a result, over the years, researchers have developed custom wireless protocol solutions with lower overhead for various applications (Agrawal et al. 2003; Nguyen and Ji 2008). In summary, it is important for the designer to understand the practical variability in throughput and account for this at the design stage. Custom solutions for specific applications are always the best way to optimize performance. However, development and implementation of custom solutions is not trivial and usually takes time and effort that make it unattractive.

3.2.3 Resource Allocations

Another important consideration in wearable systems is the means to distribute the information processing load. A typical wearable system consists of a network of sensors that measure complementary (sometimes redundant) health information. An initial vision was to have the sensors store their raw data and periodically upload those data to a central processing unit. However, with advances in embedded processing and the low cost of memory, it is now practical to perform local data processing at the sensor level prior to communication, saving battery life in the meantime. For example, a wearable pulse oximeter may decide to calculate and transmit only the blood oxygen saturation level rather than send raw photo-plethysmographic (PPG) data. This distributes the information processing task to the sensor and minimizes the amount of data communicated over the wireless link.

The key benefit of distributed signal processing is that power spent on communication is lowered at the cost of more processing power at the sensor level. Since wireless communication is the dominant drain for battery power, distributed signal processing approaches have received attention in the sensor networking arena. At the same time, it is important to remember that communicating "quantized" or "partial" information to the central processing unit limits the ability of the system and its users to detect and act on information. For example, the knowledge one can gain from postprocessing a complete ECG or PPG waveform is no longer available if only a summative health metric is communicated. Distributed processing variants can be implemented to overcome this drawback. For example, if the wearable devices detect abnormal health metrics, they can then send entire waveforms to the central processing unit for postprocessing. If the patient is healthy, devices can communicate summative health metrics periodically and sleep in the interim.

In summary, no universal solution exists with regard to resource/task allocation in a wearable system. This depends on the specific application and the goals of the monitoring system. Most importantly, implementing a centralized or smart distributed processing solution must consider cost/complexity/communication/ power constraints that vary from system to system.

3.2.4 Power Optimization

Sensor battery lifetime is a priority in wearable, wireless sensor networks. One straightforward way to increase the lifetime of a BAN is to improve the lifetimes of the batteries that power the sensors. However, the goal of this section is not to focus on battery technology but rather opportunities for power optimization and management that are unique to BANs.

Since communication consumes the bulk of the power, it is natural to first consider how to optimally manage transmit power for a specific BAN topology. The discussion of distributed signal processing in the previous section is closely related to power optimization. Effective data compression/quantization at the sensor level also reduces the amount of information to transmit, which in turn reduces the power consumed to complete the transmission. In addition, when developing custom wireless solutions, one may wish to implement adaptive power control strategies. For example, calculating the minimum power required to accomplish successful transmission and ensuring that every sensor is operating at that power level can significantly improve aggregate battery life. Power control can also be implemented in a manner similar to cellular phone architectures. Here, the base station determines the power level for every mobile phone operating in the cell and instructs (over a control channel) each mobile phone to switch to the appropriate power level. A central processing unit can similarly act as the power manager in a BAN. Additionally, power can be indirectly reduced if the sensors can be programmed to determine the best modulation and coding characteristic for each transmission. In a sense, the idea of software defined radio modules (Kenington 2005) can be used to ensure that power utilized to exchange information is always minimized. Of course, strategies such as power control and adaptive modulation come at the cost of increased complexity and additional control signaling – a typical tradeoff in communication system design.

The baseline power consumed by the sensor for its processor and idle time activities can also be managed to maximize lifetime. Researchers have assessed the impact of putting sensor nodes in sleep mode, where their power consumption is minimal (Li et al. 2006; Wu et al. 2006; Ye et al. 2004). Sensors are awakened only when their information is required. Effective scheduling of sensor measurements in time and space is an area that has received attention in a number of application domains (Chhetri et al. 2005; Gupta et al. 2004; Xiao et al. 2008).

Energy harvesting is another enabling power management scheme for wearable devices, where energy is salvaged from sources such as solar, mechanical, thermal, and chemical processes to reduce or eliminate the need for batteries that must be replaced and disposed of over time. In the case of BANs, the idea is to exploit the body and its associated activities as energy sources. For example, piezoelectric systems can convert motion from the human body into electrical power (Khaligh et al. 2010). As a result, one can harness energy from leg/arm motion and foot falls: piezoelectric materials can be embedded in shoes to recover "walking energy" (Shenck and Paradiso 2001). Additionally, miniature thermo-couples have been developed that convert body heat into electricity (Leonov et al. 2009). Another method to harvest energy is through the oxidation of blood sugars as demonstrated by researchers at Saint Louis University (Minteer et al. 2007). This approach can power implanted electronic devices (e.g., implanted biosensors for diabetics, implanted active RFID devices, etc.). In all of these cases, the harvested energy can power sensors and radios to form sensor networks. It is important to note that the energy/power levels harvested from these sources can be intermittent and quite small (on the order of microwatts). Devices conforming to traditional WPAN/WLAN standards cannot function under these constraints. Therefore, completely new PHY/MAC and higher layer protocol designs will be needed to implement a network of energy harvesting sensors. Recently, the IEEE 802.15.4f task group has started working on developing guidelines for such low power sensor systems.

3.3 Recent Applications of Wireless Technology in Wearable Health Monitoring Systems

3.3.1 Human Applications

Published applications for wireless, wearable systems applied to human health monitoring are numerous. Many of these are noted in the survey papers cited in Sect. 3.1. This broad range of monitoring applications, coupled with the fact that pervasive resources are a relatively new addition to the biomedical monitoring toolset, has led to the development of a diverse set of custom, proof-of-principle systems, as noted in Sect. 3.1.1. Table 3.2 collates some of the more recent human health monitoring applications and populations for which wearable, wireless tech-nologies show promise.

Many wearable, wireless systems have been built to accommodate one or two primary types of biomedical sensors to address a niche application. The most popular sensor category by far involves applications of electrocardiography (Bouwstra et al. 2009; Cheng et al. 2008a, 2008b; Finlay et al. 2008; Fulford-Jones et al. 2004; Isais et al. 2003; Jovanov et al. 2005; Jun et al. 2005; Kyriacou et al. 2007; Munshi et al. 2008; Rashid et al. 2008; Shen et al. 2008; Wang et al. 2010; Yoo et al. 2010). The hardware for one of these sensors, an ActiS activity sensor, is depicted in Fig. 3.2 (Jovanov et al. 2005). This sensor node consists of a daughter card [Intelligent Signal Processing Module" (ISPM)] interfaced to an off-the-shelf Moteiv Telos wireless mote (Moteiv 2010). The ISPM contains two orthogonal dual-axis ADXL202 accelerometers, a biosignal amplifier for ECG/EMG signal conditioning,

Table 3.2 Health monitoring application areas targeted by wearable, wireless systems

Application/population	Research efforts
Cardiopulmonary assessment	Cho and Asada 2002; Kim et al. 2007; Mandal et al. 2009; Mendoza and Tran 2002; Wen et al. 2008
Chronic pain and pain prevention	Hu et al. 2010; Jones et al. 2008
Elder care	Atallah et al. 2008; Dinh et al. 2009; Hong et al. 2008; Stanford 2002
Emergency, rescue, and extreme environments	Chen et al. 2003; Montgomery et al. 2004
Epilepsy	Jones et al. 2008
Fitness and rehabilitation	Fenu and Steri 2009; Jovanov et al. 2005; Li 2009; Li and Zhang 2007; Melzi et al. 2009; Waluyo et al. 2009
General/multipurpose	Ashok and Agarwal 2003; Chen et al. 2008; Cho et al. 2009; Crowe et al. 2004; Durresi et al. 2008; Fulford-Jones et al. 2004; Galeottei et al. 2008; Gialelis et al. 2008; Haahr et al. 2008; Huang et al. 2009; Hughes et al. 2007; Hung et al. 2004; Isais et al. 2003; Jones et al. 2008; Jovanov et al. 2001; Jung et al. 2009; Kim et al. 2009; Kroc and Delic 2003; Kyriacou et al. 2007; Lee et al. 2007; Massot et al. 2009; Paradiso 2008; Park and Kang 2004; Strömmer et al. 2006; Sukor et al. 2008; van de Ven et al. 2009; Wang et al. 2009a, 2009b; Wu and Xiaoming 2007; Yao and Warren 2005; Yao et al. 2005; Zito et al. 2007a, 2007b
Implantable/ingestible sensors	Fereydouni_Forouzandeh et al. 2008; Furse 2009; Guo et al. 2009; Liang et al. 2005; Mackay 1961; Simons et al. 2006; Troyk and DeMichele 2003; Zierhofer and Hochmair 1990
Inter-vehicle communication and highway use	Durresi et al. 2007
Activity/movement/motor skill assessment	Bajcsy 2007; Barth et al. 2009; Bonato et al. 2003; Hedman et al. 2009; Moy et al. 2003; Purwar et al. 2007; Shaban et al. 2009; Thiruvengada et al. 2008; Zhang et al. 2008
Parkinsons/tremors	Patel et al. 2009
Sleep apnea	Lam et al. 2009
Stress level detection	Chatterjee and Somkuwar 2008

and an MSP430F1232 microcontroller for digital signal processing and control. The host mote offers a USB interface for programming/communication, a wireless ZigBee radio transceiver, open source software support, and on-board sensors for humidity, temperature, and light level. Using this design approach, a developer can take advantage of the mote's built-in wireless networking capabilities and focus their time on custom sensor issues.

Pulse oximeter sensors are also popular candidates for these wearable wireless systems, as they offer the ability to obtain heart rate, blood oxygen saturation, respiration rate, and other important hemodynamic and biometric parameters at low power and without the need for electrical contact with tissue (Haahr et al. 2008; Han et al. 2007; Jung et al. 2008; Kim and Lee 2008; Li and Warren 2010). A new pulse oximeter design intended for such purposes is illustrated in Fig. 3.3 (Li and Warren 2010).

Fig. 3.2 ActiS activity sensor, which incorporates an acceleration/ECG daughtercard interfaced to a Moteiv Telos wireless mote. Courtesy Dr. Emil Jovanov, University of Alabama at Huntsville, USA. Originally published in (Jovanov et al. 2005)

This reflectance-mode sensor is intended for use with wrist watches, head bands, helmets, and other wearable configurations where transmission-mode devices are not sensible. It is also intended to be a "surface integratable" element that can interface with handheld platforms such as cell phones, PDAs, etc. so that users have easy access to health sensors. The device is battery operated and rechargeable through a mini-USB port. It can store many hours of data and incorporates a ZigBee wireless link on a Jennic 5139 platform. PPG data from this unit are high fidelity: they are unfiltered, exhibit thousands of peak-to-valley digitization levels, are sampled at a high sample rate (240 Hz per channel), and have a high signal-to-noise ratio. The sensor's design helps in this regard, as it incorporates large-area photodiodes and an excitation/detection separation that preferentially accepts photons, which have traveled deeper into blood-perfused tissue. The insets in Fig. 3.3 illustrate the high quality, unfiltered PPG data that can be acquired at different body locations using this sensor.

3.3.2 Animal Applications

On a final note, it is worth mentioning that wearable, wireless technologies for health monitoring in humans demonstrate potential in the animal domain. In fact, the human and animal domains exhibit tremendous similarity in terms of design challenges at the sensor and system levels. In the animal domain, the primary monitoring themes are animal tracking, traceability (the location of an animal over its life span), and state of health. Wearable, wireless animal tracking technologies speak primarily to resource management in poultry (Chansud et al. 2008; Oswald 2010) and other livestock (Laursen 2006; Salman 2003). The most

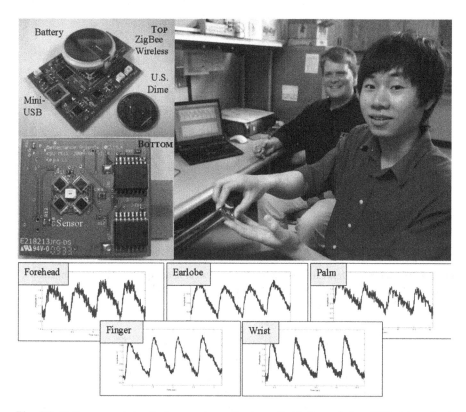

Fig. 3.3 Battery-operated reflectance pulse oximeter with ZigBee wireless/USB data upload capabilities, on-board data storage, a large-area sensor, and a filter-free design. The insets depict unfiltered PPG data obtained at different body locations. Courtesy Kejia Li, Kansas State University, Manhattan, KS, USA

common tracking and traceability approaches involve the use of RFID technologies (GrowSafe 2006; Lehpamer 2007; McAllister et al. 2000; Pereira et al. 2008; Pongpaibool 2008; Ting et al. 2007; Tuttle 1997; Ze et al. 2008). In recent years, more sophisticated systems have targeted health assessment through collections of various sensors. For example, the GrowSafe system uses low-range RF technology to ascertain feeding frequency and overall time spent at the feed bunk; parameters that can be correlated to animal health (GrowSafe 2006). Other systems use small collections of sensors to determine parameters such as animal lying time, heat stress, GPS location, and estrus (Darr and Epperson 2009; HQ 2006; Lefcourt et al. 2009; Moen et al. 1996; Redden et al. 1993; Rodgers and Rempel 1996).

Some recent systems purport to combine wireless networking technologies and wearable health monitoring sensors into more general-purpose systems. One of the first systems to be designed in such a manner is illustrated in Fig. 3.4 (Nagl et al. 2003; Smith et al. 2006; Warren et al. 2004a, 2004b, 2005a, 2005b, 2008a, 2008b). The overall embedded system design is based on the premise that each animal in a

Fig. 3.4 Wearable, wireless cattle health monitoring system that acquires core body temperature, heart rate, three acceleration axes, ambient temperature, ambient humidity, and global position (Nagl 2004; Smith et al. 2006; Warren et al. 2008a, 2008b)

herd will host a multi-parameter, wearable health monitoring device. In its current implementation, the monitoring equipment illustrated in Fig. 3.4 will periodically record an animal's heart rate (through an adapted horse heart monitor from Polar (Polar 2006)), core body temperature (ingestible pill from HQ Inc.; HQ 2006), head motion (three axes of acceleration), and absolute position [via the global positioning system (GPS)], as well as the ambient temperature and humidity of the surrounding environment. The system is designed to buffer these data for several days if needed. When an animal wanders within range of a wireless, ZigBee-enabled receiver (i.e., a base station), health data stored on the animal are uploaded for analysis and storage. Base stations can be placed in animal congregation areas, such as water troughs, feed bunks, and shelters. Given the weather and terrain challenges associated with placing wearable sensors on the exterior of an animal, KSU investigators have begun to focus their attention on ingestible pills that offer potential to acquire health parameters such as core temperature, heart rate, rumen pH, and respiration rate (Martinez 2007; Martinez et al. 2006; Warren et al. 2008a, 2008b). Because RF attenuation is significant in tissue, these pills will require the use of new long-range inductive links that can transmit data several feet through tissue from the inside of an animal to an external receiver (see Sect. 3.2.1) (Hoskins et al. 2009).

Acknowledgments The authors acknowledge Dr. Emil Jovanov, The University of Alabama in Huntsville, USA and Kejia Li, Kansas State University, USA for their contributions to the figures used in this chapter and the associated text.

References

Agrawal P, Teck TS, Ananda AL (2003) A lightweight protocol for wireless sensor networks. IEEE Wireless Communications and Networking (WCNC 2003) New Orleans, LA, 20 March, pp 1280–1285

Ashok RL, Agarwal DP (2003) Next generation wearable networks. IEEE Comput 36:31–39

Atallah L, Lo B, Yang GZ, Siegemund F (2008) Wirelessly accessible sensor populations (WASP) for elderly care monitoring. In: Second international conference on pervasive computing technologies for healthcare (PerrvasiveHealth 2008), Tampere, 30 January to 1 February 2008, pp 2–7

Bajcsy R (2007) Distributed wireless sensors on the human body. In: 7th IEEE international conference on bioinformatics and bioengineering (BIBE 2007), Boston, MA, 14–17 October 2007, pp 1448–1448

Barth AT, Hanson MA, Powell HC, Lach J (2009) TEMPO 3.1: a body area sensor network platform for continuous movement assessment. In: Sixth international workshop on wearable and implantable body sensor networks (BSN 2009), Berkeley, CA, 3–5 June 2009, pp 71–76

BluetoothSIG. (2010) Bluetooth special interest group. http://www.bluetooth.org/

Bonato P (2009) Clinical applications of wearable technology. In: 31st annual international conference of the IEEE engineering in medicine and biology society (EMBC 2009), Minneapolis, MN, 3–6 September 2009, pp 6580–6583

Bonato P, Mork PJ, Sherrill DM, Westgaard RH (2003) Data mining of motor patterns recorded with wearable technology. IEEE Eng Med Biol Mag 22:110–119

Bouwstra S, Feijs L, Chen Wei, Oetomo SB (2009) Smart jacket design for neonatal monitoring with wearable sensors. In: 6th international workshop on wearable and implantable body sensor networks (BSN 2009), Berkeley, CA, 3–5 June 2009, pp 162–167

Bunszel Chris (2001) Magnetic induction: a low-power wireless alternative. RF Des Mag 78–80

Chansud W, Wisanmongkol J, Ketprom U (2008) RFID for poultry traceability system at animal checkpoint. In: 5th international conference on electrical engineering/electronics, computer, telecommunications and information technology (ECTI-CON 2008), Krabi, 14–17 May 2008, pp 753–756

Chatterjee G, Somkuwar A (2008) Accident avoidance of fighter pilots by stress monitoring with wireless sensors in BAN. In: 3rd international conference on sensing technology (ICST 2008), Tainan, 30 November to 3 December 2008, pp 329–335

Cheng MH, Chen LC, Hung YC, Chen CN, Yang CM, Yang TL (2008a) A vital wearing system with wireless capability. In: Second international conference on pervasive computing technologies for healthcare (PervasiveHealth 2008), Tampere, Finland, 30 January – 1 February 2008, pp 268–271

Cheng MH, Chen LC, Hung YC, Yang CM, Yang TL (2008b) A real-time heart-rate estimator from steel textile ECG sensors in a wireless vital wearing system. In: 2nd international conference on bioinformatics and biomedical engineering (ICBBE 2008), Shanghai, 16–18 May 2008, pp 1339–1342

Chen SC, Chen HW, Lee A, Chao KH, Huang YC, Lai F (2003) E-vanguard for emergency – a wireless system for rescue and healthcare. In: 5th international workshop on enterprise networking and computing in healthcare industry (Healthcom 2003), Santa Monica, CA, 6–7 June 2003, pp 29–35

Chen Z, Chao H, Liao J, Liu S (2008) Protocol architecture for wireless body area network based on nrf24l01. In: IEEE international conference on automation and logistics (ICAL 2008), Qingdao, 1–3 September 2008, pp 3050–3054

Chhetri AS, Morrell D, Papandreou-Suppappola A (2005) Energy efficient target tracking in a sensor network using non-myopic sensor scheduling. In: 8th international conference on information fusion (Fusion 2005), Philadelphia, PA, 25–28 July 2005, p 8

Chirwa LC, Hammond PA, Roy S, Cumming DRS (2003) Electromagnetic radiation from ingested sources in the human intestine between 150 MHz and 1.2 GHz. IEEE Trans Biomed Eng 50:484–492

Cho KJ, Asada HH (2002) Wireless, battery-less stethoscope for wearable health monitoring. In: 28th annual northeast bioengineering conference, Philadelphia, PA, 20–21 April 2002, pp 187–188

Cho N, Yoo J, Song SJ, Lee J, Jeon S, Yoo HJ (2007) The human body characteristics as a signal transmission medium for intrabody communication. IEEE Trans Microw Theory Tech 55:1080–1086

Cho N, Bae J, Kim S, Yoo HJ (2009) A 10.8 mW body-channel-communication/MICS dual-band transceiver for a unified body-sensor-network controller. In: IEEE international solid-state circuits conference - digest of technical papers (ISSCC 2009), San Francisco, CA, 8–12 February 2009, pp 424–425,425a

Corcoran P, Iancu C, Callaly F, Cucos A (2007) Biometric access control for digital media streams in home networks. IEEE Trans Consum Electron 53:917–925

Crk I, Albinali F, Gniady C, Hartman J (2009) Understanding energy consumption of sensor enabled applications on mobile phones. In: 31st annual international conference of the ieee engineering in medicine and biology society (EMBC 2009), Minneapolis, MN, 3–6 September 2009, pp 6885–6888

Crossbow (2010) Crossbow wireless sensor networks. http://www.xbow.com/

Crowe J, Hayes-Gill B, Sumner M, Barratt C, Palethorpe B, Greenhalgh C, Storz O, Friday A, Humble J, Setchell C, Randell C, Muller HL (2004) Modular sensor architecture for unobtrusive routine clinical diagnosis. In: 24th international conference on distributed computing systems workshops, Hachioji, Tokyo, 23–24 March 2004, pp 451–454

Dağtaş S, Pekhteryev G, Şahinoğlu Z, Çam H, Challa N (2008) Real-time and secure wireless health monitoring. Int J Telemed Appl 2008

Daradimos I, Papadopoulos K, Stavrakas I, Kaitsa M, Kontogiannis T, Triantis D (2007) A physical access control system that utilizes existing networking and computer infrastructure. In: The international conference on computer as a tool (EUROCON 2007), Warsaw, 9–12 September 2007, pp 501–504

Darr M, Epperson W (2009). Embedded sensor technology for real time determination of animal lying time. Comput Electron Agric 66

DHHS (2010) Health information privacy, vol. 2010. U.S. Department of Health & Human Services, Washington, DC. http://www.hhs.gov/ocr/privacy/

Dinh A, Teng D, Li C, Shi Y, McCrosky C, Basran J, Del Bello-Hass V (2009) Implementation of a physical activity monitoring system for the elderly people with built-in vital sign and fall detection. In: Sixth international conference on information technology: new generations (ITNG 2009), Las Vegas, NV, 27–29 April 2009, pp 1226–1231

Durresi A, Paruchuri V, Kannan R, Iyengar SS (2005) Data integrity protocol for sensor networks. Int J Distrib Sens Netw 1:205–214

Durresi A, Durresi M, Barolli L (2007) Wireless communications for health monitoring on highways. In: 27th international conference on distributed computing systems workshops (ICDCSW 2007), Toronto, ON, 22–29 June 2007, p 37

Durresi A, Durresi M, Merkoci A, Barolli L (2008) Networked biomedical system for ubiquitous health monitoring. Mob Inf Syst 4:211–218

Enz CC, Baborowski J, Chabloz J, Kucera M, Muller C, Ruffieux D, Scolari N (2007) Ultra low-power MEMS-based radio for wireless sensor networks. In: 18th European conference on circuit theory and design (ECCTD 2007), Seville, 27–30 August 2007, pp 320–331

Fenu G, Steri G (2009) Two methods for body parameter analysis using body sensor networks. In: International conference on ultra modern telecommunications & workshops (ICUMT 2009), St. Petersburg, 12–14 October 2009, pp 1–5

Fereydouni_Forouzandeh F, Mohamed OA, Sawan M (2008) Ultra low energy communication protocol for implantable body sensor networks. In: Joint 6th international IEEE northeast workshop on circuits and systems and TAISA conference (NEWCAS-TAISA 2008), Montreal, QC, 22–25 June 2008, pp 57–60

Finlay DD, Nugent CD, Donnelly MP, McCullagh PJ, Black ND (2008) Optimal electrocardiographic lead systems: practical scenarios in smart clothing and wearable health systems. IEEE Trans Inf Technol Biomed 12:433–441

Fujii K, Takahashi M, Ito K, Inagaki N (2006) Study on the electric field distributions around whole body model with a wearable device using the human body as a transmission channel. In: First European conference on antennas and propagation (EuCAP 2006), Nice, 6–10 November 2006, pp 1–4

Fujii K, Takahashi M, Ito K (2007) Electric field distributions of wearable devices using the human body as a transmission channel. IEEE Trans Antennas Propag 55:2080–2087

Fulford-Jones TRF, Wei GY, Welsh M (2004) A portable, low-power, wireless two-lead ekg system. In: 26th annual international conference of the IEEE engineering in medicine and biology society (EMBC 2004), San Francisco, CA, 1–5 September 2004, pp 2141–2144

Furse CM (2009) Biomedical telemetry: today's opportunities and challenges. In: IEEE international workshop on antenna technology (iWAT 2009), Santa Monica, CA, 2–4 March 2009, pp 1–4

Galeottei L, Paoletti M, Marchesi C (2008) Development of a low cost wearable prototype for long-term vital signs monitoring based on embedded integrated wireless module. Comput Cardiol: 905–908

Gao YM, Pun SH, Du M, Mak PU, Vai MI (2009a) Simple electrical model and initial experiments for intra-body communications. In: 31st annual international conference of the IEEE engineering in medicine and biology society (EMBC 2009), Minneapolis, MN, 3–6 September 2009, pp 697–700

Gao YM, Pun SH, Min D, Vai MI, Mak PU (2009b) Quasi-static field modeling and validation for intra-body communication. In: 3rd international conference on bioinformatics and biomedical engineering (ICBBE 2009), Beijing, 11–13 June 2009, pp 1–4

Gao YM, Pun SH, Mak PU, Min D, Vai Mang I (2009c) A multilayer cylindrical volume conductor model for galvanic coupling intra-body communication. In: 7th international conference on information, communications and signal processing (ICICS 2009), Macau, 8–10 December 2009, pp 1–4

Gialelis J, Foundas P, Kalogeras A, Georgoudakis M, Kinalis A, Koubias S (2008) Wireless wearable body area network supporting person centric health monitoring. In: IEEE international workshop on factory communication systems (WFCS 2008) Dresden, 21–23 May 2008, pp 77–80

Goldstein H (2006) A touch too much [intrabody communications]. IEEE Spectr 43:24–25

GrowSafe (2006) GrowSafe Systems Ltd, vol. 2006. http://www.growsafe.com

Guo L, Wang B, Pan X (2009) The real-time wireless infrastructure for family medical care base on wearable technology. In: International conference on future biomedical information engineering (FBIE 2009), Sanya, 13–14 December 2009, pp 343–345

Gupta V, Chung T, Hassibi B, Murray RM (2004) Sensor scheduling algorithms requiring limited computation [vehicle sonar range-finder example]. In: IEEE international conference on acoustics, speech, and signal processing (ICASSP 2004), 17–21 May 2004, pp 825–828

Gu YY, Zhang Y, Zhang YT (2003) A novel biometric approach in human verification by photoplethysmographic signals. In: 4th international IEEE EMBS special topic conference on information technology applications in biomedicine, Birmingham, 24–26 April 2003, pp 13–14

Haahr RG, Duun S, Thomsen EV, Hoppe K, Branebjerg J (2008) A wearable electronic patch for wireless continuous monitoring of chronically diseased patients. In: 5th international summer school and symposium on medical devices and biosensors (ISSS-MDBS 2008), Hong Kong, 1–3 June 2008, pp 66–70

Hachisuka K, Nakata A, Takeda T, Terauchi Y, Shiba K, Sasaki K, Hosaka H, Itao K (2003) Development and performance analysis of an intra-body communication device. In: 12th international conference onsolid-state sensors, actuators and microsystems (Transducers 2003), Boston, MA, 8–12 June 2003, pp 1722–1725

Hachisuka K, Terauchi Y, Kishi Y, Hirota T, Sasaki K, Hosaka H, Ito K (2005) Simplified circuit modeling and fabrication of intrabody communication devices. In: 13th international conference on solid-state sensors, actuators and microsystems (Transducers 2005), Seoul, 5–9 June 2005, pp 461–464

Han H, Kim MJ, Kim J (2007) Development of real-time motion artifact reduction algorithm for a wearable photoplethysmography. In: 29th annual international conference of the IEEE engineering in medicine and biology society (EMBS 2007), Lyon, 22–26 August 2007, pp 1538–1541

Hedman E, Wilder-Smith O, Goodwin MS, Poh MZ, Fletcher R, Picard R (2009) icalm: measuring electrodermal activity in almost any setting. In: 3rd international conference on affective computing and intelligent interaction and workshops (ACII 2009), Amsterdam, 10–12 September 2009, pp 1–2

Hong YJ, Kim IJ, Ahn SC, Kim HG (2008) Activity recognition using wearable sensors for elder care. In: Second international conference on future generation communication and networking (FGCN 2008), Hainan Island, 13–15 December 2008, pp 302–305

Hoskins S, Sobering T, Andresen D, Warren S (2009) Near-field wireless magnetic link for an ingestible cattle health monitoring pill. In: 31st annual international conference of the IEEE engineering in medicine and biology society (EMBC 2009), Minneapolis, MN, 3–6 September 2009, pp 5401–5404

HQ (2006) HQ Inc., formerly HTI Technologies, manufactures the CorTemp temperature pill used in detecting heat stress, vol. 2006. http://www.hqinc.net

Huang ML, Park SC (2009) A WLAN and zigbee coexistence mechanism for wearable health monitoring system. In: 9th international symposium on communications and information technology (ISCIT 2009) Icheon, Korea, 28–30 September 2009, pp 555–559

Huang YM, Hsieh MY, Chao HC, Hung SH, Park JH (2009) Pervasive, secure access to a hierarchical sensor-based healthcare monitoring architecture in wireless heterogeneous networks. IEEE J Sel Areas Commun 27:400–411

Hughes E, Masilela M, Eddings P, Raflq A, Boanca C, Merrell R (2007) vmote: a wearable wireless health monitoring system. In: 9th international conference on e-health networking, application and services, Taipei, pp 330–331

Hung K, Zhang YT, Tai B (2004) Wearable medical devices for tele-home healthcare. In: 26th annual international conference of the IEEE engineering in medicine and biology society (EMBC 2004), San Francisco, CA, 1–5 September 2004, pp 5384–5387

Hu Y, Stoelting A, Wang YT, Yi Z, Sarrafzadeh M (2010) Providing a cushion for wireless healthcare application development. IEEE Potentials 29:19–23

Hyoung CH, Sung JB, Hwang JH, Kim JK, Park DG, Kang SW (2006) A novel system for intrabody communication: touch-and-play. In: IEEE international symposium on circuits and systems (ISCAS 2006), Island of Kos, 21–24 May 2006, p 4

IEEE (2004) IEEE 802.15 WPAN Task Group 1 (TG1). IEEE. http://www.ieee802.org/15/pub/TG1.html

IEEE (2009) IEEE 802.15 WPAN Task Group 3 (TG3). IEEE. http://www.ieee802.org/15/pub/TG3.html

IEEE (2010a) IEEE 802.3 Ethernet Working Group. IEEE. http://www.ieee802.org/3/

IEEE (2010b) IEEE 802.15 WPAN (Task Group 4 (TG4). http://www.ieee802.org/15/pub/TG4.html

Intel (2010) Intel motes. http://techresearch.intel.com/articles/Exploratory/1503.htm

Isais R, Nguyen K, Perez G, Rubio R, Nazeran H (2003) A low-cost microcontroller-based wireless ECG-blood pressure telemonitor for home care. In: 25th annual international conference of the IEEE engineering in medicine and biology society (EMBC 2003), Cancun, 17–21 September 2003, pp 3157–3160

Istepanian RSH, Laxminaryan S (2000) UNWIRED, the next generation of wireless and internetable telemedicine systems - editorial paper. IEEE Trans Inf Technol Biomed 4:189–194

Istepanian RSH, Jovanov E, Zhang YT (2004) Guest editorial – introduction to the special section on M-health: beyond seamless mobility and global wireless health-care connectivity. IEEE Trans Inf Technol Biomed 8:405–414

Jones VM, Huis in't Veld R, Tonis T, Bults RB, van Beijnum B, Widya I, Vollenbroek-Hutten M, Hermens H (2008) Biosignal and context monitoring: distributed multimedia applications of body area networks in healthcare. In: 10th workshop on multimedia signal processing Cairns, Qld, 8–10 October 2008, pp 820–825

Jovanov E, Raskovic D, Price J, Chapman J, Moore A, Krishnamurthy A (2001) Patient monitoring using personal area networks of wireless intelligent sensors. Biomed Sci Instrum 37:373–378

Jovanov E, Milenkovic A, Otto C, de Groen PC (2005) A wireless body area network of intelligent motion sensors for computer assisted physical rehabilitation. J Neuroeng Rehabil 2:1–10

Jun D, Miao X, Hong-Hai Z, Wei-Feng L (2005) Wearable ECG recognition and monitor. In: 18th IEEE symposium on computer-based medical systems, Dublin, 23–24 June 2005, pp 413–418

Jung SJ, Lee YD, Seo YS, Chung WY (2008) Design of a low-power consumption wearable reflectance pulse oximetry for ubiquitous healthcare system. In: International conference on control, automation and systems (ICCAS 2008), Seoul, 14–17 October 2008, pp 526–529

Jung SJ, Kwon TH, Chung WY (2009) A new approach to design ambient sensor network for real time healthcare monitoring system. IEEE Sens J: 576–580

Kenington P (2005) RF and baseband techniques for software defined radio. Artech House Publishers, London, 978–1580537933

Khaligh A, Zeng P, Zheng C (2010) Kinetic energy harvesting using piezoelectric and electro-magnetic technologies - state of the art. IEEE Trans Ind Electron 57:850–860

Kim DS, Lee SY, Kim BS, Lee SC, Chung DH (2008) On the design of an embedded biometric smart card reader. IEEE Trans Consum Electron 54:573–577

Kim HC, Chung GS, Kim TW (2009) A framework for health management services in nanofiber technique-based wellness wear systems. In: 11th international conference on e-health networking, applications and services (Healthcom 2009), Sydney, NSW, 16–18 December 2009, pp 70–73

Kim KA, Lee IK, Choi SS, Kim SS, Lee TS, Cha EJ (2007) Wearable transducer to monitor respiration in a wireless way. In: 6th international special topic conference on information technology applications in biomedicine (ITAB 2007), Tokyo, 8–11 November 2007, pp 174–176

Kim Y, Lee J (2008) A wearable health context aware system. In: International conference on consumer electronics (ICCE 2008), Las Vegas, NV, 9–13 January 2008, pp 1–2

Kroc S, Delic V (2003) Personal wireless sensor network for mobile health care monitoring. In: 6th international conference on telecommunications in modern satellite, cable and broadcasting service (TELSIKS 2003), Serbia and Montenegro, NiS, 1–3 October 2003, pp 471–474

Kyriacou E, Pattichis C, Pattichis M, Jossif A, Paraskevas L, Konstantinides A, Vogiatzis D (2007) An m-health monitoring system for children with suspected arrhythmias. In: 29th annual international conference of the IEEE engineering in medicine and biology society (EMBC 2007), Lyon, 22–26 August 2007, pp 1794–1797

Lam SCK, Wong KL, Wong KO, Wong W, Mow WH (2009) A smartphone-centric platform for personal health monitoring using wireless wearable biosensors. In: 7th international conference on information, communications and signal processing (ICICS 2009), Macau, 8–10 December 2009, pp 1–7

Laursen W (2006) Managing the mega flock (individual animal management systems). IEE Rev 52:38–42

Lee TS, Hong JH, Cho MC (2007) Biomedical digital assistant for ubiquitous healthcare. In: 29th annual international conference of the ieee engineering in medicine and biology society (EMBC 2007), Lyon, 22–26 August 2007, pp 1790–1793

Lefcourt AM, Erez B, Varner MA, Barfield R, Tasch U (2009) A noninvasive rediotelemetry system to monitor heart rate for assessing stress responses of bovines. J Dairy Sci 82:1179–1187

Lehpamer H (2007) RFID design principles. Artech House Publishers, Boston, 978–1596931947

Leonov V, Van Hoof C, Vullers RJM (2009) Thermoelectric and hybrid generators in wearable devices and clothes. In: Sixth international workshop on wearable and implantable body sensor networks (BSN 2009), Berkeley, CA, 3–5 June 2009, pp 195–200

Lewis Carol (2001) Emerging trends in medical device technology: home is where the heart monitor is. FDA Consum Mag 35:10–14

Liang CK, Chen JJJ, Chung CL, Cheng CL, Wang CC (2005) An implantable bi-directional wireless transmission system for transcutaneous biological signal recording. Physiol Meas 26:83–97

Li F, Li Y, Zhao W, Chen Q, Tang W (2006) An adaptive coordinated MAC protocol based on dynamic power management for wireless sensor networks. In: International conference on wireless communications and mobile computing (IWCMC 2006), Vancouver, BC, pp 1073–1078

Li K, Warren S (2010) A high-performance wireless reflectance pulse oximeter for photoplethysmogram acquisition and analysis in the classroom. In: Annual conference and exposition, American Society for Engineering Education, Louisville, KY, 20–23 June 2010

Li Z (2009) Exercises intensity estimation based on the physical activities healthcare system. In: WRI international conference on communications and mobile computing (CMC 2009), Yunnan, 6–8 January 2009, pp 132–136

Li Z, Zhang G (2007) A physical activities healthcare system based on wireless sensing technology. In: 13th IEEE international conference on embedded and real-time computing systems and applications (RTCSA 2007), Daegu, 21–24 August 2007, pp 369–376

Lymberis A (2003a) Smart wearable systems for personalised health management: current R&D and future challenges. In: 25th annual international conference of the IEEE engineering in medicine and biology society, Cancun, pp 3716–3719

Lymberis A (2003b) Smart wearables for remote health monitoring, from prevention to rehabilitation: current R&D, future challenges. In: 4th international IEEE embs special topic conference on information technology applications in biomedicine (ITAB 2003), Birmingham, 24–26 April 2003, pp 272–275

Mackay RS (1961) Radio telemetering from within the body. Science 134:1196–1202

Mandal S, Turicchia L, Sarpeshkar R (2009) A battery-free tag for wireless monitoring of heart sounds. In: Sixth international workshop on wearable and implantable body sensor networks (BSN 2009), Berkeley, CA, 3–5 June 2009, pp 201–206

Martinez A (2007) Acquisition of heart rate and core body temperature in cattle using ingestible sensors. Electrical & Computer Engineering. Kansas State University, Manhattan, KS, 77

Martinez A, Schoenig S, Andresen D, Warren S (2006) Ingestible pill for heart rate and core temperature measurement in cattle. In: 28th Annual Conference of the IEEE EMBS, Marriott Times Square, New York, NY, 30 August to 3 September 2006, pp 3190–3193

Massot B, Gehin C, Nocua R, Dittmar A, McAdams E (2009) A wearable, low-power, health-monitoring instrumentation based on a programmable system-on-chip™. In: 31st annual international conference of the ieee engineering in medicine and biology society (EMBC 2009), Minneapolis, MN, 3–6 September 2009, pp 4852–4855

McAllister TA et al. (2000) Electronic identification: applications in beef production and research. Can J Anim Sci 80:48–49

Meijer GAL, Westerterp KR, Verhoeven FMH, Koper HBM, ten Hoor F (1991) Methods to assess physical activity with special reference to motion sensors and accelerometers. IEEE Trans Biomed Eng 38:221–229

Melzi S, Borsani L, Cesana M (2009) The virtual trainer: supervising movements through a wearable wireless sensor network. In: 6th annual IEEE communications society conference onsensor, mesh and ad hoc communications and networks workshops (SECON Workshops 2009), Rome, 22–26 June 2009, pp 1–3

Mendoza GG, Tran BQ (2002) In-home wireless monitoring of physiological data for heart failure patients. In: 24th annual conference of the IEEE engineering in medicine and biology society (EMBS 2002), Houston, TX, pp 1849–1850

Meraki (2009) HIPAA compliance for the wireless LAN. Meraki, Inc., San Francisco, CA. http:// meraki.com/library/collateral/white_paper/meraki_white_paper_HIPAA.pdf

Microchip (1998) microID (125 kHz RFID system design guide. Microchip Technology, Inc., Chandler, AZ, http://ww1.microchip.com/downloads/en/devicedoc/51115f.pdf

Minteer SD, Liaw BY, Cooney MJ (2007) Enzyme-based biofuel cells. Curr Opin Biotechnol 8:228–234

Moen R, Pastor J, Cohen Y (1996) Interpreting behavior from activity counter in GPS collars on moose. Alces 80:101–108

Montgomery K, Mundt C, Thonier G, Tellier A, Udoh U, Barker V, Ricks R, Giovangrandi L, Davies P, Cagle Y, Swain J, Hines J, Kovacs G (2004) Lifeguard – a personal physiological monitor for extreme environments. In: 26th annual international conference of the IEEE engineering in medicine and biology society (EMBC 2004), San Francisco, CA, 1–5 September 2004, pp 2192–2195

Moteiv (2010) Moteiv. http://www.moteiv.com

Moy ML, Mentzer SJ, Reilly JJ (2003) Ambulatory monitoring of cumulative free-living activity. IEEE Eng Med Biol Mag 22:89–95

Munshi MC, Xiaoyuan X, Zou Xiaodan, Soetiono E, Teo CS, Lian Y (2008) Wireless ECG plaster for body sensor network. In: 5th international summer school and symposium on medical devices and biosensors (ISSS-MDBS 2008), Hong Kong, 1–3 June 2008, pp 310–313

Nagl LJ (2004) The design and implementation of a cattle health monitoring system. *Electrial and Computer Engineering*. Manhattan, KS, Kansas State University, 113

Nagl L, Schmitz R, Warren S, Hildreth TS, Erickson H, Andresen D (2003) Wearable sensor system for wireless state-of-health determination in cattle. In: 25th annual international conference of the IEEE engineering in medicine and biology society (EMBC 2003), Cancun, 17–21 September 2003, pp 3012–3015

Nakajima N (2009) Short-range wireless network and wearable bio-sensors for healthcare applications. In: 2nd international symposium on applied sciences in biomedical and communication technologies (ISABEL 2009), Bratislava, Slovakia, 24–27 November 2009, pp 1–6

Nguyen K, Ji Y (2008) LCO-MAC: a low latency, low control overhead MAC protocol for wireless sensor networks. In: 4th international conference on mobile ad-hoc and sensor networks (MSN 2008), Wuhan, 10–12 December 2008, pp 271–275

Nordic (2010) nRF24L01 Single Chip 2.4GHz Transceiver. Nordic Semiconductor. http://www.nordicsemi.com/files/Product/data_sheet/nRF24L01_Product_Specification_v2_0.pdf

O'Dorisio D (2003) Securing wireless networks for HIPAA compliance. The SANS Institute 8120 Woodmont Avenue, Suite 205 Bethesda, Maryland 20814. http://www.sans.org/reading_room/whitepapers/awareness/securing-wireless-networks-hipaa-compliance_1335

Ooi P, Culjak G, Lawrence E (2005) Wireless and wearable overview: stages of growth theory in medical technology applications. In: International conference on mobile business (ICMB 2005), Sydney, 11–13 July 2005, pp 528–536

Oswald T (2010) MSU studies use of wireless sensors to monitor chicken well-being. Michigan State University, East Lansing, MI. http://news.msu.edu/story/7397/

Otis BP, Chee YH, Lu R, Pletcher NM, Rabaey JM (2004) An ultra-low power MEMS-based two-channel transceiver for wireless sensor networks. In: 2004 symposium on VLSI circuits, Honolulu, HI, pp 20–23

Pantelopoulos A, Bourbakis NG (2010) A survey on wearable sensor-based systems for health monitoring and prognosis. IEEE Trans Syst Man Cybern C Appl Rev 40:1–12

Papastergiou KD, Macpherson DE (2008) Air-gap effects in inductive energy transfer. In: IEEE power electronics specialists conference (PESC 2008), Rhodes, pp 4092–4097

Paradiso J (2008) On-body wireless sensing for human-computer interfaces. In: 5th international summer school and symposium on medical devices and biosensors (ISSS-MDBS 2008), Hong Kong, 9 December 2008, pp 7–7

Park DG, Kang SW (2004) Development of reusable and expandable communication for wearable medical sensor network. In: 26th annual international conference of the IEEE engineering in medicine and biology society (EMBC 2004), San Francisco, CA, 1–5 September 2004, pp 5380–5383

Patel S, Lorincz K, Hughes R, Huggins N, Growden J, Standaert D, Akay M, Dy J, Welsh M, Bonato P (2009) Monitoring motor fluctuations in patients with Parkinson's disease using wearable sensors. *IEEE Trans Inf Technol Biomed* 13:864–873

Perakslis C, Wolk R (2005) Social acceptance of RFID as a biometric security method. In: International symposium on technology and society. weapons and wires: prevention and safety in a time of fear (ISTAS 2005), Los Angeles, CA, 8–10 June 2005, pp 79–87

Pereira DP, Dias W, Braga M, Barreto R, Figueiredo CMS, Brilhante V (2008) Model to integration of RFID into wireless sensor network for tracking and monitoring animals. In: 11th IEEE international conference on computational science and engineering (CSE 2008), Sao Paulo, 16–18 July 2008, pp 125–131

Polar (2006) Polar horse heart rate monitor S610i, vol. 2006. http://www.heartmonitors.com/horse/polar_horse_s610.htm

Pongpaibool P (2008) A study on performance of UHF RFID tags in a package for animal traceability application. In: 5th international conference on electrical engineering/electronics, computer, telecommunications and information technology (ECTI-CON 2008), Krabi, 14–17 May 2008, pp 741–744

Purwar A, Do DJ, Chung WY (2007) Activity monitoring from real-time triaxial accelerometer data using sensor network. In: International conference on control, automation and systems (ICCAS 2007), Seoul, 17–20 October 2007, pp 2402–2406

Raciti M, Pisani P, Emdin M, Carpeggiani C, Ruschi S, Kraft G, Francesconi R, Membretti G, Marchesi C (1994) Circadian dynamics of respiratory parameters from ambulatory monitoring. Comput Cardiol:581–584

Rappaport T (2002) Wireless communications: principles and practice. Prentice Hall, Upper Saddle River, NJ, 0130422320

Rashid RA, Rahim MRA, Sarijari MA, Mahalin N (2008) Design and implementation of wireless biomedical sensor networks for ECG home health monitoring. In: International conference on electronic design (ICED 2008), Penang, 1–3 December 2008, pp 1–4

Redden KD, Kennedy AD, Ingalls JR, Gilson TL (1993) Detection of estrus by radiotelemetric monitoring of vaginal and ear skin temperature and pedometer measurements of activity. J Dairy Sci 76:713–721

Reed JH (2005) An introduction to ultra wideband communication systems. Prentice Hall, Upper Saddle River, NJ, 978–0131481039

Rodgers A, Rempel RS, KF Abraham (1996) A GPS-based telemetry system. Wildl Soc Bull 24:559–566

Ruiz JA, Shimamoto S (2005) A study on the transmission characteristics of the human body towards broadband intra-body communications. In: Ninth international symposium on consumer electronics (ISCE 2005), Macau, 14–16 June 2005, pp 99–104

Sagahyroon A, Rady H, Ghazy A, Suleman U (2008) A wireless healthcare monitoring platform. In: international conference on innovations in information technology (IIT 2008), Al Ain, Abu Dhabi, 16–18 December 2008, pp 126–129

Salman MD (2003) Animal disease surveillance and survey systems: methods and applications. Wiley-Blackwell, New York, 978–0813810317

Sasaki A, Shinagawa M, Ochiai K (2004) Sensitive and stable electro-optic sensor for intrabody communication. In 17th annual meeting of the IEEE lasers and electro-optics society (LEOS 2004), Puerto Rico, 7–11 November 2004, pp 122–123

Sasaki A, Shinagawa M, Ochiai K (2009) Principles and demonstration of intrabody communication with a sensitive electrooptic sensor. IEEE Trans Instrum Meas 58:457–466

Sasamori T, Takahashi M, Uno T (2009) Transmission mechanism of wearable device for on-body wireless communications. IEEE Trans Antennas Propag 57:936–942

Shaban H, El-Nasr MA, Buehrer RM (2009) A framework for the power consumption and BER performance of ultra-low power wireless wearable healthcare and human locomotion tracking systems via UWB Radios. In: IEEE international symposium on signal processing and information technology (ISSPIT 2009), Ajman, 14–17 December 2009, pp 322–327

Sharma PK, Guha SK (1975). Transmission of time-varying magnetic field through body tissue. *J Biol Phys* vol. 3

Shenck NS, Paradiso JA (2001) Energy scavenging with shoe-mounted piezoelectrics. IEEE Micro 21:30–42

Shen TW, Hsiao T, Liu YT, He TY (2008) An ear-lead ECG based smart sensor system with voice biofeedback for daily activity monitoring. In: IEEE region 10 conference (TENCON 2008), Hyderabad, 19–21 November 2008, pp 1–6

Shinagawa M, Fukumoto M, Ochiai K, Kyuragi H (2004) A near-field-sensing transceiver for intrabody communication based on the electrooptic effect. IEEE Trans Instrum Meas 53:1533–1538

Shu-Di B, Yuan-Ting Z, Lian-Feng S (2005) Physiological signal based entity authentication for body area sensor networks and mobile healthcare systems. In: 27th annual international conference of the engineering in medicine and biology society (EMBC 2005), Shanghai, 1–4 September 2005, pp 2455–2458

Simons RN, Miranda FA (2006) Modeling of the near field coupling between an external loop and an implantable spiral chip antennas in biosensor systems. In: Joint international symposium sponsored by IEEE AP-S, USNC/URSI, and AMEREM, Albuquerque, NM, 9–14 July 2006

Simons RN, Miranda FA, Wilson JD, Simons RE (2006) Wearable wireless telemetry system for implantable bio-MEMS sensors. In: 28th international conference of the IEEE engineering in medicine and biology society (EMBC 2006), New York, 30 August to 3 September 2006, pp 6245–6248

Sklavos N, Zhang X (2007) Wireless security and cryptography: specifications and implementations. CRC Press, Boca Raton, FL, 9780849387715

Smith K, Martinez A, Craddolph R, Erickson H, Andresen D, Warren S (2006) An integrated cattle health monitoring system. In: 28th annual international conference of the IEEE engineering in medicine and biology society (EMBC 2006), New York, 30 August to 3 September 2006, pp 4659–4662

Sriram J, Shin M, Choudhury T, Kotz D (2009) Activity-aware ECG-based patient authentication for remote health monitoring. In: International conference on multimodal interfaces (ICMI-MLMI 2009), Cambridge, MA, 2–4 November 2009, pp 279–304

Stanford V (2002) Using pervasive computing to deliver elder care. IEEE Pervasive Comput 1:10–13

Stavroulakis P, Stamp M (2010) Handbook of information and communication security. Springer, New York, 978–3642041167

Strömmer E, Kaartinen J, Pärkkä J, Ylisaukko-oja A, Korhonen I (2006) Application of near field communication for health monitoring in daily life. In: 28th international conference of the IEEE engineering in medicine and biology society (EMBC 2006), New York, 30 August to 3 September 2006, pp 3246–3249

Sukor M, Ariffin S, Fisal N, Yusof SKS, Abdallah A (2008) Performance study of wireless body area network in medical environment. In: Second Asia international conference on modeling & simulation (AICMS 08) Kuala Lumpur, 13–15 May 2008, pp 202–206

Sun M, Hackworth SA, Tang Z, Gilbert G, Cardin S, Sclabassi RJ (2007) How to pass information and deliver energy to a network of implantable devices within the human body. In: 29th annual international conference of the IEEE engineering in medicine and biology society (EMBC 2007), Lyon, 22–26 August 2007, 5286–5289

Swain P (2003) Wireless capsule endoscopy. Gut 52:48–50

Teng XF, Zhang YT, Poon CC, Bonato P (2008) Wearable medical systems for p-health. IEEE Rev Biomed Eng vol. 1:pp 62–74

TexasInstruments (2010) eZ430-chronos wireless watch development tool (915 MHz US Version). http://www.ti-estore.com/merchant2/merchant.mvc?Screen=PROD&Product_Code==EZ430-Chronos-915

Thakor NV (1984) From Holter monitors to automatic defibrillators: developments in ambulatory arrhythmia monitoring. IEEE Trans Biomed Eng BME-31:770–778

Thepvilojanapong N, Motegi S, Idoue A, Horiuchi H (2008) Resource allocation for coexisting zigbee-based personal area networks. In: Seventh international conference on networking (ICN 2008), Cancun, 13–18 April 2008, pp 36–45

Thiruvengada H, Srinivasan S, Gacic A (2008) Design and implementation of an automated human activity monitoring application for wearable devices. In: IEEE international conference on systems, man and cybernetics (SMC 2008), Singapore, 12–15 October 2008, pp 2252–2258

Ting JSL, Kwok SK, Lee WB, Tsang AHC, Cheung BCF (2007) A dynamic RFID-based mobile monitoring system in animal care management over a wireless network. In: International conference on wireless communications, networking and mobile computing (WiCom 2007), Shanghai, 21–25 September 2007, pp 2085–2088

Troyk PR, DeMichele GA (2003) Inductively-coupled power and data link for neural prostheses using a class-E oscillator and FSK modulation. In: 25th annual international conference of the IEEE engineering in medicine and biology society (EMBC 2003), Cancun, 17–21 September 2003, pp 3376–3379

Tuttle JR (1997) Traditional and emerging technologies and applications in the radio frequency identification (RFID) industry. In: IEEE radio frequency integrated circuits (RFIC) symposium (IEEE), Denver, CO, 8–11 June 1997, pp 5–8

van de Ven P, Bourke A, Tavares C, Feld R, Nelson J, Rocha A, Laighin GO (2009) Integration of a Suite of Sensors in a Wireless Health Sensor Platform. IEEE Sens J: 1678–1683

Venkatasubramanian KK, Gupta SKS (2006) Security for pervasive health monitoring sensor applications. In: Fourth international conference on intelligent sensing and information processing (ICISIP 2006), Bangalore, 15–18 December 2006, pp 197–202

Venkatasubramanian KK, Venkatasubramanian AB, Gupta SKS (2008) EKG-based key agreement in body sensor networks. *INFOCOM workshops*, Phoenix, AZ, 13–18 April 2008, pp 1–6

Waluyo AB, Pek I, Yeoh WS, Kok TS, Chen X (2009) Footpaths: fusion of mobile outdoor personal advisor for walking route and health fitness. In: 31st annual international conference of the IEEE engineering in medicine and biology society (EMBC 2009), Minneapolis, MN, 3–6 September 2009, pp 5155–5158

Wang B, Wang L, Lin SJ, Wu D, Huang BY, Zhang YT, Yin Q, Chen W (2009a) A body sensor networks development platform for pervasive healthcare. In: 3rd international conference on bioinformatics and biomedical engineering (ICBBE 2009), Beijing, 11–13 June 2009, pp 1–4

Wang H, Peng D, Wang W, Sharif H, Chen H, Khoynezhad A (2010) Resource-aware secure ecg healthcare monitoring through body sensor networks. IEEE Wirel Commun 17:12–19

Wang Y, Li L, Bo W, Wang L (2009b) A body sensor network platform for in-home health monitoring application. In: 4th international conference on ubiquitous information technologies & applications (ICUT 2009), Fukuoka, 20–22 December 2009, pp 1–5

Wan Y, Sun X, Yao J (2007) Design of a photoplethysmographic sensor for biometric identification. In: International conference on control, automation and systems (ICCAS 2007), Seoul, 17–20 October 2007, pp 1897–1900

Warren S, Jovanov E (2006) The need for rules of engagement applied to wireless body area networks. In: 3rd IEEE consumer communications and networking conference (CCNC 2006), Las Vegas, NV, 8–10 January 2006, pp 979–983

Warren S, Yao Jianchu, R Schmitz, Lebak J (2004a) Reconfigurable point-of-care systems designed with interoperability standards. In: 26th annual international conference of the ieee engineering in medicine and biology society (EMBC 2004), San Francisco, CA, 1–5 September 2004, pp 3270–3273

Warren S, Andresen D, Nagl L, Schoenig S, Krishnamurthi B, Erickson H, Hildreth T, Poole D, Spire M (2004b) Wearable and wireless: distributed, sensor-based telemonitoring systems for state of health determination in cattle. In: 9th annual talbot informatics symposium, AVMA annual convention, Philadelphia Convention Center, Philadelphia, PA, 23–27 July 2004

Warren S, Lebak J, Yao Jianchu, J, Creekmore A Milenkovic, Jovanov E (2005a) Interoperability and security in wireless body area network infrastructures. In: 27th annual international conference of the IEEE engineering in medicine and biology society (IEEE-EMBS 2005), Shanghai, 1–4 September 2005, pp 3837–3840

Warren S, Nagl L, Schoenig S, Krishnamurthi B, Epp T, Erickson H, Poole D, Spire M, Andresen D (2005b) Veterinary telemedicine: wearable and wireless systems for cattle health assessment. In: 10th annual meeting of the american telemedicine association, Colorado Convention Center, Denver, CO, 17–20 April 2005

Warren S, Lebak J, Yao J (2006) Lessons learned from applying interoperability and information exchange standards to a wearable point-of-care system. In: 1st transdisciplinary conference on distributed diagnosis and home healthcare (D2H2 2006), Arlington, VA, 2–4 April 2006, pp 101–104

Warren S, Andresen D, Wilson D, Hoskins S (2008a) Embedded design considerations for a wearable cattle health monitoring system. In: International conference on embedded systems and applications (ESA 2008), Monte Carlo Resort, Las Vegas, NV, 14–17 July 2008

Warren S, Martinez A, Sobering T, Andresen D (2008b) Electrocardiographic pill for cattle heart rate determination. In: 30th annual conference of the IEEE EMBS, Vancouver Convention & Exhibition Centre, Vancouver, BC, 20–24 August 2008, pp 4852–4855

Wegmueller MS, Hediger M, Kaufmann T, Oberle M, Kuster N, Fichtner W (2007) Investigation on coupling strategies for wireless implant communications. In: IEEE instrumentation and measurement technology conference proceedings (IMTC 2007), Warsaw, 1–3 May 2007, pp 1–4

Wegmueller MS, Huclova S, Froehlich J, Oberle M, Felber N, Kuster N, Fichtner W (2009) Galvanic coupling enabling wireless implant communications. IEEE Trans Instrum Meas 58:2618–2625

Wegmueller MS, Oberle M, Felber N, Kuster N, Fichtner W (2010) Signal transmission by galvanic coupling through the human body. IEEE Trans Instrum Meas 59:963–969

Wen Y, Yang R, Chen Y (2008) Heart rate monitoring in dynamic movements from a wearable system. In: 5th international summer school and symposium on medical devices and biosensors (ISSS-MDBS 2008), Hong Kong, 1–3 June 2008, pp 272–275

WiMedia (2010) Wimedia Alliance. http://www.wimedia.org/

Wong A, McDonagh D, Omeni O, Nunn C, Silveira M, Burdett A (2009) Sensium: an ultra-low-power wireless body sensor network platform: design & application challenges. In: 31st international conference of the IEEE engineering in medicine and biology society, Minneapolis, MN, 3–6 September 2009, pp 6576–6579

Wu K, Xiaoming W (2007) A wireless mobile monitoring system for home healthcare and community medical services. In: 1st international conference on bioinformatics and biomedical engineering (ICBBE 2007), Wuhan, 6–8 July 2007, pp 1190–1193

Wu Y, Fahmy S, Shroff NB (2006) Optimal qos-aware sleep/wake scheduling for time-synchronized sensor networks. In: 40th annual conference on information sciences and systems (CISS 2006), Princeton, NJ, 22–24 March 2006, pp 924–930

Xiao S, Wei X, Wang Y (2008) Energy-efficient schedule for object detection in wireless sensor networks. In: IEEE international conference on service operations and logistics, and informatics (SOLI 2008), Beijing, China, 12–15 October 2008, pp 602–605

Xu R, Zhu H, Yuan J (2009) Characterization and analysis of intra-body communication channel. In: IEEE antennas and propagation society international symposium (APSURSI 2009), Charleston, SC, 1–5 June 2009, pp 1–4

Yao Jianchu, Schmitz Ryan, Warren Steve (2005) A wearable point-of-care system for home use that incorporates plug-and-play and wireless standards. IEEE Trans Inf Technol Biomed 9:363–371

Yao J, Warren S (2005) Applying the ISO/IEEE 11073 standards to wearable home health monitoring systems. J Clin Monit Comput 19:427–436

Yao J, Schmitz R, Warren S (2003) A wearable standards-based point-of-care system for home use. In: 25th annual international conference of the IEEE engineering in medicine and biology society (EMBC 2003), Cancun, 17–21 September 2003, pp 3732–3735

Ye W, Heidemann J, Estrin D (2004) Medium access control with coordinated adaptive sleeping for wireless sensor networks. IEEE/ACM Trans Netw 12:493–506

Yoo J, Yan L, Lee S, Kim Y, Yoo H (2010) A 5.2 mW self-configured wearable body sensor network controller and a 12 uw wirelessly powered sensor for a continuous health monitoring system. IEEE J Solid-State Circuits 45:178–188

Ze L, Haiying S, Alsaify B (2008) Integrating RFID with wireless sensor networks for inhabitant, environment and health monitoring. In: 14th IEEE international conference on parallel and distributed systems (ICPADS 2008), Melbourne, 8–10 December 2008, pp 639–646

Zhang S, Ang MH, Xiao W, Tham CK (2008) Detection of activities for daily life surveillance: eating and drinking. In: 10th international conference on e-health networking, applications and services (HealthCom 2008), Singapore, 7–9 July 2008, pp 171–176

Zierhofer CM, Hochmair ES (1990) High-efficiency coupling-insensitive transcutaneous power and data transmission via an inductive link. IEEE Trans Biomed Eng 37:716–722

ZigBeeAlliance (2010) Zigbee Alliance. http://www.zigbee.org

Zito D, Pepe D, Neri B, De Rossi D, Lanata A (2007a) Wearable system-on-a-chip pulse radar sensors for the health care: system overview. In: 21st international conference on advanced information networking and applications workshops (AINAW 2007), Niagara Falls, Ontario, 21–23 May 2007, pp 766–769

Zito D, Pepe D, Neri B, De Rossi D, Lanata A, Tognetti A, Scilingo EP (2007b) Wearable system-on-a-chip UWB radar for health care and its application to the safety improvement of emergency operators. In: 29th annual international conference of the ieee engineering in medicine and biology society (EMBS 2007) Lyon, 22–26 August 2007, pp 2651–2654

Chapter 4
Design of Wireless Health Platforms

Lawrence Au, Brett Jordan, Winston Wu, Maxim Batalin,
and William J. Kaiser

4.1 System Architecture Requirements for Wireless Health Platforms

Wireless embedded platforms play a significant role in Wireless Health: For many hidden medical conditions, symptoms may not reveal during traditional clinical visits. *Cumulative, free-living monitoring*, where individuals are monitored continuously with the use of wireless electronics and sensors, is considered a potential solution for capturing additional physiological data (Bonato 2003). From chronic disease management to physical rehabilitation, these platforms have demonstrated their potential uses (Moy et al. 2003; Bonato 2005).

Designing *wireless health platforms* for *continuous subject monitoring* entails multiple requirements and tradeoffs. For clinical use, physicians should be able to specify diagnostic parameters and have complete control of the platforms based on real-time diagnostic needs. In addition to accurate acquisition of sensor data, information and tools that are significant or valuable to physicians include:

1. Event triggers: alerting physicians when any physiological sensing variables exceeds a threshold value or when an event of concern is derived from sensor fusion (Winters and Wang 2003; Wu et al. 2007).
2. Context snapshots: selectively collecting physiological data during specific user contexts (e.g., jogging) (Park and Jayaraman 2003).
3. Evaluation of quality of life: recognizing various activities of daily living (ADLs) and evaluating a subject's overall well-being (Tu et al. 1997).

Design tradeoffs often pose challenges to the deployment of these platforms. The combination of sensor data acquisition and signal processing algorithms must provide adequate diagnostic and classification information to the subject or physician. For

W.J. Kaiser (✉)
Electrical Engineering Department, University of California, Los Angeles, CA, USA
e-mail: kaiser@ee.ucla.edu

A. Bonfiglio and D. De Rossi (eds.), *Wearable Monitoring Systems*,
DOI 10.1007/978-1-4419-7384-9_4, © Springer Science+Business Media, LLC 2011

battery-powered sensor platforms, they must be sufficiently compact and lightweight to be worn without inconvenience (Anliker et al. 2004; Bharatula et al. 2006).

The rest of the chapter describes major design considerations and applications related to wireless health platforms for continuous subject monitoring.

4.2 System Design

Figure 4.1 illustrates the general system architecture of a wireless health platform.

4.2.1 Sensors

Sensors are essential for all monitoring applications. For continuous monitoring, the most prominent sensors for activity recognition include accelerometers and gyroscopes (Bao and Intille 2004; Jovanov et al. 2005; Stager et al. 2004). Additionally, physiological signals are important in continuous monitoring. Typical physiological sensors include heart rate monitor,[1] ECG/EMG sensor (Wu et al. 2007), pulse oximeter,[2] and respiratory sensor.[3] Table 4.1 shows the typical power consumptions of various sensors. For activity re-cognition, motion sensors (accelerometers and gyroscopes) can provide accurate information (Bao and Intille 2004), and they should be integrated into the main printed circuit board to reduce the overall size of the platform. External connectors for sensor board extension should

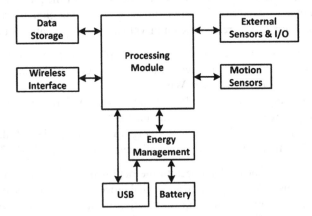

Fig. 4.1 System architecture of wireless health platform for continuous monitoring

---

[1] Polar T31 Transmitter Belt; http://www.polarusa.com

[2] Nonin Pulse Oximeter; http://www.nonin.com

[3] MLT1132 Piezo Respiratory Belt Transducer; http://www.adinstruments.com

Table 4.1 Typical power consumption of various sensors (Au et al. 2007)

Sensor	Power
Piezoelectric sensor	Passive
Temperature	$\sim$0.1 mW
Accelerometer (tri-axial)	$\sim$1 mW
ECG/EMG sensor	3–20 mW
Pulse oximeter	$\sim$30 mW
Gyroscope (bi-axial)	$\sim$30 mW

TMP35 Low Voltage Temperature Sensor (http://www.analog.com); Polar T31 Transmitter Belt (http://www.polarusa.com); ANT Alliance (http://www.thisisant.com/); Zigbee Alliance (http://www.zigbee.org/); WT12 Data Sheet (http://www.bluegiga.com)

be made available to accommodate additional sensors and provide sufficient flexibility for more specific applications.

4.2.2   Signal Acquisition

Analog signals from sensors must be acquired and preprocessed appropriately to obtain useful information. The preprocessing stage includes *amplification* and *filtering*. The type of circuits necessary depends on the type of sensor used (piezoelectric, voltage-output, current-output, etc.), dynamic range of signal output, and the sampling requirements (Webster 1997; Wu 2008). Sampling requirements are important in establishing whether a signal is adequately acquired to achieve sufficient accuracy to enable event detection, motion classification, or sensor fusion (Au et al. 2009). Typical preprocessing stages include a low-noise, high-impedance amplifier, and a second anti-aliasing filter for the analog-digital (A/D) conversion. A practical example is the electrocardiogram (ECG) sensor circuit (Medical Applications Guide 2007).

Signal filtering presents critical requirements. For instance, low-pass filters for accelerometers must be designed properly to reduce high-frequency spectral components that are outside of the band of interest (Steele et al. 2003). The A/D converter is an essential component for all signal acquisition system, and should be selected for low-power operation. The resolution of A/D converters should have a minimum of 12 bits for typical applications, but for some physiological signals such as ECG, 16 bits or higher is needed.

Certain microelectromechanical systems (MEMS) sensors provide digital outputs by embedding A/D conversion on the same MEMS packages. Future platforms can take advantage of these sensors, further reducing the number of external components, size of platform, and the overall power consumption. Digitized signals are usually acquired through serial communication interfaces, such as SPI and I2C.

4.2.3 Processing Module and Data Storage

In terms of processing, embedded processors are used extensively in wireless health applications: The CodeBlue project uses Crossbow's MICA2 to implement a wearable sensor with capabilities for physiological monitoring (Malan et al. 2004). Jovanov et al. have implemented an ECG sensor board, oximeter sensor, and other signal processing modules on the Moteiv's Tmote Sky platform (Jovanov et al. 2005). The BTnode provides a dual-radio (CC1000 and Bluetooth) architecture for various applications in Body Sensor Networks (Beutel 2006). The iMote provides an ARM7-based embedded processor and a Bluetooth radio (Nachman et al. 2005). Other notable platforms include SHIMMER (Patel et al. 2007), eWatch (Maurer et al. 2006), and e-AR (Pansiot et al. 2007).

For wireless health platforms, an embedded processor should be capable of efficiently performing simple feature extraction algorithms (e.g., averaging, signal peak detection). Most importantly, it requires multiple low-power modes to enable energy-efficient operations and extend battery life (McIntire et al. 2006). Furthermore, sensor data are either preprocessed and stored locally on the platform (Kamijoh et al. 2001; Anliker et al. 2004), or transmitted to another location (e.g., mobile devices) wirelessly (Jovanov et al. 2005; Gao et al. 2005; Wu et al. 2007). Consequently, additional data storage, such as external flash memory cards, provides the capability to store raw sensor data for future analyses.

4.2.4 Wireless Interface

Prevalent wireless standards most suitable for wireless health platforms include WiFi (IEEE 802.11), Bluetooth (IEEE 802.15.1),[4] IEEE 802.15.4 (usually associated with Zigbee protocol[5]), and the ANT protocol.[6] WiFi is most appropriate for high data rate applications (e.g., up to 54 Mbit/s with IEEE 802.11 g). It is also extremely power-hungry, and is not used in many platforms with limited battery capacity. Bluetooth is widely supported in commercial mobile devices, such as cell phones. It provides moderate data rates (3 Mbit/s in Bluetooth 2.0 + EDR), sufficient for many real-time continuous monitoring applications. IEEE 802.15.4 is a low-power wireless standard designed for automated industrial control, and is suitable for applications that require low data rates (250 kbit/s) and intermittent data transfer. The ANT protocol is similar to IEEE 802.15.4 with comparable power consumption and higher data rate (1 Mbit/s), and it is primarily a proprietary protocol.

---

[4] Bluetooth Special Interest Group; http://www.bluetooth.org/

[5] Zigbee Alliance; http://www.zigbee.org/

[6] ANT Alliance; http://www.thisisant.com/

Selection of the wireless interface hinges on application needs, and the designer must study the application carefully to determine the worst-case scenario (e.g., maximum data rate required).

4.2.5 Energy Management

To improve energy efficiency, special energy management features should be taken into consideration. A particularly important feature of the system architecture is that it should be equipped with real-time energy accounting capability (Au et al. 2007; McIntire et al. 2006). The energy information derived from energy accounting capability can be relayed to applications in software, and applications can enable or disable subsystems on the platform through software-enabled power switches based on real-time conditions. This hardware–software system approach is critical for energy management because the real-time energy consumption of the platforms translates to how effectively energy can be managed in applications (Au et al. 2007; McIntire et al. 2006).

4.3 MicroLEAP: A Wireless Health Platform with Integrated Energy Accounting

MicroLEAP (ULEAP) provides a new architectural solution for wireless health platforms (Au et al. 2007). Primarily focusing on continuous subject monitoring, its design emphasizes the unique requirements in wireless health applications. It provides the required sensing resolution and a Bluetooth interface for compatibility with mobile devices with which the platforms must be interoperable. Most importantly, MicroLEAP provides real-time energy accounting and management; it can be placed in low-power modes to conserve battery energy, based on real-time application needs.

4.3.1 Hardware

The MicroLEAP platform consists of three subsystems: processing, wireless, and sensor. A simplified hardware architecture is shown in Fig. 4.2. A photo of the MicroLEAP platform is shown in Fig. 4.3.

The processing subsystem contains the Texas Instruments's MSP430 embedded processor,[7] an external 16-bit ADC, a microSD card slot, and energy accounting circuitry. MSP430 provides multiple low-power modes to reduce energy

---

[7] MSP430F1611 Data Sheet; http://www.ti.com/msp430

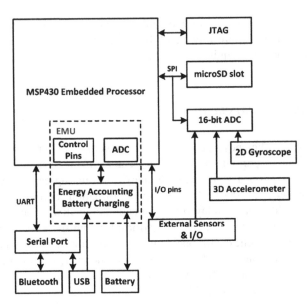

Fig. 4.2 MicroLEAP: hardware architecture

Fig. 4.3 MicroLEAP

consumption during idle periods. The external 16-bit ADC supports an effective sampling throughput of 100 ksps, sufficient for typical sampling requirements in wireless health applications. UART, SPI, and I2C are available for serial communications between subsystems.

The Energy Management Unit (EMU) is responsible for measuring energy consumption and power-cycling subsystems. Current-sensing technique is used to capture energy consumption. Charge samples are acquired with the on-chip ADC and accumulated in memory, and they can be fetched through software. Corresponding charge and energy values can be calculated using (4.1) and (4.2).

$$Q = \frac{V_{\text{ADC},range} \text{ sum}}{\left(2^B - 1\right) A_v R_{sense} f_{sample}} \tag{4.1}$$

$$E = Q V_{supply}, \tag{4.2}$$

where, $V_{ADC,range}$ = input dynamic range of on-chip ADC (V); sum = cumulative sum of sample in memory; B = number of bits available in on-chip ADC; A_v = voltage gain of the amplifying stage (V/V); R_{sense} = current-sense resistor (Ω); V_{supply} = supply voltage of the component (V); f_{sample} = sampling frequency of EMU (Hz); Q = total change accumulated (C); E = total energy consumption (J).

MicroLEAP is powered by a rechargeable Li-Polymer battery. A mini-USB port is available for both platform programming and battery recharge.

The wireless subsystem contains a Class-2 Bluetooth module that can be controlled via an ASCII-based protocol.[8] The protocol stack of the Bluetooth module includes a serial port profile (SPP), which allows raw bytes to be transmitted via UART.

The sensor subsystem comprises a tri-axial accelerometer[9] and a bi-axial gyroscope.[10] Additional analog channels are available for external sensor circuits. MicroLEAP accommodates an extension board for custom sensor development. The external connector provides three to five additional analog channels, SPI and UART interfaces, and an external voltage source. The extension board can be mounted either perpendicularly or horizontally onto the main module. For instance, a second gyroscope can be mounted perpendicularly onto the board to capture the yaw angle (Fig. 4.3).

4.3.2 Software

MicroLEAP uses microC/OS, a multitasking, preemptive real-time operating system, due to its small kernel footprint and relaxed constraints on hardware (Labrosse 2002). It contains all primary operating system features, such as a scheduler, semaphores, mutexes, memory management, and task synchronization. Device drivers and user APIs are designed in a layered approach to abstract any hardware details when developing new applications. A simplified software architecture is shown in Fig. 4.4.

Access to common features on MicroLEAP is done through a remote command interface via Bluetooth's SPP. Basic operations, such as selecting sampling rate and sensor channels, can be enabled through sending ASCII-based commands from a mobile device.[11]

---

[8] WT12 Data Sheet; http://www.bluegiga.com

[9] ADXL330 Tri-axial Accelerometer; http://www.analog.com

[10] IDG-300 Integrated Dual-Axis Gyroscope; http://www.invensense.com

[11] Au LK. MicroLEAP documentation; http://www.ascent.ucla.edu/wiki/index.php/Uleap

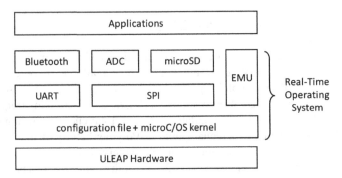

Fig. 4.4 MicroLEAP: software architecture

Table 4.2 Average power consumption

	Operating mode	Power (mW)
Processing module	Active @ 1 MHz Accelerometer = On Sampling rate = 250 Hz	9.84
Bluetooth	Tx @ 115200 bps	72.74
Gyroscopes	On	53.0
EMU overhead	~0.8 mW (current-sensing circuitry)	

4.3.3 Performance

Table 4.2 shows the average power consumption of MicroLEAP. As expected, Bluetooth consumes a significant portion of the total power. Furthermore, the gyroscopes are also power-hungry, suggesting that it should be used only when needed.

4.4 MicroLEAP Application: SmartCane

Falls are particularly serious among the elderly (Sattin and Nevitt 1993) and disabled (Rubenstein and Josephson 2006), where the number of individuals with fall-induced injury has been steadily increasing at a rate greater than accounted for in terms of demographic changes (Kannus et al. 1999). Canes provide the required biomechanical support for mobility and are used by over several millions of individuals in the United States (Bateni and Maki 2005). Traditional technique of analyzing cane usage typically involves sensor data collection in the laboratory with customized equipment, such as a set of imagers that capture the walking motion. While this may provide information on the initial progress of a particular subject's cane usage, various events may cause one to vary from the proper usage in the long run.

Consequently, real-time feedback becomes increasingly essential in training cane users. Regular cane users may receive initial training instructions, but they tend to deviate from proper use after long hours. This type of "open-loop" training does not provide an individual any type of feedback regarding their present state of cane usage at any given time. Continuous training within a controlled environment (such as an instrumented laboratory) is impractical mainly because the results obtained from such training may not reflect the actual cane usage in real life. Furthermore, commercialization of such equipment is often financially prohibitive.

The primary objective of SmartCane is to provide real-time guidance to the cane user regarding the present state of cane usage with low-cost wireless health platforms (Wu et al. 2008; Au et al. 2008).

4.4.1   System Implementation

The SmartCane architecture (Fig. 4.5) consists of the MicroLEAP platform, sensors for measuring motion, rotation and pressure, and a compact piezoelectric speaker that provides feedback to the cane user, alerting any improper cane usage.

The set of sensors on SmartCane consists of a tri-axial accelerometer,[12] two bi-axial gyroscopes,[13] and pressure sensors.[14] The accelerometer captures the linear accelerations, and the gyroscopes are mounted orthogonally to one another, thus providing the ability to capture angular rotations in all three dimensions. One of the pressure sensors detects the downward force applied on the cane tip, while the other pressure sensor captures the pressure applied on the cane handle. Using these sensor data, one can compute the orientation with respect to the gravity and swing characteristic of the cane.

Audio is an effective method for providing feedback information. The Smart-Cane uses a piezoelectric speaker that is controlled by MicroLEAP to produce different audio tones based on its usage. Subjects can adjust their movement based on the tones and guide themselves during usage.

4.4.2   Real-Time Feedback

Benefits associated with real-time feedback include immediate prevention of certain improper cane usage that can potentially lead to falls and various injuries, longer battery life (due to reduced energy consumption through local data processing), and elimination of mobile devices during usage.

---

[12] ADXL330 Tri-axial Accelerometer; http://www.analog.com

[13] IDG-300 Integrated Dual-Axis Gyroscope; http://www.invensense.com

[14] FlexiForce Force Sensor; http://www.tekscan.com

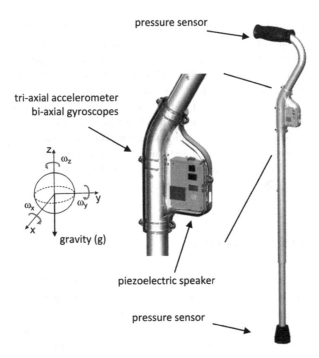

pressure sensor

tri-axial accelerometer
bi-axial gyroscopes

gravity (g)

piezoelectric speaker

pressure sensor

Fig. 4.5 The SmartCane architecture

To provide real-time feedback, SmartCane must be able to recognize a proper stride. A proper stride involves exerting a certain amount of force on both the cane and the opposite leg. The overall walking motion should also exhibit some periodic behavior. SmartCane can collect all these information and classify proper and improper strides.

The classification algorithm used on SmartCane is shown in Fig. 4.6. One main advantage of this approach is that cane users and medical professionals can interpret visually how SmartCane determines a proper stride. For instance, if the feedback indicates insufficient weight dependence on the cane, the subject can immediately remedy the problem by placing more weight on the cane.

Thresholds are determined through manual training for each subject. Specifically, the threshold values are the different features obtained from the proper cane usage. Based on these thresholds, the algorithm classifies, in real time, whether the subject is using the cane properly.

4.4.3 Proper Strides

Correct detection of proper strides corresponds to the sensitivity of the algorithm. The experiment involving such analysis includes a subject making four hundred proper strides with SmartCane on a treadmill at a specific pace. Sensor data are collected in real time, and the classification algorithm determines if the subject has

Fig. 4.6 Classification
algorithm

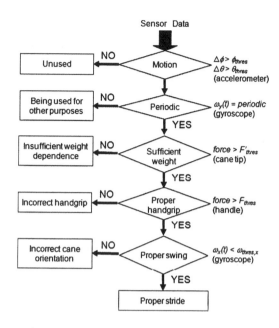

Table 4.3 Classification of proper strides with separate thresholds

	Thresholds		
	Subject 1	Subject 2	Subject 3
$\phi_{thres}(^\circ)$	20	20	20
θ_{thres} $(^\circ)$	30	30	30
Period of $\omega_y(t)$ (s)	(0.5, 2.1)	(0.5, 2.1)	(0.5, 2.1)
F'_{thres} (average, lbs/stride)	7.1	6.3	2.0
F_{thres} (maximum, lbs/stride)	7.1	7.1	4.0
ω_x $(^\circ/s)$	13.0	21.0	21.0
Classification rates (%)	99.2	97.0	99.0

made a stride, and if so, whether it is a proper one. The same experiment is
performed on three different subjects.

Table 4.3 lists the corresponding classification rates. Most of the threshold
values for the three individuals are relatively close, except those related to pressure
sensors. The threshold values of the pressure sensors from the handle and cane tip
are much lower for the third subject, and that can be attributed to the gender
(female) and her relatively lighter body weight.

4.4.4 Continuous Monitoring

One of the key objectives of the SmartCane system is to process real-time sensor
data locally and provide appropriate feedback to the cane user regarding the state of

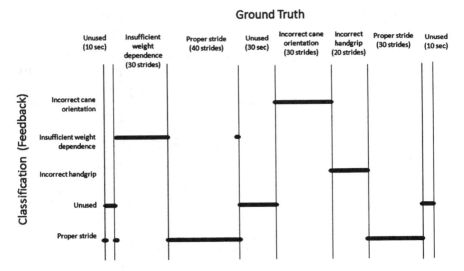

Fig. 4.7 Classification result: continuous monitoring

the subject's cane usage. Figure 4.7 illustrates the classification result of an experiment that involves using the cane through a series of events that include both proper and improper strides. The horizontal axis represents the ground truth, and the vertical axis corresponds to the classification result from SmartCane. When the SmartCane is unused, an LED lights up. SmartCane will show a blinking LED if it detects a proper stride. Audio tones with different frequencies indicate various improper cane use.

4.5 MicroLEAP Application: Episodic Sampling

Episodic sampling is a context-aware technique where context classification occurs only episodically, and the time for the next episode depends on prior context information (Au et al. 2009). This takes advantage of the observation that human activities typically vary slowly and constant state classification is not necessary in most circumstances (Krause et al. 2005). Figure 4.8 compares episodic sampling with the traditional continuous sampling. In continuous sampling, the system performs constant classifications in the active state. In episodic sampling, the system only executes classification at specific episodes and remains in a sleep state between episodes. T_{episode} represents the duration of an episode, and it mainly depends on the amount of sensor data required for context classification. T_{sleep} represents the sleep time between successive episodes; this is the time interval when the platform spends in sleep mode.

The EMU in MicroLEAP provides a software-based interface for real-time energy information and power control. Sensor data are transmitted to a mobile

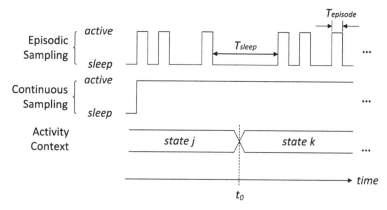

Fig. 4.8 Episodic sampling and continuous sampling

device, which performs feature extraction and classification in a Linux-based, multi-threaded environment.

MicroLEAP incorporates various sensors for continuous monitoring:

4.5.1 Motion Sensors

The platform is integrated with both accelerometers and gyroscopes. These sensors provide measurements of motions, capable of capturing various body movements related to ADLs.

4.5.2 Physiological Sensors

The respiratory sensor is ADInstrument's MLT1132,[15] a piezoelectric belt that generates a small voltage in response to mechanical stretching. The sensor is completely noninvasive. MicroLEAP amplifies and conditions the voltage generated by the respiratory sensor.

For heart rate monitoring, MicroLEAP includes the commercially available heart rate monitor from Polar.[16] MicroLEAP collects the heart beats through a separate receiver and locally computes the corresponding heart rate.

---

[15] MLT1132 Piezo Respiratory Belt Transducer; http://www.adinstruments.com
[16] Polar T31 Transmitter Belt; http://www.polarusa.com

4.5.3   Feature Extraction and Classification

For many ADLs, the N-point fast Fourier transform (FFT) is an effective algorithm because all the activities of interest exhibit certain periodicity. Specifically, the peak magnitude, F_{peak}, and the energy, F_{energy} of the N-point FFT, $X(k)$, from the accelerometer signals form the feature vector, $\mathbf{F}$:

$$F_{peak} = \max_k \| X(k) \| \qquad (4.3)$$

$$F_{energy} = \sum_{k=1}^{N/2} X(k) \times X(k). \qquad (4.4)$$

Both of these features are highly correlated with body movements (Bao and Intille 2004), and a sampling rate of 128 Hz captures major human motions with sufficient accuracy. For each N-point FFT operation, a sliding window of size 256 data points ($N = 256$) is used to ensure sufficient frequency resolution. A new $\mathbf{F}$ can be derived every second; each operation takes the newest 128 samples and discards the oldest 128 samples.

4.5.4   Control Algorithm

Intuitively, classification should be performed when a significant variation in activity pattern appears. On the other hand, if the activity pattern does not vary frequently, classification can be minimized. By adaptively adjusting the frequency of context classification, one can reduce the amount of time the platform spends in active mode performing constant context classification. The apparent tradeoff is the classification accuracy.

Algorithm 1 Episodic Sampling

$T_{sleep} = 0$
while in each episode **do**
    Collect Sensor Data
    Extract Features, $\mathbf{F} = (F_1, F_2, ...)$
    $state \Leftarrow Classification(\mathbf{F})$
    if $state$ has changed since last classification **then**
        $T_{sleep} \Leftarrow a \cdot T_{sleep}$, where $0 \leq a \leq 1$
    else
        $T_{sleep} \Leftarrow T_{sleep} + T_{incr}$
        if $T_{sleep} > T_{sleep,max}$ **then**
            $T_{sleep} \Leftarrow T_{sleep,max}$
        end if
    end if
end while

Episodic sampling is based on additive-increase/multiplicative-decrease (AIMD). Consider Fig. 4.8: when the state (context) remains the same, the algorithm gradually increases T_{sleep}. When a state change is detected, the algorithm aggressively decreases T_{sleep} to avoid future misses. Algorithm 1 illustrates the pseudocode. During each episode, sensor data are collected for classification. If the pattern exhibits constant state changes (e.g., during interval training), the algorithm decreases T_{sleep} by a constant factor, a. In the limiting case (i.e., when T_{sleep} equals zero), the algorithm reduces to continuous sampling. When the state remains identical for an extended period, the algorithm increases T_{sleep} by an amount T_{incr}. To ensure T_{sleep} does not become unbounded, the algorithm caps T_{sleep} at some maximum value, $T_{sleep,max}$.

4.6   Conclusion and Next Generation Platforms

Designing wireless health platforms requires application-dependent tradeoffs: sensors, wireless interfaces, processors, size and weight, and convenience. This chapter describes MicroLEAP, a wireless health platform for continuous subject monitoring. Applications are also outlined with two examples.

Next generation of platforms will likely demand even better wireless connectivity due to the constant effort of digitizing health records (also known as Electronic Health Records). Real-time processing may also become desirable for applications that can benefit from providing feedback training. Furthermore, wireless health platforms should leverage server-side computing power for more complex signal processing.

References

Anliker U, Beutel J, Dyer M, Enzler R, Lukowicz P, Thiele L, Trster G (2004) A systematic approach to the design of distributed wearable systems. IEEE Trans Comput 53(8):1017–1033

Anliker U, Ward J, Lukowicz P, Troster G, Dolveck F, Baer M, Keita F, Schenker E, Catarsi F, Coluccini L, Belardinelli A, Shklarski D, Alon M, Hirt E, Schmid R, Vuskovic M (2004) Amon: a wearable multiparameter medical monitoring and alert system. IEEE Trans Inf Technol Biomed 8(4):415–427. doi:10.1109/TITB.2004.837888

Au L, Wu W, Batalin M, McIntire D, Kaiser W (2007) MicroLEAP: energy-aware wire-less sensor platform for biomedical sensing applications. In: Biomedical circuits and systems conference, 2007. BIOCAS 2007. IEEE, pp 158–162. doi:10.1109/BIOCAS.2007.4463333

Au L, Wu W, Batalin M, Kaiser W (2008) Active guidance towards proper cane usage. In: 5th International summer school and Symposium on medical devices and biosensors, 2008. ISSS-MDBS 2008., pp. 205–208. doi:10.1109/ISSMDBS.2008.4575054

Au L, Batalin M, Stathopoulos T, Bui A, Kaiser W (2009) Episodic sampling: towards energy-efficient patient monitoring with wearable sensors. In: 31st Annual international conference of the IEEE engineering in medicine and biology society (EMBC'09)

Bao L, Intille SS (2004) Activity recognition from user-annotated acceleration data. Lect Notes Comput Sci 3001/2004:1–17. doi:10.1007/b96922

Bateni H, Maki BE (2005) Assistive devices for balance and mobility: benefits, demands, and adverse consequences. Arch Phys Med Rehabil 86:134–145

Beutel J (2006) Fast-prototyping using the BTnode platform. In: Proceedings of the conference on design, automation and test in Europe (DATE'06)

Bharatula NB, Anliker U, Lukowicz P, Trster G (2006) Architectural tradeoffs in wearable systems. In: Architecture of Computing Systems - ARCS 2006, ser. Lecture Notes in Computer Science, Grass W, Sick B, and Waldschmidt K, eds., vol. 3894. Springer Berlin/Heidelberg, pp 217–231

Bonato P (2003) Wearable sensors/systems and their impact on biomedical engineering. IEEE Eng Med Biol Mag 22(3):18–20. doi:10.1109/MEMB.2003.1213622

Bonato P (2005) Advances in wearable technology and applications in physical medicine and rehabilitation. J Neuroeng Rehabil 2(2) http://www.jneuroengrehab.com/content/2/1/2

Gao T, Greenspan D, Welsh M, Juang R, Alm A (2005) Vital signs monitoring and patient tracking over a wireless network. 27th annual international conference of the IEEE engineering in medicine and biology society, 2005, pp 102–105. doi:10.1109/IEMBS.2005.1616352

Jovanov E, Milenkovic A, Otto C, Groen PD, Johnson B, Warren S, Taibi G (2005) A WBAN system for ambulatory monitoring of physical activity and health status: applications and challenges. In: Proceedings of the international conference on engineering in medicine and biology society (IEEE-EMBS), pp 3810–3813

Kamijoh N, Inoue T, Olsen C, Raghunath M, Narayanaswami C (2001) Energy trade-offs in the ibm wristwatch computer. In: Proceedings of the fifth international symposium on wearable computers, 2001, pp 133–140. doi:10.1109/ISWC.2001.962115

Kannus P, Parkkari J, Koskinen S, Palvanen SNM, Jrvinen M, Vuori I (1999) Fall-induced injuries and deaths among older adults. JAMA 281:1895–1899

Krause A, Ihmig M, Rankin E, Leong D, Gupta S, Siewiorek D, Smailagic A, Deisher M, Sengupta U (2005) Trading off prediction accuracy and power consumption for context-aware wearable computing. In:. Proceedings of the ninth IEEE international symposium on wearable computers, 2005, pp 20–26. doi:10.1109/ISWC.2005.52

Labrosse JJ (2002) MicroC/OS II: the real-time kernel. CMP Books, Burlington, MA

Malan D, Fulford-Jones TRF, Welsh M, Moulton S (2004) CodeBlue: an ad hoc sensor net-work infrastructure for emergency medical care. In: Proceedings of the MobiSys 2004 workshop applications mobile embedded systems (WAMES 2004), Boston, MA, pp 12–14

Maurer U, Rowe A, Smailagic A, Siewiorek D (2006) eWatch: a wearable sensor and notication platform. In: International workshop on wearable and implantable body sensor networks, 2006, pp 4–145. doi:10.1109/BSN.2006.24

McIntire D, Ho K, Yip B, Singh A, Wu W, Kaiser WJ (2006) The low power energy aware processing (LEAP) embedded networked sensor system. In: IPSN'06: Proceedings of the fifth international conference on information processing in sensor networks. ACM Press, New York, pp 449–457. doi:http://doi.acm.org/10.1145/1127777.1127846

Medical Applications Guide (2007), TI's Medical Applications Guide, Texas Instruments. URL: www.ti.com/medical

Moy M, Mentzer S, Reilly J (2003) Ambulatory monitoring of cumulative free-living activity. IEEE Eng Med Biol Mag 22(3):89–95. doi:10.1109/MEMB.2003.1213631

Nachman L, Kling R, Adler R, Huang J, Hummel V (2005) The intel mote platform: a bluetooth-based sensor network for industrial monitoring. In: Proceedings of the international conference on information processing in sensor networks (IPSN), Los Angeles, CA

Pansiot J, Stoyanov D, McIlwraith D, Lo BP, Yang GZ (2007) Ambient and wearable sensor fusion for activity recognition in healthcare monitoring systems. In: 4th international workshop on wearable and implantable body sensor networks (BSN 2007)

Park S, Jayaraman S (2003) Enhancing the quality of life through wearable technology. IEEE Eng Med Biol Mag 22(3):41–48. doi:10.1109/MEMB.2003.1213625

Patel S, Lorincz K, Hughes R, Huggins N, Growdon J, Welsh M, Bonato P (2007) Analysis of feature space for monitoring persons with parkinson's disease with application to a wireless wearable sensor system. In: Engineering in Medicine and Biology Society, 2007. 29th annual international conference of the IEEE, pp 6290–6293. doi:10.1109/IEMBS.2007.4353793

Rubenstein LZ, Josephson KR (2006) Falls and their prevention in elderly people: what does the evidence show? Med Clin North Am 90:807–824

Sattin RW, Nevitt MC (1993) Injuries in later life: epidemiology and environmental aspects. Oxford Textbook of Geriatric Medicine. Oxford University Press, New York

Stager M, Lukowicz P, Troster G (2004) Implementation and evaluation of a low-power sound-based user activity recognition system. In: Proceedings of the eighth international symposium on wearable computers, 2004. ISWC 2004. 1, pp 138–141. doi:10.1109/ISWC.2004.25

Steele BG, Belza B, Hunziker J, Holt L, Legro M, Coppersmith J, Buchner D, Lak- shminaryan S (2003) Monitoring daily activity during pulmonary rehabilitation using a triaxial accelerometer. J Cardiopulm Rehabil 23:139–142

Tu SP, McDonell MB, Spertus JA, Steele BG, Fihn SD (1997) A new self-administered questionnaire to monitor health-related quality of life in patients with COPD. Chest 112(3):614–622. doi:10.1378/chest.112.3.614. http://www.chestjournal.org/content/112/3/614.abstract

Webster JG (1997) Medical instrumentation: application and design. Wiley, New York

Winters J, Wang Y (2003) Wearable sensors and telerehabilitation. IEEE Eng Med Biol Mag 22 (3):56–65. doi:10.1109/MEMB.2003.1213627

Wu W (2008) MEDIC: an end-to-end biomedical system based on active sensor fusion. Ph.D. Thesis, University of California, Los Angeles

Wu W, Batalin M, Au L, Bui A, Kaiser W (2007) Context-aware sensing of physiological signals. In: Engineering in Medicine and Biology Society, 2007. EMBS 2007. 29th annual international conference of the IEEE, pp 5271–5275. doi:10.1109/IEMBS.2007.4353531

Wu W, Bui A, Batalin M, Liu D, Kaiser W (2007) Incremental diagnosis method for intelligent wearable sensor systems. IEEE Trans Inf Technol Biomed 11(5):553–562. doi:10.1109/TITB.2007.897579

Wu W, Au L, Jordan B, Stathopoulos T, Batalin M, Kaiser W, Vahdatpour A, Sarrafzadeh M, Fang M, Chodosh J (2008) The SmartCane system: an assistive device for geriatrics. In: Body-Nets'08: proceedings of the ICST 3rd international conference on body area networks, pp 1–4. ICST (Institute for Computer Sciences, Social-Informatics and Telecommunications Engineering)

Chapter 5
Lightweight Signal Processing for Wearable Body Sensor Networks

Hassan Ghasemzadeh, Eric Guenterberg, and Roozbeh Jafari

Use of mobile sensor-based platforms for human action recognitionis an ever-growing area of research. Recent advances in this field allowpatients to wear several small sensors with embedded processors and radios.Collectively, these sensors form a body sensor network (BSN). AlthoughBSNs have the potential to enable many useful applications [1], limitedprocessing power, storage and energy make efficient use of these systemscrucial. Moreover, user comfort is a major issue, which can cause patients tobecome frustrated and stop wearing the sensor nodes. The interactionbetween the human body and these wearable nodes here is defined as wearability.

This chapter will start by examining wearability constraints in greater detail in Sect. 5.1, then look at a general signal processing scheme for physical movement monitoring under these constraints in Sect. 5.2. Finally, two methods are examined which use this general system for specific movement monitoring problems. The first technique, which centers around the problem of optimal node placement, aims to minimize the number of sensor nodes required to classify a predefined set of human movements Sect. 5.3. The second one is a temporal parameter extraction technique that leads to an efficient human movement recognition algorithm in terms of accuracy and simplicity for implementation Sect. 5.4.

5.1 Wearability Issues

Design for wearable BSNs focuses on specific and important issues for developing wearable computing systems that take into account the physical shape of the sensors and their active relationship with the human form. In this section, we have outlined several design guidelines for the creation of wearable BSNs.

Design for wearability requires unobtrusive sensor node placement on the human body based on application-specific criteria. Criteria for placement can vary with the needs of functionality and convenience. Functionality criteria

R. Jafari (✉)
University of Texas at Dallas, TX, USA
e-mail: rjafari@utdallas.edu

A. Bonfiglio and D. De Rossi (eds.), *Wearable Monitoring Systems*,
DOI 10.1007/978-1-4419-7384-9_5, © Springer Science+Business Media, LLC 2011

constrains node placement to regions where relevant data can be sensed. The number of nodes required to capture all relevant data can vary based on the quality of information sensed at individual locations. Convenience criteria include: (1) physical interference with movement, (2) difficulty in removing and placing nodes, (3) social and fashion concerns, (4) frequency and difficulty of maintenance (charging and cleaning) [2].

For example, in continuous healthcare monitoring, patients will be expected to charge the sensors or replace the batteries on a regular basis, as they do with cell phones and other electronics. However, the frequent need to charge and the bulk of the battery can frustrate the users, causing them to no longer wear the sensors. Furthermore, batteries are the heaviest component in the system. By decreasing power usage, the size and weight of each sensor node can decrease, thus increasing patient comfort and device wearability.

This makes energy usage a primary constraint in designing BSNs, limiting everything from data sensing rates and link bandwidth, to node size and weight. Thus, one of the important goals in designing BSNs is to minimize energy consumption while preserving an acceptable quality of service.

Energy consumption can be decreased by lower sampling frequency, decreasing processing power, and simplifying signal processing. Another effective technique is deactivating nodes that are unnecessary for specific tasks that is investigated in details in Sect. 5.3.

5.2 System Architecture and Signal Processing Flow

The purpose of an action recognition system is to classify transitional movements into pre-defined actions. Given a set of movements, the system must distinguish between every pair of motions. These sensors capture inertial information from physical movements. An example platform used to generate data and achieve results in this chapter is shown Fig. 5.1a. The system consists of several sensor units; each has a tri-axial accelerometer, a bi-axial gyroscope, a microcontroller, and a radio. In this specific setup, the accelerometers are LIS3LV02DQ with $1,024$ LSb/g sensitivity. The IDG-300 gyroscopes have $2\,mV/^{\circ}/s$ sensitivity. The processing unit of each node, or mote, can sample sensor readings at certain rate and transmit the data wirelessly to a base station. A common protocol for this transmission is TDMA. For the purpose of action recognition, authors of [3] use the TelosB motes [4], which are commercially available from XBow®. The sensor board they use is predominantly a custom designed with an integrated Li-Ion battery that powers both sensor board and mote as shown in Fig. 5.1a. The sensor nodes can be placed on different locations on the body to capture movements of their subjects. Figure 5.1b shows a subject with nine nodes placed on different body segments.

Figure 5.2 shows several processing tasks typically used for signal processing and action recognition in BSNs. Each processing block is described as follows.

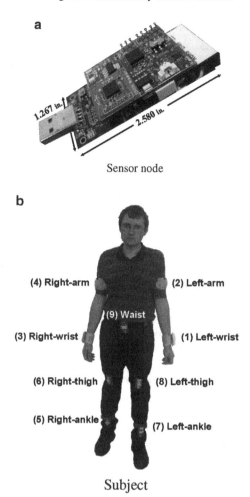

Fig. 5.1 (**a**) A sensor node composed of processing unit and custom-designed sensor board. The motion sensor board has a triaxial accelerometer and a biaxial gyroscopeSensor node and subject (**b**) A subject wearing nine sensor nodes

Sensor data collection: Data is collected from all of the sensors on each of the nodes at a specified frequency. The sampling rate can be empirically chosen to provide sufficient resolution while compensating for bandwidth constraints of the system [5], or it can be determined to satisfy the Nyquist criterion [6]. Usually, 20 samples per seconds would provide fine details of human movements [7].

Preprocessing: Data is filtered with a small window moving average to remove high frequency noise. The number of points used to average the signal can be chosen by examining the power spectral density of the signals. The filter is required to remove unnecessary motions (e.g., tremors) while maintains significant data.

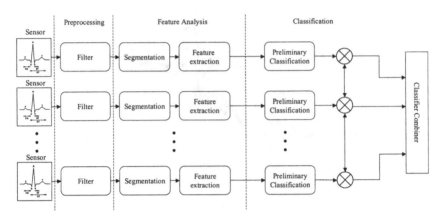

Fig. 5.2 Signal processing flow

With these objectives, authors in [8] test several moving average filters with varying window sizes and choose the filter that best satisfies the aforementioned requirements.

Segmentation: It determines the portion of the signal that represents a complete action. For action recognition, usually the start and the end of signal are determined in segmentation. To enable real-time movement monitoring, an automated method is required.

Feature extraction: Statistical and morphological features are extracted from the signal segment. For example, sensor readings can be transformed into a set of informative attributes, including Mean, Start-to-End, Standard Deviation, Peak-to-Peak Amplitude, RMS power, Median and Maximum value.

Per-node classification: Each node uses the feature vector generated during feature extraction to determine the most likely action. Due to its simplicity and scalability, k-Nearest Neighbor (k-NN) [9] is a widely used classifier [10].

Final classification: The final decision can be made using either a data fusion or a decision fusion scheme. In the data fusion, features from all sensor nodes are fed into a central classifier. The classifier then combines the features to form a higher dimensional feature space and classifies movements using the obtained features. In the decision fusion, however, each sensor node makes a local classification and transmits the result to a central classifier where a final decision is made according to the received labels.

5.3 Action Coverage for Node Placement

Action coverage aims to select the smallest number of sensor nodes that can adequately distinguish among all expected activities. This selection can be altered dynamically to disperse power load, route around a failed node, or cover a diverse set of activities. To address coverage problem in BSNs, a model of local knowledge provided by individual sensor nodes is required. A compatibility graph is a powerful model that representns capability of a sensor node in discriminating between movements.

5.3.1 Compatibility Graph

The amount of knowledge presented by each node determines the node's ability in action recognition. An example is shown in Fig. 5.3. Figure 5.3a is an example of two feature spaces with associated distributions drawn for four classes. The ellipses represent classification boundaries. In reality, the shapes are not perfect ellipses. For example, each node in the test system may have five data streams (x, y, z acceleration, and x, y angular velocity) and multiple features per data stream, forming a high-dimensional feature space per node.

Regions where the ellipses overlap represent potential misclassifications. Any point in the intersection of A and B or B and C cannot be confidently assigned to either class. In Fig. 5.3b, overlapping vs. well-separated classes is translated into a conflict graph. The vertices represent classes, and the edges represent ambiguities between the classes. Finally, Fig. 5.3c, the so-called compatibility graph, is generated by complementing the conflict graph of Fig. 5.3b. If a compatibility graph is not complete, then there exist some movements that the node cannot correctly classify. A complete graph is equivalent to the capability of distinguishing between every pair of classes.

One of the most popular class separability measures in the field of pattern recognition is the Bhattacharyya distance [11, 12]. It is theoretically sound because it relates directly to the upper bound of classification error the probabilities.

One of the most popular class separability measures in the field of pattern recognition is the Bhattacharyya distance [4, 17]. This measure is related to the well-known Chernoff bound and therefore has an explicit expression for a generalized Gaussian distribution. The Transformed Divergence is another common empirical measure of class separability, which is computationally simpler

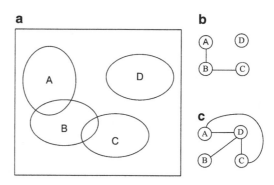

Fig. 5.3 Evolving toward a compatibility graph

than the Bhattacharyya distance. However, the Bhattacharyya distance is more theoretically sound because it relates directly to the upper bound of classification error the probabilities [18]. Both the Transformed Divergence and Bhattacharyya distance measures are real values between 0 and 2, where 0 indicates complete overlap between the signatures of two classes, and 2 indicates a complete separation between the two classes. Both measures are monotonically related to classification accuracies. The larger the separability value, the better the final classification result.

For action coverage, the Bhattacharyya distance can be exploited as a measure of separability between pairs of classes. The distance between two distributions i and j is represented by $\beta(i,j)$ in (5.1), where μ_i and Σ_i denote the mean vector and the covariance matrix associated with distribution i, respectively.

$$\beta(i,j) = 2(1 - e^{-\alpha(i,j)})$$

$$\alpha(i,j) = \frac{1}{8}(\mu_i - \mu_j)'\left(\left(\frac{\Sigma_i + \Sigma_j}{2}\right)^{-1}(\mu_i - \mu_j) + \frac{1}{2}\ln\left(\frac{\frac{|\Sigma_i| + |\Sigma_j|}{2}}{\sqrt{|\Sigma_i||\Sigma_j|}}\right)\right) \quad (5.1)$$

5.3.2 Problem Definition

Given a set of sensor nodes $S = \{s_1, s_2, \ldots, s_n\}$ placed in a body sensor network to detect a set of movements $M = \{1, 2, \ldots, m\}$, the action coverage problem can be formulated as follows.

Definition 1. *Two movements j_1 and j_2 are said to be* compatible *if they have complete separability based on Bhattacharyya metric indicated by (5.1).*

Definition 2. *A* compatibility graph *is an undirected graph $G_i = (V, E_i)$ constructed for a sensor node s_i, where V is a set of vertices identical to the set of movements M, and E_i is a set of undirected edges such that edge $(u, v) \in E_i$ if movements u and v are compatible at node s_i.*

The action coverage problem is used to find a minimal set of nodes that still encompasses full coverage within their capacity. The idea behind action coverage is that a subset of sensor nodes is sufficient to provide accurate detection of every target action. This subset is referred to as complete set and is defined as follows.

Definition 3. *A simple graph $G = (V, E)$ is a* complete graph *if for every pair of distinct vertices u and v, there is an edge $(u, v) \in E$.*

Definition 4. *A subset S' of sensor nodes ($S' \subseteq S$) is a* complete set*, if the compatibility graphs derived from S' altogether form a complete graph. That is, the graph G' computed by $G' = \cup_{i:s_i \in S'} G_i$ is a complete graph.*

Definition 5. *Given a finite set of sensor nodes $S = \{s_1, s_2, \ldots, s_n\}$ and a set of movements $M = \{1, 2, \ldots, m\}$,* Minimum-Cost Action Coverage (MCAC) *is the* problem of finding a subset $S' \subseteq S$ in which every pair $j_1, j_2 \in M$ are compatible and $|S'|$ is minimized.

The action coverage problem defined in Sect. 5.3.2 can be shown to be NP-hard by exact reduction from the well-studied Minimum Set Cover (MSC) problem. The following subsections are aimed to provide an ILP formulation of the MCI problem as well as a greedy approximation. The ILP can be used to obtain the lower bound of the solution, while the greedy approach provides a fast approximation.

5.3.3 ILP Approach

In this section, an integer linear programming formulation for the action coverage problem is presented. Since each node is represented by a graph, this problem can be stated as follows.

Problem 1. *Given compatibility graphs $G_1 = (V, E_1)$, $G_2 = (V, E_2), \ldots, G_n = (V, E_n)$, and a complete set of all edges $E = \bigcup_{i=1}^{n} E_i$, select a subset of graphs $G'_1, G'_2, \ldots, G'_m$ taken from $G_1, G_2, \ldots, G_n$, such that $\bigcup_{i=1}^{m} E'_i = E$ and the number of selected graphs (m) is minimized.*

The corresponding ILP formulation is presented as follows.

$$x_i = \begin{cases} 1, & \textit{if graph } G_i \textit{ is selected} \\ 0, & \textit{otherwise} \end{cases} \tag{5.2}$$

$$Min \sum_{i=1}^{n} x_i \tag{5.3}$$

subject to:

$$\sum_{i: e_j \in G_i} x_i \geq 1 \qquad \forall e_j \in E \tag{5.4}$$

$$x_i \in \{0, 1\} \tag{5.5}$$

The variables x_i ($i = 1, 2, \ldots, n$) indicate whether graph G_i is selected to form a complete graph. The inequality constraint (5.4) ensures that for each edge e_j in the complete graph, at least one of the compatibility graphs that contains that edge is selected. The objective function (5.3) attempts to minimize the number of graphs selected to form a complete graph. This is equivalent to minimizing the number of active nodes, which suitably leads to energy reduction in the system.

5.3.4 Greedy Approach

The greedy approach selects the compatibility graphs as follows: at each stage, it picks a compatibility graph G_i that covers the most uncovered edges; next it picks

---

Algorithm 1 Greedy Solution for Action Coverage

Require: Set of compatibility graphs $G_1 = (V, E_1)$, $G_2 = (V, E_2)$,..., $G_n = (V, E_n)$

Ensure: Target complete graph $G = (V, E)$

$\quad CG = G_1 \cup G_2 \cup ... \cup G_n$

$\quad G = \emptyset$

$\quad$ **while** $G \neq CG$ **do**

$\quad\quad$ **for all** uncovered graphs G_i **do**

$\quad\quad\quad \alpha_i = |G_i \cap (CG - G)|$

$\quad\quad$ **end for**

$\quad\quad$ Find uncovered graph G_i s.t. $G_i = \mathrm{argmax}_i \{\alpha_i\}$

$\quad\quad G = G \cup G_i$

$\quad\quad$ Add G_i to the list of covered graphs

$\quad$ **end while**

---

the next graph that covers the most remaining edges; this continues until all edges are covered. At the end of the algorithm, graph G will be a complete graph. A detailed description of this approach is shown in Algorithm 1.

5.3.5 Dynamic Design Decision

Static action coverage for a movement monitoring system finds the minimum number of active nodes that cover all actions. However, the model can be used for the dynamic deactivation of nodes. Once the action has occurred, each node classifies it individually. The final classification involves some notion of collaboration between the nodes in real-time. The dynamic sensor selection tends to find even smaller set of nodes based on current classification results obtained by individual nodes. In the following, details of this technique are explained through an example, where the system consists of three sensor nodes with compatibility graphs shown in Fig. 5.4. This system monitors subjects for the five movements A, B, C, D, and E. Sensor nodes I, II, and, III classify the movement as A, A, and E, respectively. These classified movements are shown as shaded vertices. The compatibility graph

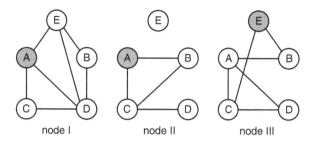

Fig. 5.4 Compatibility graphs for dynamic design decisions

for node I indicates that the target movement could be A or B as this node is not able to distinguish between A and B. The graph for node II indicates that the movement could be A, D, or E; and for node III, target movement could be one of E, D, or A. By intersecting these possibilities, global classification would result movement A. However, not both nodes II and III are required to determine this; one or the other is sufficient. Therefore, power can be potentially reduced by eliminating one of the nodes before initiating communication.

Hence, we propose the following approach: First, select a master node. This is done by selecting the node whose target movement vertex has the highest out degree. In this case, in the compatibility graph for *node I*, movement *A* has an out-degree of three, for *node II*, movement *B* has an out-degree of two; and for *node III*, movement *D* also has an out-degree of two. Therefore, *node I* should be the master. Next, add the master node to the solution space. Then, apply the action coverage problem from the master node's point of view and find the minimum number of nodes that will achieve full coverage of the target movement. In this case, only the edge (A, B) is missing from the master node, which can be covered by either of the remaining nodes. Finally, obtain the set of possible classifications from each of the remaining nodes (including the master), and intersect them to achieve final classification. Assume the action coverage allows *nodes I* and *II* to be the active nodes. The results issued are $\{A, B\}$ and $\{B, E, D\}$, leaving B as the final target movement.

5.3.6 Experimental Analysis

The results discussed in this section are based on an experiment reported in [3]. Eight sensor nodes are placed on a subject using TelosB with the custom-designed sensor board in Fig. 5.1a. The location of each sensor node is listed in Table 5.1. For each of the five data streams (x, y, z acceleration and x, y angular velocity), seven features including Mean, Start-to-End, Standard Deviation, Peak-to-Peak Amplitude, RMS power, Median and Maximum value are extracted. Three male test

Table 5.1 Mote locations

No.	Description
1	Waist
2	Left-forearm
3	Left-arm
4	Right-forearm
5	Right-ankle
6	Right-thigh
7	Left-ankle
8	Left-thigh

Table 5.2 Movements for experimental analysis

No.	Description	Category
1	Stand to sit	Full
2	Sit to stand	Full
3	Stand to sit to stand	Full
4	Sit to lie	Full
5	Lie to sit	Full
6	Sit to lie to sit	Full
7	Bend and Grasp	Upper
8	Kneeling, right leg first	Lower
9	Kneeling, left leg first	Lower
10	Turn clockwise 90°	Turning
11	Turn counter clockwise 90°	Turning
12	Turn clockwise 360°	Turning
13	Turn counter clockwise 360°	Turning
14	Look back clockwise	Upper
15	Move forward (1 step)	Full
16	Move backward (1 step)	Full
17	Move to the left (1 step)	Full
18	Move to the right (1 step)	Full
19	Reach up with one hand	Upper
20	Reach up with two hands	Upper
21	Grasp an object with right hand, turn 90° and release	Turning
22	Grasp an object with two hands, turn 90° and release	Turning
23	Jumping	Full
24	Going upstairs (2 stairs)	Lower
25	Going downstairs (2 stairs)	Lower

subjects between the ages of 25 and 35 perform the 25 movements listed in Table 5.2, for ten trials each.

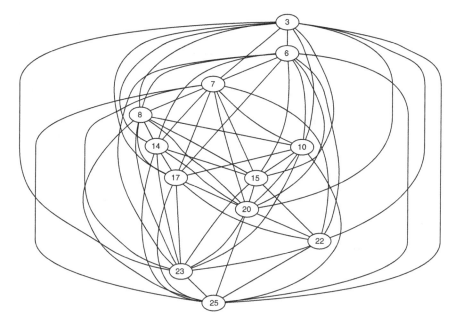

Fig. 5.5 Compatibility graph based on data from the waist node for 12 movements

5.3.6.1   Compatibility Graphs

For each sensor node, the Bhattacharyya distance is calculated between all movement pairs, and compatibility graphs are generated. A compatibility graph generated from data collected from the "waist" node is shown in Fig. 5.5. For this figure,
a subset of movements is shown for simplicity. Each vertex corresponds to a
movement as labeled in Table 5.2. The edges represent pairs of compatible movements. For example, there are edges between vertex 10 and all other vertices except
14 and 22. This means the action "turn clockwise 90°" can be distinguished with a
high level of confidence from all actions except "look back clockwise" and "grasp
an object with two hands."

5.3.6.2   Static Design Coverage

In this section, the ILP and greedy approaches are compared using the experimental data mentioned earlier. Using both algorithms, the number of nodes needed to
distinguish between all 25 movements is determined. For a more comprehensive
study, the movements are divided into four mutually exclusive subsets, shown
under the "Category" label in Table 5.2. The split is performed intuitively and is

Table 5.3 Solutions to action coverage problem

Movements Mote #	ILP solution 1 2 3 4 5 6 7 8	Greedy solution 1 2 3 4 5 6 7 8	#Nodes	Power saving (%)	Execution time (s)
All	0 0 1 1 0 1 1 0	1 1 0 1 0 1 1 0	4	%50	6.761290
Upper body	0 1 0 0 0 0 0 0	0 1 0 0 0 0 0 0	1	87	0.082245
Lower body	0 0 0 0 1 0 0 0	0 0 0 0 1 0 0 0	1	87	0.060753
Turning	0 1 0 1 0 0 0 0	0 1 0 1 0 0 0 0	2	75	0.065421
Full body	0 0 0 0 0 1 1 0	1 1 0 0 0 1 0 0	2	75	0.074850

Table 5.4 Classification analysis

K (No reduction)	Accuracy (%) (static)	Accuracy (%) (dynamic)	Accuracy (%) (dynamic)	#Nodes
1	97.5	96.0	92.2	1.84
2	96.6	95.0	89.9	1.85
3	96.6	94.3	88.9	1.85
4	94.8	95.2	89.1	1.85
5	95.4	92.9	88.2	1.86
6	91.2	89.5	81.1	1.86
7	90.3	90.3	80.8	1.85
8	89.5	88.0	78.9	1.86

based on the level of involvement of body segments in each movement. This categorization provides meaningful information for designing a system that is restricted to monitoring particular movements. The idea comes from the fact that in many medical applications, only a subset of actions are valid movements with respect to the temporal and spatial conditions. In the temporal case, the set of actions that are addressed changes from time to time while in the spatial case, movements of interest are updated when the subject moves to a new geographical area. Physicians need to quantify the level of daily activities with respect to certain movements for their patients. Furthermore, the movements a person might perform can significantly change when cooking in the kitchen compared to going to the gym. We present results for a few categories, but the approach can be used for any subset of interest. Table 5.3 compares the performance of the two methods on the full set of movements and on each subset. The nodes that are selected to be active nodes are expressed by a "1" in the resulting pattern. For example, to provide distinguishability information for all 25 movements, the ILP technique chose nodes 3, 4, 6, and 7 while the greedy algorithm chose nodes 1, 2, 4, 6, and 7. As expected, the ILP generated a slightly smaller set of nodes compared to the greedy approach. As the results show, the system is capable of providing full coverage of the movements using only four sensor nodes placed on four different segments of the body. To test the effectiveness of action coverage, the classification accuracy before and after node reduction is compared. The results are shown in Table 5.4 where the second and third columns represent the accuracy using original data and the data collected from active nodes respectively. The very small differences between two cases (e.g., 1.5% for k = 1) demonstrate the capability of action coverage to reduce the number of active nodes while maintaining an acceptable quality of service.

Table 5.3 also presents the amount of power saving along with the running time of the ILP algorithm. The power saving shows the total percentage of the power preserved in the system compared to the case when no power reduction technique is applied. When running the linear programming optimization problem in MATLAB on a Dell Laptop with a 1.6 GHz Core 2 Duo processor, the execution time for 25 movements is about 6.8 s.

5.3.6.3 Dynamic Design Coverage

Throughout the classification, three of the trials for each subject and movement are used for training, and the remaining trials were used for validation. This gives a good division of data set into training and test sets. Since the experiments are carried out in a controlled environment, this split between training and test data provides good classification results as will be demonstrated later in this section. Per-node feature extraction is performed to obtain seven features for each action across five sensor readings. A data fusion strategy is employed to integrate data from different nodes at the feature level. Therefore, the final classifier would work on the same training and test sets, whereas the feature space has been extended by the sensor nodes. Only the four nodes selected by ILP for all movements are used for the dynamic analysis (see Table 5.4). Compatibility graphs are generated from the training set. Classification is performed using a k-NN classifier, where k varies from 1 to 8. This dynamic technique further reduces the number of active nodes to an average of 1.84 nodes (for k = 1) per classification.

5.3.6.4 Classifier Accuracy

Classification accuracy exhibits how confident 25 movements can be recognized. Therefore, accuracy can be defined as follows:

$$A = \frac{TP + TN}{N}, \tag{5.6}$$

where TP is the number of true positive samples, TN represents the number of true negative samples and N is the total number of test points.

As reported in [3], by feeding all the features from all eight motes into a k-NN classifier with k = 1, an accuracy reading of 97.5% can be obtained. Repeating this test using data from only the four nodes selected by the ILP (and shown in Table 5.3) gives an accuracy of 96.0%. The results based on dynamic design coverage solution provides an accuracy of 92.2%. The complete results are shown in Table 5.4.

5.4 Efficient Temporal Parameter Extraction

Human movement models often divide movements into parts. In walking the stride can be segmented into four different parts, and in golf and other sports, the swing is divided into sections based on the primary direction of motion. These parts are often divided based on key events, also called temporal parameters. When analyzing a movement, it is important to correctly locate these key events, and so automated techniques are needed. There exist many methods for dividing specific actions using data from specific sensors, but for new sensors or sensing positions, new techniques must be developed. To address this problem, this section introduces a generic method for temporal parameter extraction called the Hidden Markov Event Model based on Hidden Markov Models. This method can be quickly adapted to new movements and new sensors/sensor placements. This method is validated on a walking dataset using inertial sensors placed on various locations on a human body. The technique is designed to be computationally complex for training, but computationally simple at runtime to allow deployment on resource-constrained sensor nodes.

5.4.1 HMEM Training and Use

The Hidden Markov Event Model (HMEM) is the name of the introduced key event labeling system, which uses an HMM with a specific state structure and a modified training procedure designed to find key events. The model also adds a feature selection and model parametrization system based on Genetic Algorithms (GAs). The HMEM makes several assumptions about the underlying data: (1) there are a number of different event types, (2) the events always occur in a specific order and for cyclical movements they repeat, (3) every single event type is represented in every action, and (4) there are a number of unlabeled samples between two adjacent events.

A traditional pattern recognition technique used for time-varying signals is the Hidden Markov Model (HMM). A basic HMM describes a discrete-time Markov process. At a particular moment, the process is in just one state. At fixed time intervals, the process produces an output and then transitions to another state (or remains in the current state). The transitions and outputs are probabilistic and based exclusively on the present state. The process generates a sequence of outputs, and a corresponding sequence of states. The states cannot be directly observed, and are thus hidden. An HMM is completely described by initial state probabilities, state transition probabilities, and output probabilities. Algorithms exist to (1) build an

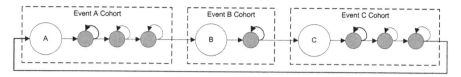

Fig. 5.6 HMEM model and structure

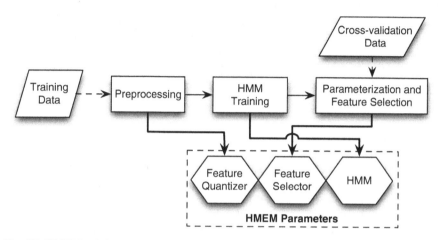

Fig. 5.7 HMEM training procedure

HMM to describe a given set of output sequences, (2) choose which of several models best describe an output sequence, (3) find the most likely current state of an HMM given the output sequence up until now, and (4) the most likely state sequence associated with a particular output sequence for a specified HMM [13, 14].

For a general HMM, it is possible for any state to transition to any other state. This is called an ergodic model. Another variant is to enforce a specific ordering for the states: each state can only transition to itself or state to the "right" of it in the ordering. This is called a left–right model [14].

Each key event can be represented by a unique state. Ideal events occur at a specific time but have no duration. However, given the idea that the key event might be associated with unique features in the observation sequence, the key event state should have a one sample duration. The HMEM encodes this concept into an HMM by removing the self-transition from states associated with key events, forcing a transition after one sample. The samples between key events are represented by transition states, which support both self-transitions and forward-transitions as seen in Fig. 5.6. States are grouped into cohorts which start with a key event state and end with the last transition state before the next key event state. For any observation sequence in the training data, the positions of the key event states are known. This means that training each cohort independently is identical to training the whole system at once.

5.4.2   Overview

There are several stages required to train the HMEM as shown in Fig. 5.7.

5.4.2.1   Preprocessing and Feature Extraction

The signal is filtered with a moving average filter to remove high frequency noise. Then it is normalized by subtracting a large-window mean and dividing by a large-window standard deviation. Several parameters representing the action data inside the signal, referred to as features, are extracted at each sample. These features are further quantized with a ten-level uniform quantizer based on the range of the features in the training data.

5.4.2.2   HMM Training

The HMM is effectively a finite state machine with probabilistic transitions and certain emission probabilities. These probabilities must be specified as part of the mathematical model defining an HMM. The exact locations of all key events states and what observations they emit are known from the annotations in the training data. However, the number of transition states and their transition and emission probabilities are unknown and must be trained. There are several well-known techniques for training HMMs, including Baum-Welch and Viterbi Path Counting [13, 15].

The training data is segmented using the canonical annotations. Each cohort is trained independently using a set of segments that start with a sample that should be labeled with the cohorts event and end just before the next labeled event. According to this model, the first state must be the cohort's event state, and the last sample must be associated with the last transition state in the cohort. During training, it is important to make sure that all considered paths meet this constraint. Viterbi Path Counting produces a single path for each event that can be edited to meet the constraints if necessary, while Baum-Welch can also be constrained in this way, VPC is much faster, which is important given the already high training times. This feature is one of the primary reasons for choosing VPC over Baum-Welch. The details of the training process are discussed in Sect. 5.4.3.

5.4.2.3   Parametrization and Feature Selection

HMMs are trained to represent a process, not to minimize segmentation error. It is possible to explicitly attempt to increase classification accuracy by choosing model parameters with that goal in mind. A genetic algorithm with uniform crossover is used to train the model. The population fitness is evaluated using the training model, and then at the end, the model that gives the best results for the cross-validation data is selected. Further discussion follows in Sect. 5.4.4.

5.4.3 HMM Training and the Viterbi Algorithm

A Markov process has N states $S = \{ s_1, s_2, \cdots, s_N \}$, and can emit M observations $X = \{ x_1, x_2, \cdots, x_M \}$. For a given observation sequence $O_T = (o_1, o_2, \cdots, o_T)$ with

---

Algorithm 2 HMM Training Procedure

Require: λ_0, $\mathscr{O} = \{ O^1_{(T_1)}, O^2_{(T_2)}, \cdots, O^Y_{(T_Y)} \}$

1:   $\lambda \leftarrow \lambda_0$
2:   **for** $i \leftarrow 1$ to K **do** {estimates from $\mathscr{Q}$}
3:     $\mathscr{Q} \leftarrow \emptyset$
4:     **for all** $O_{(T)} \in \mathscr{O}$ **do**
5:       $Q_{(T)} \leftarrow$ collect weighted state sequences using λ
6:       $\mathscr{Q} \leftarrow \mathscr{Q} \cup Q_{(T)}$
7:     **end for**
8:     $\bar{\pi}_i = \dfrac{\text{number of times in state } s_i \text{ at time 1}}{\text{number of times at time 1}}$
9:     $\bar{a}_{ij} = \dfrac{\text{number of transitions from state } s_i \text{ to state } s_j}{\text{number of transitions from state } s_i}$
10:    $\bar{b}_j(k) = \dfrac{\text{number of times in state } s_j \text{ observing symbol } x_k}{\text{number of times in state } s_j}$
11:    $\lambda \leftarrow \{ \bar{\pi}, \bar{A}, \bar{B} \}$
12: **end for**

T observations, there is a corresponding state sequence $Q_{(T)} = (q_1, q_2, \cdots, q_T)$. The Hidden Markov Model $\lambda = \{ \pi, A, B \}$ is defined by three sets of probabilities: initial state probabilities $\pi = \{ \pi_i | \pi_i = P(q_o = s_i) \}$, state transition probabilities $A = \{ a_{ij} | a_{ij} = P(q = s_j | q_{prev} = s_i) \}$, and observation probabilities $B = \{ b_j(k) | b_j(k) = P(o = x_k | q = s_j) \}$.

The most common training algorithm is the Baum-Welch algorithm [13]; however, a newer algorithm, the Viterbi Path Counting (VPC) algorithm is more appropriate for this work [15]. Both algorithms follow the training procedure shown in Algorithm 2. They start with a fixed number of states and an initial set of model parameters, then extract probabilistically weighted state sequences using the model parameters. Next, the transition and emission probabilities are updated based on the transitions and observations associated with each state sequence. The initial model is updated with these new probabilities. This process repeats until some desired level of convergence is reached. It will implicitly converge to a local minima.

The key to Viterbi Path Counting is extracting the most likely state sequence.

$$Q_{(T)\,\max} = \arg_{Q_{(T)} \in S^T} \max P\left(Q_{(T)}, O_{(T)}\right) \tag{5.7}$$

A dynamic programing algorithm, called the Viterbi algorithm [13], solves this problem. Using the most likely state sequence extracted using the Viterbi algorithm, the transition and emission probabilities are found simply by counting the occurrences in all the most likely state sequences for each observation sequence in the training set. Since we are training one cohort at a time, each observation sequence is a sequence starting on the key event and ending right before the next key event.

Maximizing the probability is equivalent to maximizing the log probability, therefore we use log probabilities to prevent numerical underflow and facilitate faster computing. The order of the Viterbi algorithm for a left–right model with independent features is $O(\text{Viterbi}) = O(TxN) \cdot O(Pr_{est})$, where $O(Pr_{est})$ is the order of algorithm required to estimate probability. This order is constant time for the state transition probability, but based on the number of features for the observation probability estimation. This means $O(Viterbi) = O(TxNx|F|)$.

5.4.4 Feature Selection and Model Parametrization Using Genetic Algorithms

The choice of whether or not to include each feature and the choice of the number of transition states for the cohorts are all tunable parameters of the HMEM. The feature selection $\Psi = \{\psi_k | \psi_k \in \{0, 1\}, i = 1, \ldots, |F|\}$ represents a choice of which features out of an exhaustive list are to be included and which are to be discarded. The number of selected features is $|\Psi| = \sum_k \psi_k$. Feature k is selected if $\psi_k = 1$ and is discarded if $\psi_k = 0$. The other parameter is the number of transitions states for each cohort, $\Omega = \{\omega_e | \omega_e \in \{1, \ldots, 5\}, e = 1 \ldots E\}$, where E is the number of key events. The full HMEM model is represented by $\lambda_{\text{HMEM}} = \{\lambda_{\text{HMM}}, \Psi, \Omega\}$.

In essence, parametrization consists of choosing several good models based on one or more objective functions applied to the training set. These models are then

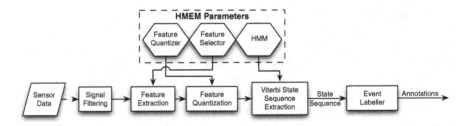

Fig. 5.8 Application of the HMEM as described in Sect. 5.4.5

compared against each other using the same objective function(s) applied to the cross-validation set. The best model on the cross-validation set is chosen. Because the cross-validation set exclusively contains data from subjects not in the training set, generalizability of the models to new subjects is improved. Genetic algorithms are used to generate the list of "good" models.

5.4.5 HMEM Application Procedure

After the HMEM is trained, it can be used to find key events in a data stream for the movement it has been trained on. The data flow for the algorithm is shown in Fig. 5.8. First, the data is filtered using the procedure described above, then features are extracted and quantized. Next, the feature selection is applied, and finally the most likely state sequence is extracted using the Viterbi algorithm. The annotation converter finds all the event states in this sequence and outputs an ordered set where each element consists of a time and event label.

5.4.6 Collaborative Segmentation

One of the main properties of a body sensor network is the capability of distributed sensing and collaboration. The design of the HMEM so far has been based on individual sensor nodes labeling the key events. An interesting question is "how much do the segmentation results improve by allowing collaboration between nodes?" A simple form of collaboration splits the nodes into a coordinator node and several operator nodes. It requires each operator node to extract the key events independently, then transmit the annotations to the coordinator node for final event labeling. This scheme has a relatively low communication overhead, and the star-topography also introduces less delay than a more general DAG collaboration scheme might.

HMEM already provides a framework for extracting key events that can be extended for use in collaborative segmentation. All the operator nodes independently label the events using an HMEM, which can observe the features extracted from their sensors. The coordinator node has an HMEM which not only observes the data from its own sensors, but also the labels assigned by the operator nodes. Since the HMEM requires an observation for each sample, samples unlabeled by a specific operator node receive a label of '0'.

5.4.7 Experimental Analysis

The results discussed in this section are based on an experiment reported in [16] and the node placement as described in Sect. 5.3.6. The results are reported with

Table 5.5 Subject error for R thigh with Accel and TP

Subject	Default (132 features)			GA (38 features)		
	P	R	RMSE	P	R	RMSE
Sub 2 train	100	100	1.44	100	100	1.62
Sub 3 train	100	100	1.33	100	100	1.11
Sub 4 cross	100	100	1.36	100	100	1.38
Sub 5 train	100	100	3.54	100	100	3.43
Sub 6 cross	100	100	1.03	100	100	1.02
Sub 7 cross	100	100	3.25	100	100	3.09
Sub 8 test	100	100	1.74	100	100	1.76
Sub 9 test	100	100	1.55	100	100	1.66

Table 5.6 Cross-validated subject error (R thigh)

Subject	P	R	RMSE	Fsel
Sub 2	100	100	1.47	6.7
Sub 3	100	100	1.38	8.7
Sub 4	100	100	1.68	4.7
Sub 5	99.8	99.8	3.55	4.7
Sub 6	100	100	1.20	6.7
Sub 7	100	100	3.09	28.7
Sub 8	100	100	1.52	10.7
Sub 9	100	100	1.59	16.0

precision (P), recall (R), Quality (RMSE). The first and last annotations were ignored because the algorithm needs context to determine annotations, and we are interested in the steady-state performance only. We show error for each mote using just the accelerometer readings, look at per-subject error for a poor performing sensor node and a well-performing sensor node.

5.4.7.1 Examination of Per-Subject Error

One of the goals of the HMEM system is good generalization to new subjects. Table 5.5 shows per-subject error for the sensor on the right thigh. Initially subjects two to four were in training, five to seven in cross validation, and eight and nine in test. However, subjects five and seven have walking patterns that differ significantly from the others, but are similar to each other. Therefore, subjects four and five were exchanged. It is likely that with a larger dataset the system could generalize better to such subjects. All the results are shown from the portion of the subjects' data that was in the test dataset.

The sensor on the right thigh, as shown in Table 5.5, performs well. Subjects eight and nine perform a little worse than subjects in the training and cross-validation sets. The worst per-subject error comes from subjects five and seven. The reason for this

Table 5.7 Sensor types with TP features on R thigh

Set	Default			Genetic algorithm			
	P	R	RMSE	P	R	RMSE	Fsel
Accel (132)	100	100	1.97	100	100	1.92	38
Gyro (88)	100	100	2.14	100	100	2.14	9
All (220)	100	100	1.84	100	100	1.84	33
Acc Mag (44)	99.4	99.8	2.02	99.9	100	1.97	5

is not entirely clear. The use of the GA does not significantly reduce the discrepancy in per-subject error. Since the final selection criteria for the solution is minimum total error, not minimum worst-case subject error, this is not surprising.

Manual partitioning of the data into training, cross validation, and testing sets can artificially bias the results. Therefore, we performed an experiment for each subject where the subject was placed exclusively in the testing set, and the training and cross-validation sets were selected randomly from the remaining subjects. The results shown in Table 5.6 are the average of three tests after the GA. These results demonstrate that the model has good generalization to many subjects, but performs poorly on some. It would be interesting as a future work to investigate the features of those subjects that cause the model to perform poorly.

5.4.7.2 Exploration of Different Sensor Types

Another goal for HMEM is the ability to use new sensors and combination of sensors without having to develop new methods to extract the key events. To simulate this, we examine the HMEM trained with different subsets of sensors, as shown in Table 5.7. The sensor types considered were (a) Accelerometer only, (b) Gyroscope only, (c) All sensors, and (d) just the magnitude from the accelerometer.

The accelerometer performs better than the gyroscope and a combination of all the sensors has the best performance. However, the most interesting test is using just the magnitude of the accelerometer. The magnitude of the accelerometer would be invariant for any rotation of the sensor. This could be especially important if the sensor is a cellphone in the subject's pocket. Even though exclusive reliance on the magnitude of acceleration gives the worse results, the performance is still reasonable.

5.4.7.3 Explicit Feature Reduction

Feature selection can be used to explicitly reduce the number of features. The algorithmic time for the HMEM event annotation algorithm increases linearly with the number of features, so feature reduction can improve performance

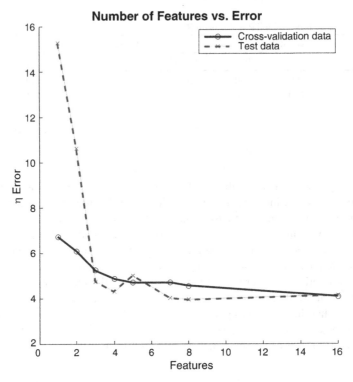

Fig. 5.9 Number of features vs. error for the right thigh

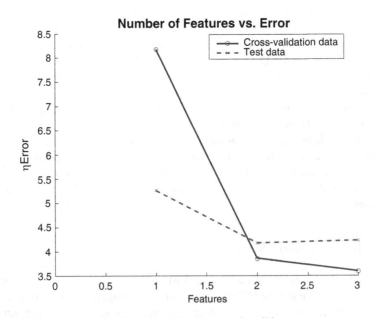

Fig. 5.10 Features vs. error for the right thigh with expansive elitism

considerably. The NSGA framework can be used as described above with the objectives of global error minimization and feature minimization.

The two objectives imply a two-dimensional pareto front. We take the population at the final generation and select the pareto front when the population error is judged using the cross-validation set. The same trained models are then judged against the test set. The results from both the test and cross-validation sets are shown in Figs. 5.9 and 5.10. The error is reported using the training error η, which is approximately equal to the square of the RMSE. Figure 5.9 shows the results where each generation saves just the pareto front for the training data. The GA used to generate Fig. 5.10 saved several successive fronts so that at least twenty of the best were saved each generation, resulting in lower error. Further, using just the two features found in Fig. 5.10 results in performance almost as good as with no feature reduction. The starting number of features was 132, so this results in a performance increase of approximately 66 times.

Moreover, while the performance on the cross-validation set and the test set are different, both have an "elbow" at the same place, where the error increases drastically with an increased number of features. This suggests that an effective way of picking the best HMEM is to pick the HMEM right before the cross-validation error starts increasing significantly.

5.5 Summary

The focus of this chapter was on signal processing platform for the creation of wearable BSNs. In this chapter, we investigated the use of BSNs for physical movement monitoring where body-mounted inertial sensors are embedded in mobile nodes to detect the movements of the subject wearing the system. In Sect. 5.1, we started with addressing wearability issues in BSNs. Sect. 5.2 discussed about system architecture and signal processing components required for human movement recognition. Since batteries are the heaviest components of the system, reducing power usage leads to smaller batteries and improves wearability. Keeping this in mind, the last two sections discussed two recently discovered techniques for power optimization in BSNs. The first technique, which centered around the problem of optimal node placement, aimed to minimize the number of sensor nodes required to classify a predefined set of human movements. The second one was a temporal parameter extraction technique for accurate movement recognition that is simple to implement.

References

1. Akyildiz I, Su W, Sankarasubramaniam Y, Cayirci E (2002) A survey on sensor networks. IEEE Commun Mag 40(8):102–114

2. Gemperle F, Kasabach C, Stivoric J, Bauer M, Martin R (1998) Design for wearability. In: Proceedings of the 2nd IEEE International Symposium on Wearable Computers, Citeseer, p 116
3. Ghasemzadeh H, Guenterberg E, Jafari R (2009) Energy-efficient information-driven coverage for physical movement monitoring in body sensor networks. IEEE J Sel Area Commun 27:58–69
4. Polastre J, Szewczyk R, Culler D (2005) Telos: Enabling ultra-low power wireless research. Information processing in sensor networks, 2005. Fourth International Symposium on IPSN 2005, pp 364–369, doi:10.1109/IPSN.2005.1440950
5. Ghasemzadeh H, Barnes J, Guenterberg E, Jafari R (2008) A phonological expression for physical movement monitoring in body sensor networks. In: Mobile ad hoc and sensor systems, 2008. 5th IEEE International Conference on MASS 2008, pp 58–68
6. Stergiou N (2003) Innovative analyses of human movement: analytical tools for human movement research. Human kinetics
7. Ghasemzadeh H, Guenterberg E, Gilani K, Jafari R (2008) Action coverage formulation for power optimization in body sensor networks. In: Design automation conference, 2008. Asia and South Pacific ASPDAC 2008, pp 446–451
8. Ghasemzadeh H, Loseu V, Jafari R (2009) Structural action recognition in body sensor networks: Distributed classification based on string matching. IEEE Trans Inform Technol Biomed
9. Duda R, Hart P, Stork D (1973) Pattern classification and scene analysis. Wiley, NY
10. Jafari R, Bajcsy R, Glaser S, Gnade B, Sgroi M, Sastry S (2007) Platform design for healthcare monitoring applications. High Confidence Medical Devices, Software, and Systems and Medical Device Plug-and-Play Interoperability, 2007 HCMDSS-MDPnP Joint Workshop on pp 88–94
11. Bhattacharyya A (1943) On a measure of divergence between two statistical populations defined by their probability distributions. Bull Calcutta Math Soc 35:99–109
12. Duda RO, Hart PE, Stork DG (2000) Pattern classification. Wiley, NY
13. Rabiner L, Juang B (1986) An introduction to hidden Markov models. IEEE ASSP Mag 3(1 Part 1):4–16
14. Rabiner L (1989) A tutorial on hidden Markov models and selected applications inspeech recognition. Proc IEEE 77(2):257–286
15. Liu N, Lovell B, Kootsookos P (2003) Evaluation of HMM training algorithms for letter hand gesture recognition. In: Signal processing and information technology, 2003. Proceedings of the 3rd IEEE International Symposium on ISSPIT 2003, pp 648–651
16. Guenterberg E, Ghasemzadeh H, Jafari R (2009) A distributed hidden Markov model for fine-grained annotation in body sensor networks. In: Proceedings of the 2009 Sixth International Workshop on Wearable and Implantable Body Sensor Networks-Volume 00, IEEE Computer Society, pp 339–344

Chapter 6
Signal Data Mining from Wearable Systems

Francois G. Meyer

6.1 Definition of the Subject

6.1.1 Introduction

Sensors from wearable systems can be analyzed in real-time on-site, or can be transmitted to a central hub to be analyzed off-line. In both cases, the goal of the analysis is to extract from the measurements information about the state of the user, and identify anomalous behavior to alert the person. The notion of state depends obviously on the particular application, but in general characterizes a high-level function: the user is awake (as opposed to asleep), the user is falling, the user is going to have a heart attack, etc. Data analysis relies on sophisticated statistical machine learning methods to learn from existing training examples the association between sensors and high-level states [1, 2]. The first stage of the analysis consists in extracting meaningful features, and in removing confounding artifacts. This stage can be achieved using various time-frequency or multiscale methods. The second stage consists in reducing the dimensionality of the data. Indeed, we know that it becomes extremely difficult to learn a function of the data, when the data are in very high dimension. Linear methods to reduce dimensionality include principal component analysis (PCA) and independent component analysis (ICA). Recently, nonlinear methods, such as Laplacian eigenmaps, offer powerful alternatives to traditional linear methods. Finally, one is ready to learn a function of the measurements that describes the state of the person wearing the devices. The sensors only provide very coarse and indirect measurements about the state of the user. For instance, one may be interested in classifying the state of the user into "normal states" or "anomalous states" (the user fell asleep, or is going to have a heart attack). It is therefore necessary to learn the state of the user as a function of the measurements using statistical learning methods.

F.G. Meyer (✉)
University of Colorado at Boulder, Boulder CO 80309, USA
e-mail: fmeyer@colorado.edu

A. Bonfiglio and D. De Rossi (eds.), *Wearable Monitoring Systems*,
DOI 10.1007/978-1-4419-7384-9_6, © Springer Science+Business Media, LLC 2011

This chapter is organized as follows. The current section provides an overview of the type of data analysis questions associated with wearable systems. Section 6.2 contains a description of the various feature extraction techniques used in wearable devices. The real-time analysis of data requires that an efficient dimension reduction be performed. Methods that can provide a faithful representation of the data with much fewer parameters are described in Sect. 6.3. Finally, statistical machine learning methods that are used to characterize the state of the user are described in Sect. 6.4. A glossary of the terms used in this chapter can be found in Sect. 6.6.

6.1.2 Shape of the Data

Sensor data can be one-dimensional (e.g., accelerometer) or two-dimensional (e.g., video). In this chapter, we will model sensor data as time series of scalars or vectors. Formally, a wearable system can generate a vector X of measurements in $\mathbb{R}^P$ (to simplify the notations and use the same formalism, we can think of an image as a vector properly reorganized). As time evolves, we can index each measurement with a time index, and we denote by $x_{i,j}$ the measurement that sensor j generates at times $t_i, i = 1, \ldots, N$. Clearly, the temporal dimension plays a different role than the sensor index j in this dataset, and methods of analysis usually take advantage of this distinction.

6.1.3 Scientific Questions

The data generated by the sensors of a wearable system can be used to answer questions about the state of an individual or the state of a population of individuals.

6.1.3.1 At the Level of the Individual

Wearable computers can provide information about the environment surrounding the user, as well as information about the state (e.g., activity and health monitoring) of the user. In both cases, the information is centered around the individual user and does not involve a network of wearables. The surrounding, or context, can be characterized by the location of the user [3], the amount of noise [4], and the output of several cameras [5]. The detection of the activity of the user [6, 7] is a process that is intrinsically dynamic and requires real-time computations. Finally, a wearable system can be used to monitor the health of the user [2, 8] and diagnose and detect anomalous events [9].

6.1.3.2 At the Level of a Group of Individuals

Wearable systems can also be used to analyze social interaction between different users and monitor social networks [10].

6.1.4   Local vs. Remote Analysis

The analysis of the data collected by the sensors can either be performed locally, using the limited computation and power resources available on the wearable system, or be sent to a hub, where remote computations can be performed.

6.1.4.1   Local or On-Site Analysis

This type of analysis can provide an immediate feedback without requiring any communication to a central hub. An on-site analysis would, for instance, be useful for the continuous health monitoring of individuals living in remote areas without access to wireless networks [11]. Wearable devices can be designed around micro-controllers [12], DSP chips [13, 14, 15], or Field Programmable Gate Array (FPGA) [16, 17]. In all cases, the complexity of the algorithms that can be programmed is limited by the computational and battery power available on the device. Such limitations may prevent the usage of classification methods that require computationally expensive algorithms and massive amounts of training data. Finally, the degree of integration of the technology (handheld vs. wearable) may further limit the amount of computational power.

6.1.4.2   Remote Analysis

The on-site analysis may be supplemented with, or replaced by, a remote analysis. In this scenario, the data harvested by the sensors may be pre-processed on site to eliminate artifacts, and then sent to a central computer, where a more complex analysis is performed. Typically, a wireless connection to a centralized computer allows the wearable device to send the data to a central hub, where the remote analysis is performed. For instance, in the case of medical monitoring, medical data can be sent to a health care center where diagnostic testing and monitoring are performed [18, 19, 2]. This type of processing makes it possible to use machine learning algorithms that are computationally intensive and require large amount of training data [20, 21]. The wireless connection can take advantage of wireless personal area networks standards such as the ZigBee specification: a suite of high level communication protocols using small, low-power digital radios based on the IEEE 802.15.4-2006 standard [20].

6.2   Feature Extraction

The very large size of the time series collected by wearable sensors is a basic hurdle to any attempt at analyzing the data. Consequently, the analysis of sensors is usually performed on a smaller set of features extracted from the data [22]. The extraction of features serves two purposes: first it reduces the dimensionality by replacing the time-series with a more compact representation in a transformed domain (e.g., Fourier, or wavelet); second, it separates the artifact and noise from the signal.

6.2.1 Time-Frequency Analysis

Many of the transforms that are used to extract features operate in the frequency domain. This can be justified by a theoretical argument: if the signals are realizations of a wide sense stationary process, then the Fourier transform provides the optimal representation [23]. In practice, many physiological signals are oscillatory and can be decomposed as a sum of a small number of sinusoidal functions. For instance, [11] use Fourier coefficients of motion sensors to study gait.

However, in many applications the signals of interest are not stationary: they contain sudden changes and the local statistical properties of the signals vary as a function of time. The ability to detect sudden changes in the local frequency content is essential (e.g., prediction of seizures [24]). One alternative to the Fourier transform consists in dividing the time series into overlapping segments within which the signal can be expanded using a Fourier transform. This local Fourier analysis requires windowing functions to isolate the different time segments. Formally, we consider a cover of the time axis $\bigcup_{n=-\infty}^{n=+\infty}[a_n, a_{n+1})$, and we write $I_n = [a_{n-1}, a_n)$. The time intervals I_n can be of fixed or adaptive sizes. To localize f around the interval of interest I_n, one can construct a projection of f, $P_{[a_n,a_{n+1}]}f$ (see Fig. 6.1). The simplest example of $P_{[a_n,a_{n+1}]}f$ is obtained by multiplying f by a smooth window function ρ_n, whose support is approximately I_n,

$$P_{[a_n,a_{n+1}]} : f \rightarrow \rho_n f. \tag{6.1}$$

We can then compute the Fourier transform of $P_{[a_n,a_{n+1}]}f$ using a fast Fourier transform.

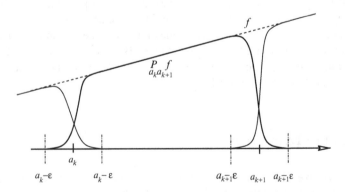

Fig. 6.1 Localization of a time interval before computing a Fourier transform

Features are then extracted by computing the energy (estimated from the magnitude squared of the Fourier transform) present in specific frequency bands. These frequency bands are determined from prior experiments, or from a priori physiological knowledge. For instance, [25] use Fourier analysis to extract relevant features from raw tremor acceleration data around a small set of frequencies of interest (3 Hz, 4 Hz, 5 Hz, and 6 Hz).

6.2.1.1   Application to Wearable Systems

The authors in [13] use a spectral representation of ECG and respiratory effort signals to estimate sleepiness and distinguish sleep from wake activity. The approach relies on the computation of the energy of the Fourier transform (spectrogram) over time intervals of length 40 s. The computation of the Fourier transform can be performed using DSP chips or FPGA, and can be performed on site [26].

6.2.1.2   Available Software

- The Time-Frequency Toolbox, http://tftb.nongnu.org/, is a collection of MATLAB functions for the analysis of nonstationary signals using time-frequency distributions.
- See also WaveLab in the next section.

6.2.2   Multiscale Analysis

While time-frequency methods can provide good energy compaction, they are not suited for analyzing phenomena that occur at very different scales (from a second to a day). One of the main limitations of Fourier-based algorithms is their inability to exploit the multiscale structure that most natural signals exhibit. As opposed to the short-time Fourier transform, the wavelet transform performs a multiscale, or

Algorithm 1: Fast wavelet transform

$$\text{Initialization}: \quad s_k^J = x_{k+1} \quad k = 0, \ldots, N-1 \tag{6.2}$$

$$\text{Iterate}: \text{for scale} \quad j = J \quad \text{down to} \quad 1: \tag{6.3}$$

$$\text{Low pass filter}: \quad s_k^{j-1} = \sum_n h_{n-2k} \, s_n^j \qquad k = 0, \ldots, 2^{j-1-J}N-1 \tag{6.4}$$

$$\text{High pass filter}: \quad d_k^{j-1} = \sum_n g_{n-2k} \, s_n^j \qquad k = 0, \ldots, 2^{j-1-J}N-1 \tag{6.5}$$

Fig. 6.2 Fast wavelet transform

multiresolution, analysis of the signal. The wavelet transform is an orthonormal transform that provides a very efficient decorrelation of many physical signals [23]. Excellent references in the mathematical theory of wavelets and their application to image compression include [23, 27]. We describe briefly the fast wavelet transform algorithm. We consider the time series $x_n, n = 1, \ldots, N$ generated from one specific sensor from time t_1 until t_N. For simplicity, we assume that $N = 2^J$. The following algorithm was discovered by [23], and is called the *fast wavelet transform*. The wavelet transform of X at scale J is given by the vector of coefficients

$$\left[s_0^0, d_0^0, d_0^1, d_1^1, \ldots \ldots, d_0^j, \ldots d_{2^{j-J}N-1}^j, \ldots \ldots, d_0^{J-1}, \ldots \ldots \ldots \ldots, d_{2^{-1}N-1}^{J-1} \right]. \quad (6.6)$$

The reconstruction formula is given by the following iteration:

$$s_k^{j+1} = \sum_{n \in \mathbf{Z}} h_{k-2n} s_n^j + \sum_{n \in \mathbf{Z}} g_{k-2n} d_n^j. \quad (6.7)$$

The fast wavelet transform has an overall complexity of $O(N)$ operations. As shown in Fig. 6.3, the algorithm organizes itself into a binary tree, where the shaded nodes of the tree represent the subspaces W_j.

One important parameter in the wavelet analysis concerns the choice of the filters (h_n, g_n). In general, biorthogonal wavelet filters with linear phase introduce less distortion than orthonormal wavelets. While longer filters provide better out of band rejection with a sharp frequency cutoff, such filters may become a computational burden if a real-time on-site application is required. Finally, the number of vanishing moments of the wavelet filter controls the number of significant coefficients that one should expect when processing smooth signals. Indeed, polynomials of degree $p - 1$ will have a very sparse representation in a wavelet basis with p vanishing moments: all the d_k^j are equal to zero, except for the coefficients located

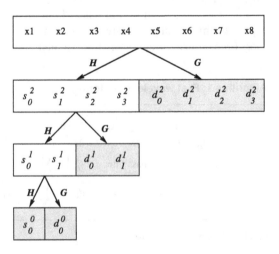

Fig. 6.3 Pyramidal structure of the fast wavelet transform

at the border of the dyadic subdivision $(k = 0, 1, 2, 4, \cdots, 2^{J-1})$. Unfortunately, wavelet with p vanishing moments have at least $2p$ coefficients, thereby increasing the computational load.

In a manner similar to Fourier analysis, one can decide a priori that certain frequency bands, which are associated with specific wavelet coefficients, should beused for the subsequent analysis of the data. Alternatively, one can remove small coefficients, which are typically generated by the noise, and keep only the largest coefficients that come from the signal. In both cases, the selected wavelet coefficients become the features representing the time series.

6.2.2.1 Application to Wearable Systems

A wavelet transform is used in [28, 29] to extract important features from ECG recordings. Similarly, data from inertial sensors are processed with filter banks in [25] to quantify tremor frequency and energy. Finally, [26] use a wavelet transform to remove the noise and smooth accelerometers time series. Wavelet analysis can be performed on site, since it can be implemented efficiently on a DSP chip or an FPGA.

6.2.2.2 Available Software

- WaveLab at Stanford University, http://www-stat.stanford.edu/~wavelab/, is a very comprehensive library of MATLAB software that implement wavelet transforms, wavelet packets, and other various time-frequency transforms.
- Wavelet toolbox in MATLAB.

6.3 Dimensionality Reduction

The extraction of features from the data effectively replaces long time series with shorter feature vectors (e.g., Fourier or wavelet coefficients). Often, this dimension reduction is not significant, and one needs to further reduce the dimension of the feature vectors. Alternatively, there are cases where standard features are not available, and one needs to apply methods to reduce dimensionality directly on the rawsensor measurements.

Methods to reduce dimension exploit the intrinsic correlations that exist in the feature vectors, or in the sensors' measurements. Principal components analysis finds the set of orthogonal components that can best explain the variance in the observations, while independent component analysis can decompose the observations into components that are statistically independent. Finally, Laplacian eigenmaps can provide a faithful parametrization of the sensor data when the data organize themselves in a nonlinear manner.

6.3.1   Principal Component Analysis

Instead of using a fixed transform, such as the Fourier or the wavelet transform, one can often construct a sparser representation by adaptively computing an optimal transform. Principal component analysis is one such adaptive transform.

We consider a wearable system equipped with several sensors. Let $x_{i,j}$ be the measurement that sensor j generates at times t_i, $i = 1, \ldots, N$. We can organize the sensor measurements as an $N \times p$ matrix

$$\mathbf{X} = \begin{bmatrix} x_{1,1} & \cdots & x_{1,p} \\ x_{2,1} & \cdots & x_{2,p} \\ \vdots & & \vdots \\ x_{N,1} & \cdots & x_{N,p} \end{bmatrix}. \tag{6.8}$$

We can think of the measurement generated by the sensors at time t_i as a vector $X_i = [x_{i,1}, \ldots, x_{i,p}]^T$ in $\mathbb{R}^p$, and we can write the matrix $\mathbf{X}$ by stacking horizontally the time series of sensor vectors,

$$\mathbf{X} = \begin{bmatrix} X_1^T \\ \hline \vdots \\ \hline X_N^T \end{bmatrix}. \tag{6.9}$$

As time evolves, the vectors $X_1, X_2, \ldots X_N$ form a trajectory in $\mathbb{R}^p$ (see Fig. 6.4). If the temporal sampling of the sensor is sufficiently fast, we expect that the discrete trajectory will be smooth, and that the points X_i will be highly correlated. The goal of principal component analysis (PCA) is to compute a low dimensional affine approximation to the set $X_1, X_2 \ldots X_N$. First, the set of measurements is centered around the center of mass

$$\overline{X} = \frac{1}{N} \sum_{i=1}^{N} X_i.$$

Then the optimal subspace of dimension r is computed from the singular value decomposition (SVD) [30] of the centered matrix

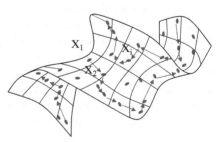

Fig. 6.4 Trajectory of the sensor vector X_i as a function of time

$$\begin{bmatrix} \dfrac{X_1^T - \overline{X}^T}{\vdots} \\ \overline{X_N^T - \overline{X}^T} \end{bmatrix} = \sum_{i=1}^{p} \sigma_i U_i V_i^T, \qquad (6.10)$$

where the vectors $U_1, \ldots, U_p$ (respectively $V_1, \ldots, V_p$) are mutually orthogonal [30]. In summary, the optimal affine approximation of rank r is given by

$$\overline{X} + \sum_{k=1}^{r} \sigma_k U_k V_k^T. \qquad (6.11)$$

This low dimensional approximation is guaranteed to maximize the dispersion (variance) of the projected points on the subspace formed by the vectors $U_1, \ldots, U_r$. Geometrically, the vectors U_i will be aligned with the successive orthogonal directions along which the data changes the most (see Fig. 6.5). From a linear algebra perspective, the affine model (6.7) results in the minimum approximation error – in the mean-squared sense – among all affine models of rank r.

Remark 1. *The eigenvalues σ_i determine the shape of the distribution of sensor measurements. If all the σ_i are equal, then all vectors U_i are equivalent. This is a degenerate case where PCA offers no gain. If the σ_i are very different, the distribution is shaped as an ellipsoid. PCA provides a gain by aligning the coordinate axes with the main axes of the ellipsoid (see Fig. 6.5).*

Remark 2. *If we include all the p components, then we have an exact decomposition of the measurements,*

$$X = \overline{X} + \sum_{k=1}^{p} \sigma_k U_k V_k^T = \overline{X} + \mathbf{U}\Sigma\mathbf{V}^T, \qquad (6.12)$$

where $\mathbf{U} = [U_1, \ldots, U_p]$, $\mathbf{V} = [V_1, \ldots, V_p]$ and $\Sigma = \mathrm{diag}(\sigma_1, \cdots, \sigma_p)$.

Remark 3. *PCA can also be used to "whiten" the data; a step which is often required before further analysis [14].*

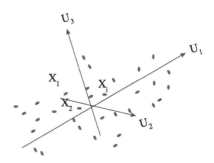

Fig. 6.5 Principal components U_1, U_2 and U_3

6.3.1.1 Application to Wearable Systems

PCA cannot be implemented on a DSP chip or an FPGA, but can be implemented on a small laptop, as described in [4], where a real-time analysis of the sensors using a time-varying PCA is computed. A clustering algorithm combined with self-organizing maps is also used to identify the main sensor clusters. The clusters are then updated in real-time. In [21], the authors combine feature extraction and PCA to analyze ECG recordings.

6.3.1.2 Available Software

The main ingredient of the PCA algorithm is the SVD decomposition of the matrix $\mathbf{X}$. There exist several implementations of SVD in Fortran, and in MATLAB.

6.3.2 Independent Component Analysis

Independent component analysis (ICA) seeks to decompose the data into a linear combination of statistically independent components [31]. The method assumes the following mixing model, where the vector of measurements X generated by the sensors at any given time can be decomposed in terms of p independent scalar "sources"

$$X = \begin{bmatrix} x_1 \\ \vdots \\ x_p \end{bmatrix} = \begin{bmatrix} a_{1,1} & \cdots & a_{1,p} \\ \vdots & & \vdots \\ a_{p,1} & \cdots & a_{p,p} \end{bmatrix} \begin{bmatrix} s_1 \\ \vdots \\ s_p \end{bmatrix}, \tag{6.13}$$

or equivalently in matrix form

$$X = \mathbf{A}S. \tag{6.14}$$

Because the variance of the sources $s_1, .., s_p$ cannot be estimated, we can assume that the sources are decorrelated (their covariance matrix is the identity) and the matrix $\mathbf{A}$ is orthogonal. A numerical solution to the estimation of the sources and the mixing matrix $\mathbf{A}$ can be obtained by maximizing the departure from Gaussianity of the vector $S = \mathbf{A}^{-1}X = \mathbf{A}^T X$. This can be achieved by maximizing the negentropy of each coordinate s_i of S [32].

It has been noted [31] that in practice, if the observations are noisy, then it becomes impossible to separate the components from the noise. In fact, if the analysis is performed on real data, then the components are not even approximately independent [31]. Finally, a common problem associated with the usage of ICA is the interpretation of the components. The interpretation usually relies on post hoc heuristics, such as visual inspection of the source time-series.

6.3.2.1 Application to Wearable Systems

ICA suffers from the same computational load as PCA, and cannot be easily implemented on small portable devices at the moment. Motion artifacts were identified in pulse oximeter signals using ICA [33]. But see [34] for a discussion of the validity of the assumption of independence of the motion artifact signals and the signal generated by arterial volume variations.

6.3.2.2 Available Software

- The implementation of Bell and Sejnowski can be found at http://www.cnl.salk. edu/~tony/ica.html.
- A fast ICA MATLAB package is available from the Laboratory of Computer and Information Science (CIS) at the Helsinki University of Technology, http:// www.cis.hut.fi/projects/ica/fastica/

6.3.3 Laplacian Eigenmaps

The methods of reduction of dimensionality described in the previous sections are linear: each vector of sensor data X_i is projected onto a set of components U_k. The resulting coefficients serve as the new coordinates in the low dimensional representation. However, in the presence of nonlinearity in the organization of the X_i in $\mathbb{R}^p$ (see Fig. 6.4), a linear mapping may distort local distances. These distortions will make the analysis of the dataset more difficult. We describe in this section a method to construct a nonlinear map Ψ to represent the dataset $\mathbf{X}$ in low dimensions. Because the map Ψ is able to preserve the local coupling between sensors vectors, low dimensional coherent structures can easily be detected with a clustering or classification algorithm.

We describe a low dimensional embedding of the set of sensor measurements $X_1, \ldots, X_N$ into $\mathbb{R}^m$, where $m \ll p$. The embedding is constructed with the eigenfunctions of the graph Laplacian [35]. First, we represent the measurements by a graph that is constructed as follows. Each sensor vector X_i becomes the node (or vertex) i of the graph. Edges between vertices quantify the similarity of sensor vectors. Each node i is connected to its n_n nearest neighbors according to the Euclidean distance $\|X_i - X_j\|$. Finally, a weight $W_{i,j}$ on the edge $\{i, j\}$ is defined as follows,

$$W_{i,j} = \begin{cases} e^{-\|X_i - X_j\|^2/\sigma^2}, & \text{if } i \text{ is connected to } j, \\ 0 & \text{otherwise.} \end{cases} \quad (6.15)$$

The weighted graph G is fully characterized by the $N^2 \times N^2$ weight matrix $\mathbf{W}$ with entries $W_{i,j}$. Let $\mathbf{D}$ be the diagonal degree matrix with entries $d_i = \sum_j W_{i,j}$.

The map Ψ is designed to preserve short-range (local) distances, as measured by $W_{i,j}$. The map is constructed one coordinate at a time. Each coordinate function ψ_k is the solution to the following minimization problem

$$\min_{\psi_k} \frac{\sum_{X_i \sim X_j} W_{i,j}\left(\psi_k(X_i) - \psi_k(X_j)\right)^2}{\sum_i D_{i,i} \Psi_k^2(X_i)}, \qquad (6.16)$$

where ψ_k is orthogonal to the previous functions $\{\psi_0, \psi_1, \cdots, \Psi_{k-1}\}$,

$$\langle \psi_k, \psi_j \rangle = \sum_{i=1}^{N} D_{i,i} \Psi_k(X_i) \Psi_j(X_i) = 0 \qquad (j = 1, \ldots, k-1). \qquad (6.17)$$

The numerator of the Rayleigh ratio (6.12) is a weighted sum of the gradient of ψ_k measured along the edges $\{i, j\}$ of the graph; it quantifies the average distortion introduced by the map Ψ_k. The denominator provides a natural normalization. The constraint of orthogonality to the previous coordinate functions (6.17) guarantees that guarantees that the coordinate ψ_k describes the dataset with a finer resolution: Ψ_k oscillates faster on the dataset than the previous Ψ_j if $\langle \Psi_k, \Psi_j \rangle = 0$. Intuitively, Ψ_k plays the role of an additional digit that describes the location of X_i with more precision. It turnsout [35] that the solution of (6.12 and 6.13) is the solution to the generalized eigenvalue problem,

$$(\mathbf{D} - \mathbf{W})\psi_k = \lambda_k \mathbf{D}\psi_k, \qquad k = 0, \ldots \qquad (6.18)$$

The first eigenvector ψ_0, associated with $\lambda_0 = 0$, is constant, $\Psi_0(X_i) = 1, i = 1, \ldots, N$; it is therefore not used. Finally, the new parametrization Ψ is defined by

$$X_i \mapsto \Psi(X_i) = [\psi_1(X_i) \; \psi_2(X_i) \ldots \psi_m(X_i)]^T. \qquad (6.19)$$

The idea of parametrizing a manifold using the eigenfunction of the Laplacian was first proposed in [36]. Recently, the same idea has been revisited in the machine-learning literature [37, 38]. The construction of the parametrization is summarized in Fig. 6.6. Unlike PCA, which yields a set of vectors on which to project each X_i, this nonlinear parametrization constructs the new coordinates of $X_i X_i$ by concatenating the values of the $\psi_k, k = 1, \cdots, mk = 1, \cdots, m$ evaluated at X_i, as defined in (6.15). The embedding obtained with the Laplacian eigenmaps is in fact very similar to a parametrization of the dataset with a kernelized version of PCA, known as kernel-PCA [39].

6.3.3.1   Application to Wearable Systems

The authors in [40] embed multidimensional sensor data using Laplacian eigenmaps and cluster the dataset using the new coordinates. The analysis is not performed locally, since the Laplacian eigenmaps require computationally intensive algorithms: nearest neighbor search and eigenvalue problems.

Algorithm 2: Laplacian eigenmaps

Input:

set of N sensor vectors, $X_1, \ldots, X_N$,
σ: width of the kernel for the graph; n_n: number of nearest neighbors of each X_i
m: dimension of the embedding

Algorithm:

1. construct the graph defined by the n_n nearest neighbors of each X_i
2. compute $\mathbf{W}, \mathbf{D}$
3. compute the m eigenvectors $\psi_1, \ldots, \psi_m$ of $\mathbf{D}^{-\frac{1}{2}} \mathbf{W} \mathbf{D}^{-\frac{1}{2}}$

Output: m coordinates $\Psi(X_i) = \left[\psi_1(X_i)\ \psi_2(X_i)\ \ldots\ \psi_m(X_i) \right]^T$.

Fig. 6.6 Construction of the embedding

6.3.3.2 Available Software

- The original MATLAB code of Beilkin is available here:http://manifold.cs.uchicago.edu/
- A suite of classification algorithms based on Laplacian eigenmaps is available here: http://manifold.cs.uchicago.edu/manifold_regularization/software.html
- A method related to Laplacian eigenmpas and known as diffusion geometry is available here: http://www.math.duke.edu/~mauro/code.html
- Finally, the Statistical Learning Toolbox of Dahua Lin available here:http://www.mathworks.com/matlabcentral/fileexchange/12333 contains code for the Laplacian eigenmaps.

6.4 Classification, Learning of States, and Detection of Anomalies

We are now concerned with the final stage of the analysis: the extraction of high-level information from the wearable system. This abstract information can be an alarm in the case of a natural catastrophe [41], or a diagnostic for a subject with a risk of heart disease [24]. We assume that each sensor vector is represented by a p-dimensional vector $X \in \mathbb{R}^p$. This vector may be composed of the raw measurements of the sensors, or may be a vector of features (see Sect. 6.2), or the outcome of a dimensionality reduction algorithm (see Sect. 6.3). At each time t_i, we have access to the sensor vector X_i (the subscript i is a time index). The question becomes: what is the state y_i that characterizes the user at time t_i given the sensor vector X_i? Depending on the application, the state y_i could, for instance, encode the presence of an alarm, or the likelihood of a heart attack. There are two broad types of approaches to answer this question.

The first approach assumes that there exists a large collection of training data composed of sensors values $X_1, \ldots, X_l$ that have been carefully labeled with the corresponding state $y_1, \ldots, y_l$ of the user. The labeling is a very time-consuming process that needs to be performed off-line by the user himself/herself, or by an expert (e.g., in the case of biomedical data). Machine-learning algorithms can use the training data to learn the association between the state y_i and the sensor measurement X_i. Support vector machines is an important example of supervised classification methods; it is discussed in Sect. 6.4.2. This supervised approach may require significant amounts of training data to achieve good performances.

An alternative approach consists in "letting the data speak for themselves" using clustering methods that identify similarities in the sensor vectors, and group the measurements into coherent clusters. Such methods are called unsupervised and do not require any training data. While they may not provide an answer to our question, namely, "what is the state y_i?", unsupervised methods can rapidly organize the data into coherent states that can then be further analyzed. We described in the next section unsupervised methods.

6.4.1 Unsupervised Methods

6.4.1.1 K-Means Clustering

The K-means clustering algorithm is a method to divide a set of vectors into homogeneous groups, within which vectors are similar to one another. The goal is to find the optimal number of clusters, K, the optimal cluster centroids, $C_1, \ldots, C_K$, and the optimal assignments of each vector X_i to a cluster k to minimize the total within cluster scatter [30]

$$\sum_{k=1}^{K} N_k \sum_{X_i \in \text{cluster k}} \|X_i - C_k\|^2, \tag{6.20}$$

where N_k is the total number of vectors assigned to cluster k. Indeed, the term

$$\sum_{X_i \in \text{cluster k}} \|X_i - C_k\|^2,$$

quantifies the scatter of the vectors in cluster k around the centroid C_k. Therefore, (6.20) quantifies the total amount of scatter within all clusters. Given a specified number of clusters, K, algorithm 3 (see Fig. 6.7) iteratively partitions the data, and computes the centroids $C_1, \ldots, C_K$ of all the clusters (see Fig. 6.8). The output of the algorithm depends on K, which can be further optimized [30], and depends also on the initial assignment of vectors to clusters. While the algorithm will eventually converge, there is no guarantee that it reaches the global minimum of (6.16). One should therefore repeat the algorithm of Fig. 6.7 with several different initial conditions, and retain the solution that minimizes (6.20).

Algorithm 3: K-means clustering

Input:

- $X_1, \ldots, X_N$
- K: number of clusters

Algorithm:

Repeat
1. assign each vector X_i to the cluster k with the closest centroid C_k.
2. recompute all the cluster centroids $C_1, \ldots, C_K$.
Until convergence

Output: The cluster label for each X_i

Fig. 6.7 K-means clustering

Fig. 6.8 K-means clustering of the dataset, with $K = 2$. The cluster centroids are *circled* in *black*. The true boundary is shown as a *dashed lined*

Fig. 6.9 Mixture of Gaussian model fitted to the dataset, with $K = 2$. The mean of each Gaussian distribution is *circled* in *black*. The true boundary is shown as a *dashed lined*

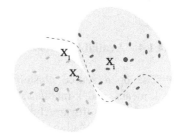

The assignment of a sensor vector to a cluster results in a hard decision. A soft version of the same idea is provided by the mixture of Gaussian densities model discussed in the next section (Fig. 6.9).

6.4.1.2   Mixture of Gaussian Densities

The Gaussian mixture model is a generative probabilistic model that assumes that the joint distribution of the vector of sensor measurements X is a finite mixture of multivariate Gaussian densities,

$$P(X) = \sum_{k=1}^{K} \pi_k \phi(X, \mu_k, \Sigma_k). \tag{6.21}$$

The mixing parameters π_k are positive weights that add up to 1. The density ϕ is the p-multivariate normal density function. The maximum likelihood estimates $\hat{\mu}_k$, $\hat{\Sigma}_k$ and $\hat{\pi}_k$ of the mixture parameters can be computed from the measurements using the expectation minimization (EM) algorithm [42]. The estimation of the number of components K (which plays the same role as the number of clusters) is often difficult [42] but can be addressed using model selection criteria [43]. We can use the estimates $\hat{\mu}_k$, $\hat{\Sigma}_k$, and $\hat{\pi}_k$ to compute the posterior probability given by

$$P(k|X) = \frac{\hat{\pi}_k \phi(X, \hat{\mu}_k, \hat{\Sigma}_k)}{\sum_{l=1}^{K} \hat{\pi}_l \phi(X, \hat{\mu}_l, \hat{\Sigma}_l)}, \tag{6.22}$$

which provides an estimate of the probability that measurement X be generated by component k.

6.4.1.3   Application to Wearable Systems

Mixture of Gaussians have been used in [44] to segment sensor time series into intervals associated with distinct activities. We note that clustering algorithms can be implemented on a personal digital assistant (PDA) [45], where a clustering algorithm learns to identify the context associated with the usage of the PDA.

6.4.1.4   Available Software

- The open source clustering software contains clustering routines such as k-means and k-medians, and is available at: http://bonsai.ims.u-tokyo.ac.jp/~mdehoon/software/cluster/
- PRTools is a Matlab based toolbox for pattern recognition, and is freely available at http://www.prtools.org/

6.4.2   *Support Vector Machine*

Among all classification techniques that have been used to analyze sensor data from wearable systems, support vector machines (SVMs) is one of the most popular methods [46, 47]. SVM can construct a nonlinear separating boundary between two classes by implicitly mapping the features to a high-dimensional space, and performing a linear classification in that space. Our discussion of SVM follows [30].

We consider the problem of classifying sensor vectors into two classes defined by the labels $y = -1$ for class 0 and $y = 1$ for class 1. Extension to more than two classes can be easily obtained by using a "one-versus-all strategy," where each class is compared to the other remaining classes, and X is assigned to the class that is most often selected by the different classifications.

6.4.2.1 Support Vector Classifier

Given a training set composed of sensor data $X_1, \ldots, X_l$ and the associated class labels, $y_1, \ldots, y_l$, our goal is to construct a classifier $f(X)$ that assigns a label -1 if X belongs to class 0, and 1 if X belongs to class 1. We assume that the training samples are linearly separable: one can find a hyperplane that divides the training dataset according to the class membership. The classifier is defined as follows,

$$f(X) = \begin{cases} -1 & \text{if } \langle W, X \rangle + b < 0 \\ 1 & \text{if } \langle W, X \rangle + b > 0. \end{cases} \tag{6.23}$$

where $W \in \mathbb{R}^p$, $b \in \mathbb{R}$ and $\langle W, X \rangle = \sum_{i=1}^{p} x_i w_i$ is the inner product between the vectors X and W. The hyperplane (defined by W and b) that optimally separates the two classes can be found by maximizing the margin between the two classes for the training data (see Fig. 6.10). Specifically, we choose W and b such that

$$\begin{cases} \frac{2}{\|W\|}, \text{ the margin width, is maximized, and} \\ \text{for all the vectors } X_i \text{ in the training data we have, } y_i(\langle W, X_i \rangle + b) \geq 1. \end{cases}$$

This quadratic programming optimization problem can be solved using Lagrange multipliers and results in the following classifier

$$f(X) = sign\left(\sum_i \alpha_i y_i \langle X_i, X \rangle + b \right), \tag{6.24}$$

where

$$W = \sum_i \alpha_i y_i X_i, \tag{6.25}$$

and the Lagrange multipliers α_i are nonzero only for those vectors that lie at the boundary of the margin (see Fig. 6.10). Such training vectors are called the "support vectors". As shown in Fig. 6.10, the width of the margin is constrained by the support vectors. The support vectors all satisfy

$$y_i(\langle W, X_i \rangle + b) = 1. \tag{6.26}$$

Fig. 6.10 Maximum margin
linear classifier

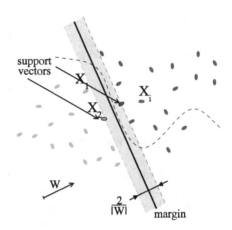

Finally, the offset b can be computed from any such support vector X_i,

$$b = y_i - \langle W, X_i \rangle. \tag{6.27}$$

If the two classes cannot be linearly separated, it is still possible to apply the
same classification method with the introduction of additional "slack variables".
These variables allow for some training vectors to be on the wrong side of the
separating hyperplane [30].

6.4.2.2 Support Vector Machines

The linear classifier can be extended to a nonlinear classifier, where the separating
boundary between the two classes is no longer a hyperplane, but can be a surface of
arbitrary geometry. Rather than defining the surface in the original domain $\mathbb{R}^p$, the
vectors are implicitly mapped to a higher dimensional space (see Fig. 6.11), where
distances and inner products are measured using a kernel $\mathbf{K}$, different from the usual
inner product [30]. The classifier becomes

$$f(X) = sign\left(\sum_i \alpha_i y_i \mathbf{K}(X_i, X) + b \right). \tag{6.28}$$

Popular choices for $\mathbf{K}$ include the polynomial kernel $\mathbf{K}(X_i, X_j) = (\langle X_i, X_j \rangle + 1)^d$
and the Gaussian kernel $\mathbf{K}(X_i, X_j) = \exp(-\|X_i - X_j\|^2 / \sigma^2)$.

6.4.2.3 Application to Wearable Systems

Dinh et al. [20] use SVM to classify various sensors measurements to detect near-fall
events in older adults. In [46] the severity of tremor in patients with Parkinson

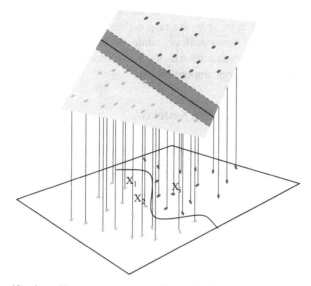

Fig. 6.11 Classification with support vector machines: the original dataset is lifted up into a high-dimensional space, where a maximal margin linear classifier can separate the two classes

disease is quantified using a multiclass (one-versus-all) classifier based on SVM. The sensors provide accelerometer data. The computational complexity of the SVM algorithm requires that the training and the classification be performed remotely on a central computer.

6.4.2.4   Available Software

Many software packages provide implementations of the SVM. Some of the main public domain packages are listed below.

- SVM light: http://svmlight.joachims.org
- LIB SVM: http://www.csie.ntu.edu.tw/~cjlin/libsvm
- Additional implementations can be found at: http://www.support-vector-machines.org/SVM_soft.html

6.4.3   Semi-Supervised

An alternative to supervised learning and unsupervised learning is "semi-supervised" learning [48]. Semi-supervised can take advantage of unlabeled sensor vectors to learn the organization of the sensor data in $\mathbb{R}^p$. Most semi-supervised learning algorithm assume that the data organize themselves smoothly, and that the geometry of the data will help construct the classifier. In the case of sensors from a wearable system, it is possible to assume that the data from a sensor lie close to a low dimensional manifold (see Sect. 6.3.3). In this case, it becomes possible to use partially labeled data to discover the geometry of the manifold, while at the same time constructing the classifier. In the context of wearable sensors, this approach is very appealing since it does not require to label a large amount of data.

6.4.3.1   Application to Wearable Systems

Ali et al. [49] combine a PCA decomposition with a semi-supervised learning approach to construct a classifier that can recognize activities. Similarly, Mahdaviani and Choudhury [50] and Stikic and Schiele [7] propose an activity recognition method that only requires very few labeled data. As with SVM, the method can only be applied remotely on a computer with enough computational power.

6.4.3.2   Available Software

*   The group of Partha Niyogi at the University of Chicago has developed some MATLAB software to perform semi-supervised classification using manifold regularization. The software is available at: http://manifold.cs.uchicago.edu/manifold_regularization/manifold.html

6.5   Conclusion and Future Directions

We have reviewed in this chapter some of the dimension reduction and statistical machine-learning methods to mine and extract information from wearable systems. Many of these techniques have been borrowed from the existing statistical and machine-learning literature. In comparison with general data analysis problems, the on-site, or local analysis of wearable data has a number of well-defined constraints: limited computational power and limited bandwidth. These constraints require efficient and fast algorithms for on-line analysis or off-line transmission to a central hub. As the presence of wearables becomes ubiquitous, and as the sensors become more integrated, we expect that there will be a need for more efficient data analysis algorithms. Increasing the speed of current algorithms is clearly not the answer. Rather, we expect to see that completely new ideas will be required to tackle the amount of data generated by wearable systems. For instance, it may come as a surprise that only reading randomly a very small subset of the sensor values (the principle underpinning "compressive sampling" [51]) yields the same accuracy as uniform sampling [52], with an enormous saving in power consumption. Other directions include the development of tailored statistical models [53] that can be estimated with fewer sensors and less computations than generic probabilistic models. Clearly, the area of data analysis for wearable systems promises to be exciting and challenging.

6.6 Glossary

- **Independent component analysis (ICA)** A linear decomposition of the data into statistically independent sources. The sources and the mixing weights are estimated by the algorithm.
- **Karhunen-Loève transform** See Principal component analysis.
- **Kernel PCA** See Laplacian Eigenmaps.
- **Laplacian eigenmaps** A nonlinear method to parametrize a dataset using the eigenvectors of the graph Laplacian defined on the dataset. The method optimally preserves the short-range distance while assembling a global parametrization of the data.
- **Multiscale analysis** A method to decompose a dataset into signals that have well defined characteristic scales. An example is provided by a wavelet analysis.
- **Principal component analysis (PCA)** A decomposition of a dataset into linear components that best capture the variance in the data.
- **Semi-supervised learning** A classification method that combines unlabeled data with label data to construct a classifier. The method exploits the underlying smoothness (e.g., the data lie on a manifold) of the data to estimate the geometry of the data and compensate for the lack of labels.
- **Short time Fourier transform** See time-frequency analysis.
- **Singular value decomposition** See principal component analysis.
- **Support vector machine (SVM)** A classification algorithm that combines a maximum margin classifier with a measure of similarity using a kernel.
- **Time-frequency analysis** A decomposition of a signal in terms of localized oscillatory patterns with well defined frequency and position.
- **Wavelet analysis** See multiscale analysis.

References

1. Ahola T, Korpinen P, Rakkola J, Ramo T, Salminen J, Savolainen J (2007) Wearable fpga based wireless sensor platform. In: Engineering in Medicine and Biology Society, 2007. 29th Annual International Conference of the IEEE EMBS 2007, pp 2288–2291
2. Ali A, King R, Yang G (2008) Semi-supervised segmentation for activity recognition with Multiple Eigenspaces. In: Medical Devices and Biosensors, 2008. 5th International Summer School and Symposium on ISSS-MDBS 2008, pp 314–317
3. Ali R, Atallah L, Lo B, Yang G (2009) Transitional activity recognition with manifold embedding. In: Proceedings of the 2009 Sixth International Workshop on Wearable and Implantable Body Sensor Networks, 1, pp 98–102
4. Atallah L, Lo B, Ali R, King R, Yang G (2009) Real-time Activity Classification Using Ambient and Wearable Sensors. IEEE Transactions on Information Technology in Biomedicine
5. Baheti PK, Garudadri H (2009) An ultra low power pulse oximeter sensor based on compressed sensing. In: Proceedings – 2009 6th International Workshop on Wearable and Implantable Body Sensor Networks, BSN 2009, pp 144–148

6. Belkin M, Niyogi P (2003) Laplacian eigenmaps for dimensionality reduction and data representation. Neural Computations 15:1373–1396
7. Bérard P, Besson G, Gallot S (1994) Embeddings Riemannian manifolds by their heat kernel. Geomet Funct Anal 4(4):373–398
8. Bonato P, Mork P, Sherrill D, Westgaard R (2003) Data mining of motor patterns recorded with wearable technology. IEEE engineering in medicine and biology magazine 22(3):110–119
9. Bonfiglio A, Carbonaro N, Chuzel C, Curone D, Dudnik G, Germagnoli F, Hatherall D, Koller J, Lanier T, Loriga G et al (2007) Managing catastrophic events by wearable mobile systems. In: Mobile Response, vol 4458/2007. Springer, Berlin, pp 95–105
10. Candes EJ, Wakin MB (2008) An introduction to compressive sampling: A sensing/sampling paradigm that goes against the common knowledge in data acquisition. IEEE Signal Process Mag 25(2):21–30
11. Chapelle O, Schölkopf B, Zien A (eds) (2006) Semi-supervised learning. MIT, MA
12. Chung F (1997) Spectral Graph Theory, 92(92). American Mathematical Society
13. Coifman R, Lafon S (2006) Diffusion maps. Appl Comput Harmonic Anal 21:5–30
14. Davrondzhon G, Einar S (2009) Gait Recognition Using Wearable Motion Recording Sensors. EURASIP Journal on Advances in Signal Processing pp 1–16
15. Dinh A, Shi Y, Teng D, Ralhan A, Chen L, Dal Bello-Haas V, Basran J, Ko S, McCrowsky C (2009) A fall and near-fall assessment and evaluation system. Open Biomed Eng J 3:1
16. Eagle N, Pentland A (2006) Reality mining: sensing complex social systems. Personal and Ubiquitous Computing 10(4):255–268
17. Flanagan J (2005) Unsupervised clustering of context data and learning user requirements for a mobile device. Lecture Notes in Computer Science, vol 3554, pp 155–168
18. Giansanti D, Maccioni G, Cesinaro S, Benvenuti F, Macellari V (2008) Assessment of fall-risk by means of a neural network based on parameters assessed by a wearable device during posturography. Medical Engineering and Physics 30(3):367– 372
19. Glaros C, Fotiadis D (2005) Wearable Devices in Healthcare. Studies in Fuzziness and Soft Computing 184:237
20. Han D, Park S, Lee M (2008) THE-MUSS: Mobile u-health service system. In: Biomedical Engineering Systems and Technologies, vol 25. Springer, Berlin, pp 377–389
21. Hastie T, Tibshinari R, Freedman J (2009) The elements of statistical learning. Springer, Berlin
22. Hu F, Jiang M, Xiao Y (2008) Low-cost wireless sensor networks for remote cardiac patients monitoring applications. Wireless Comm Mobile Comput 8(4):513–530
23. Huynh T, Blanke U, Schiele B (2007) Scalable recognition of daily activities with wearable sensors. In: Location-and context-awareness: Third international symposium, LoCA 2007, Oberpfaffenhofen, Germany, September 20–21. 2007 Proceedings, p 50
24. Hyvärinen A (1999) Survey on independent component analysis. Neural Comput Surv 2:94–128
25. Hyvärinen A, Oja E (2000) Independent component analysis: Algorithms and applications. Neural Network 13(4–5):411–430
26. Jaffard S, Meyer Y, Ryan R (2001) Wavelets: tools for science & technology. Society for Industrial and Applied Mathematics
27. Karlen W, Mattiussi C, Floreano D (2009) Sleep and wake classification with ECG and respiratory effort signals. IEEE Trans Biomed Circ Syst 3(2):71–78
28. Ko L, Tsai I, Yang F, Chung J, Lu S, Jung T, Lin C (2009) Real-time embedded EEG-based brain-computer interface. In: Advances in neuro-information processing. Springer, Berlin, pp 1038–1045
29. Krause A, Smailagic A, Siewiorek D (2006) Context-aware mobile computing: Learning context-dependent personal preferences from a wearable sensor array. IEEE Transactions on Mobile Computing pp 113–127

30. Mahdaviani M, Choudhury T (2008) Fast and scalable training of semi-supervised crfs with application to activity recognition. In: Platt J, Koller D, Singer Y, Roweis S (eds) Advances in Neural Information Processing Systems 20,MIT Press, Cambridge, MA, pp 977–984
31. Mallat S (1999) A wavelet tour of signal processing. Academic, NY
32. Matsuyama T (2007) Ubiquitous and wearable vision systems. In: Imaging beyond the pinhole camera, Springer, pp 307–330
33. McLachlan G, Krishnan T (1997) The EM algorithm and extensions. Wiley, NY
34. Minnen D, Starner T, Essa M, Isbell C (2006) Discovering characteristic actions from on-body sensor data. In: 10th IEEE International Symposium on Wearable computers 2006, pp 11–18
35. Pantelopoulos A, Bourbakis N (2010) Design of the new prognosis wearable system-prototype for health monitoring of people at risk. In: Advances in biomedical sensing, measurements, instrumentation and systems. Springer, Berlin, pp 29–42
36. Paoletti M, Marchesi C (2006) Discovering dangerous patterns in long-term ambulatory ECG recordings using a fast QRS detection algorithm and explorative data analysis. Comput Meth Programs Biomed 82(1):20–30
37. Patel S, Lorincz K, Hughes R, Huggins N, Growdon J, Standaert D, Akay M, Dy J, Welsh M, Bonato P (2009) Monitoring motor fluctuations in patients with Parkinson's disease using wearable sensors. IEEE Trans Inform Technol Biomed 13(6):864
38. Poh MZ, Kim K, Goessling AD, Swenson NC, Picard RW (2009) Heartphones: Sensor earphones and mobile application for non-obtrusive health monitoring. Wearable Computers, IEEE International Symposium pp 153–154
39. Powell Jr H, Hanson M, Lach J (2009) On-body inertial sensing and signal processing for clinical assessment of tremor. IEEE Trans Biomed Circ Syst 3(2):108–116
40. Preece S, Goulermas J, Kenney L, Howard D (2009) A comparison of feature extraction methods for the classification of dynamic activities from accelerometer data. Biomedical Engineering, IEEE Transactions on 56(3):871–879
41. Schölkopf B, Smola A, Müller K (1999) Kernel principal component analysis. In: Advances in kernel methods: Support vector learning. MIT, MA
42. Stetson P (2004) Independent component analysis of pulse oximetry signals based on derivative skew. Lecture notes in computer science pp 1072–1078
43. Stikic M, Schiele B (2009) Activity recognition from sparsely labeled data using multi-instance learning. In: Proceedings of the 4th International Symposium on Location and Context Awareness, Springer, p 173
44. Subramanya A, Raj A, Bilmes J, Fox D (2006) Recognizing activities and spatial context using wearable sensors. In: Proc. of the Conference on Uncertainty in Artificial Intelligence
45. Sun Z, Mao X, Tian W, Zhang X (2009) Activity classification and dead reckoning for pedestrian navigation with wearable sensors. Meas Sci Technol 20:015203
46. Tanner S, Stein C, Graves S (2009) On-board Data Mining. Scientific Data Mining and Knowledge Discovery pp 345–376
47. Tsai D, Morley J, Suaning G, Lovell N (2009) A wearable real-time image processor for a vision prosthesis. Comput Meth Program Biomed 95(3):258–269
48. Verbeek J, Vlassis N, Krose B (2003) Efficient greedy learning of Gaussian mixtures. Neural Comput 15:469–485
49. Viswanathan M (2007) Distributed data mining in a ubiquitous healthcare framework. In: Proceedings of the 20th conference of the Canadian Society for Computational Studies of Intelligence on Advances in Artificial Intelligence. Springer, Berlin, p 271
50. Yao J, Warren S (2005a) A short study to assess the potential of independent component analysis for motion artifact separation in wearable pulse oximeter signals. In: 27th Annual International Conference of the Engineering in Medicine and Biology Society, IEEE-EMBS, pp 3585–3588
51. Yao J, Warren S (2005b) Applying the ISO/IEEE 11073 standards to wearable home health monitoring systems. J Clin Monit Comput 19(6):427–436

52. Zinnen A, Blanke U, Schiele B (2009) An analysis of sensor-oriented vs. modelbased activity recognition. In: IEEE International Symposium onWearable Computers, pp 93–100
53. Zhang F, Lian Y (2009) QRS Detection Based on Multiscale Mathematical Morphology for Wearable ECG Devices in Body Area Networks. IEEE Transactions on Biomedical Circuits and Systems 3(4):220–228

Chapter 7
Future Direction: E-Textiles

Danilo De Rossi and Rita Paradiso

7.1 Introduction

Body worn systems, endowed with autonomous sensing, processing, actuation, communication and energy harvesting and storage are emerging as a solution to the challenges of monitoring people anywhere and at anytime in applications such as healthcare, well-being and lifestyle, protection and safety.

Textiles, being a pervasive and comfortable interface, are an ideal substrate for integrating miniaturized electronic components or, through a seamless integration of electroactive fibres and yarns, they have the potentiality to become fully functional electronic systems. The idea of e-textiles being a viable solution to implement truly wearable, smart platforms as bidirectional interfaces with human body and functions has emerged as a result of the work of independent groups around the world almost at the same time. A group at MIT-Media Lab (Post and Orth 1997) reported about early results using embroidery to realize different interfaces from human–computer interactions; researchers at Georgia Tech presented (Luid et al. 1997) a shirt using fibre optics for monitoring soldiers conditions in the battlefield; researchers in Tokyo (Ishijima 1997) reported experimental results on the use of textile electrodes for cardiopulmonary monitoring and researchers at the University of Pisa (De Rossi et al. 1997) disclosed the use of thermo and piezoresistive fabrics for recording thermal and biomechanical parameters in ergonomics and rehabilitation. Since then, diverse commentaries and critical reviews have been made available in the literature further exploring the viability and potentials of smart fabrics and interactive textiles (SFITs) (Marculescu et al. 2003; Service 2003). Although additional progress has been made (Carpi and De Rossi 2005; Coyle et al. 2007), the realization of viable textile fibres and yarns endowed with active electronic functions has proved to be more elusive and major scientific, technical and economic limitations still need to be overcome to bring the technology to fulfil the initial expectations.

D. De Rossi (✉)
Interdepartmental Research Centre "E. Piaggio", University of Pisa, Pisa, Italy
e-mail: d.derossi@ing.unipi.it

A. Bonfiglio and D. De Rossi (eds.), *Wearable Monitoring Systems*,
DOI 10.1007/978-1-4419-7384-9_7, © Springer Science+Business Media, LLC 2011

Work performed along this direction for most of the textile-based devices is, with the sole exception of textile electrodes, fragmented at a very early stage of development. Besides the great technical challenges to be faced, it is nowadays difficult to predict if issues such as compatibility with industrial textile technology, durability, water resistance, cost and user acceptance can be successfully dealt with.

In the following paragraphs, we report on research work disclosed in the literature aimed at developing microsystem components (sensor, actuators, logics, energy storage and harvesting devices, keyboards, displays and antennas) needed for the implementation of full e-textile wearable systems. Finally, we briefly illustrate some e-textile-based wearable platforms made available in recent years. As it will be shown, all these platforms rely on integration at various levels of conventional off-the-shelf microelectronics components with fabrics.

7.2   Fibres and Textiles for Bioelectrodes

Monitoring diagnostics, therapy and rehabilitation in the medical domain are often based on methodologies and techniques exploiting bioelectrical phenomena. Non-invasive monitoring of vital signs such as electrocardiogram (ECG), electrodermal response (EDR), electromyogram (EMG), stimulus-evoked potentials (SEPs) and others biosignals need electrodes to convert body-generated ionic currents into electronic signals for further processing.

Injecting electrical currents into the body for therapeutic purposes is also widely used; functional electrical stimulation (FES) and electrotherapy are good examples. Supporting the extension and expansion of health services outside the classical healthcare establishments through wearable systems has evidenced the need for long-term use of bioelectrodes. Textile electrodes have been developed and have reached the level of commercial development. Textile-grade fibres made of stainless steel, carbon, silver and silver-coated polyester have all been weaved or knitted to realize fabric electrodes. The inappropriateness of using gel coupling in long-term monitoring, the difficulty in proper positioning and fixing electrodes on skin and the susceptibility to motion artefacts still pose problems for the development of dry textile electrodes, in particular when intended to monitor non-sweating or elderly subjects.

Most textile electrodes have been realized using commercial stainless steel threads wound around a standard cotton textile yarns. They have been realized either through a flat weaving process or knitted using the tubular intarsia technique to get double face apparels (see Fig. 7.1), where the external part is realized with the basal yarn, non-conductive, to isolate the electrode from the external environment (Paradiso et al. 2005; Ctrysse et al. 2004).

Adhesive hydrogel membranes shown in Fig. 7.2, or integrated pushing cushions have been used to improve electrode–skin coupling. Fabric electrodes have been proved to offer very good performances in recording either EMG or ECG signals even also after repeated washing (Scilingo et al. 2005; Xu et al. 2008).

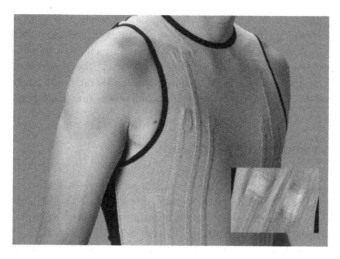

Fig. 7.1 Knitted intarsia electrodes and textile connections, particularly on the skin side on the right

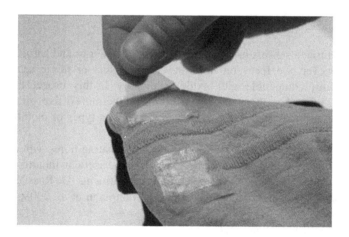

Fig. 7.2 Hydrogel membrane coupled with textile electrode

Non-contact capacitive electrodes have also been developed and disclosed, in particular, for EMG recording (Gourmelon and Langereis 2006). Their susceptibility to motion artefacts and power line interference, however, is very high.

The use of active electrodes, which implies the integration of miniaturized buffer amplifiers (impedance converters) to suppress noise, is well known; a new embodiment has been recently developed in the form of non-woven fabric, adding complexity and cost but being more effective in term of front-end electronics signal-to-noise ratio (Kang et al. 2008).

Existing fabric electrodes still suffer limitations when skin is dry and when the wearer performs intense physical activity. Minimizing electrical contact imped-ance between the electrode and the skin, maximizing adhesive properties without the use of covalent bonding, and mechanical stabilizing the ionic–electronic inter-face are crucial interventions needed to improve textile electrode performances in terms of signal-to-noise ratio, motion artefact and power line interference rejection.

7.3   Fibres and Textiles for Sensing

Sensing and monitoring physical and chemical parameters is finding useful appli-cation in e-textiles. The signals can be endogenous (related to the person wearing the garment) or exogenous (related to the environment). As an example, vital signs monitoring can be useful in the wellness, sport or medical domains, while chemi-cally sensitive fabrics can be exploited for alerting from dangerous chemical or biological substances as needed by people working in noxious environments or, in the longer term, to monitor metabolites and biomarker in sweat or in wounds.

7.3.1   Physical Sensing

Many miniaturized sensors have been integrated into fabrics and garments. In this chapter, however, we focus on fibre and yarn sensors or fabric sensors being potentially fully compatible with textile technology. In this respect, thermal and mechanical sensors have been disclosed, relying on thermoresistive, piezoresistive, piezoelectric and piezocapacitive effects or on various types of modifications of optical fibres.

Thermo- and piezoresistive fabrics, prepared by coating fibres with conducting polymers (Gregory et al. 1989), have been proved to be useful in monitoring surface temperature and surface strain in using body-fitting garments (De Rossi et al. 1999). Similar work has been subsequently reported (Mattmann et al. 2008; Gibbs and Asada 2005).

A different technique and methodology for developing textile wearable posture and gesture monitoring systems has been disclosed, using redundant arrays of stretch sensors made of piezoresistive rubber, screen printed onto fabrics and garments (Lorussi et al. 2004). Very recently, textile-based electrogoniometers have been conceived and tested to further increase the performance of 3D recon-struction of body kinematics through sensorized garments (Lorussi et al. 2009). Experimental work to enable clinical applications, mostly related to neurorehabil-itation, is in progress (Tognetti et al. 2005; Pacelli et al. 2006; Paradiso et al. 2006; Giorgino et al. 2009). See Fig. 7.3.

Fibres, films and patches of piezoelectric polymers such as polyvinylidene fluoride (PVDF) and its copolymers may expand the range of biomechanical and physiological parameters detected by electronic textiles. Compared to piezoresistive materials,

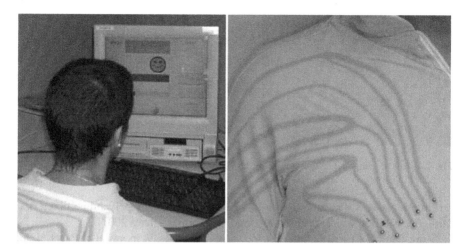

Fig. 7.3 System developed in the frame of MyHeart project for neuro-rehabilitation therapy

piezoelectrics offer higher sensitivity and wider bandwidth in transducing mechanical to electrical signals. They do not, however, consent the acquisition of static or slowly varying signals because of the intrinsic nature of the physical principle they exploit. PVDF narrow strips have been incorporated into fabrics matrices (Edmison et al. 2002), and multifunctional monitoring of cardiopulmonary parameters through a very wide bandwidth PVDF sensor has been reported (Lanatà et al. 2010).

Piezocapacitive stress-sensitive mats have also been realized (Sergio et al. 2002), which might be useful for body–object contact pressure sensing.

Textile optical sensors making use of fibre optics have also been proposed (El-Sherif 2000) for sensing a wide range of parameters either through the modification of the cladding or by incorporating metallic or semiconducting nanoparticles into the core of the optical fibre (Dhawan et al. 2006).

7.3.2   Chemical Sensors and Biosensors

Synthetic textile fibres (nylon and polyesters) coated with conductive polymers, polypirrole or polyaniline, woven into fabric mesh have been used for toxic vapour sensing (Collins and Buckley 1996). Detection limits of parts per million of vapours such as ammonia and nitrous oxide were obtained.

Conducting polymer-based electrospun polyaniline fibre sensors have also been shown (Koncar 2005) to enable pathogens detection through exhaled breath analysis. The large specific surface of this kind of nanofibres may offer potentiality for high

Fig. 7.4 Fabric pump and sweat rate sensor developed in the frame of Biotex project

sensitivity and low detection threshold. Humidity sensing by polyaniline fibres is a further possibility, which has been reported (Norris and Mattes 2007).

More complex textile platforms for sensing chemicals in sweat (Morris et al. 2009) or for monitoring sweat rate (Salvo et al. 2010) have been described, also including fabric pumps and valves (Coyle et al. 2010). Examples are shown in Fig. 7.4 www.biotex.it

7.4   Active Fibre Electronics and Woven Logics

A very challenging task resides in the realization of viable textile fibres and fabrics endowed with active electronic and logic functions.

Organic field effect transistors (OFETs) are commonly used in flexible electronics. The use of OFETs in a textile substrate has been proposed and different configurations examined (Bonfiglio et al. 2005). OFETs on fibres have also been disclosed (Lee and Subramanian 2005; Maccioni et al. 2006), but their realization onto fibres is a complex process, since precise micropatterning of source, drain and gate electrodes is needed. A different approach has been disclosed (Hamedi et al. 2007, 2009) making use of micrometre-sized organic electrochemical transistors (OECTs). Wiring fibre OECTs may prove to be an easier route to weaving electronics directly into fabrics to implement universal logic operations (De Rossi 2007), but very major obstacles exist in making these prototypes working in the real world as all these procedures are incompatible with current textile technology. In particular, to realize fabric microcomponents, i.e. fibres and yarns endowed with even simple electronic functions, as electrical conductors, insulators, semiconductors, material properties must be deeply modified and this modification must not invalidate other fundamental properties as mechanical flexibility and robustness, not mentioning the fact that these treatments must be done on large scale, to pave the way for the future industrial exploitation of these technologies.

Some encouraging reports have recently appeared in the literature: for instance, carbon nanotubes modified cotton yarns (Shim et al. 2008) and cotton yarns modified by a metal nanoparticles and conductive polymers (Mattana et al. 2010).

Both these examples are extremely interesting as they pave the way to the use of modified natural fibres as a basis for new active devices, with the tremendous advantage of preserving their favourable textile properties while modifying their electrical behaviour in a stable way.

7.5   Fibres and Textiles for Energy Harvesting and Storage

Energy harvesting and storage for wireless autonomous systems is a crucial issue in particular for wearables when intended to work continuously. Harvesting energy from motion and vibrations, from thermal gradients, from environmental RF or solar irradiation is nowadays actively investigated. Besides MEMS and silicon technology, some effort is also devoted to flexible plastic electronics and textile devices. Similarly, flexibles, organic batteries and, more recently, fabric batteries and supercapacitors for energy storage have been disclosed. As long as in the previous paragraphs, we confine our discussion to truly fibre and textile-based devices.

7.5.1   Textile-Based Solar Cells

Lightweight, portable solar panels are commercially available using polymer-based organic photovoltaics (www.konarka.com).

The development of dye-sensitized photovoltaic fibres has been recently reported (Fan et al. 2008; O'Connor et al. 2007) and their mechanical compatibility with textile manufacturing and weaving operation documented (Ramier et al. 2008). Several technical challenges have still to be solved, but the potentiality of textile photovoltaic might be substantial, in particular when covering large area and because fibre devices are less sensitive to light source orientation (O'Connor et al. 2008). Power conversion efficiencies are quite low at present (no more than 0.5%); the use of more efficient dyes-on-fibres (dyes of this kind are already available) may well bring efficiency to 5% in the near future (Yadav et al. 2008).

7.5.2   Electronic Textile Batteries

Thin, flat batteries using solid electrolytes are already commercially available covering the need of flexible, lightweight power sources for smart garments in which they can be easily integrated. The realization of charge storage devices directly onto a textile substrate is a more recent endeavour, and limited experimental work is available in the literature. Early work on the use of polypirrole and derivatized polytiophene on non-woven polyester as cathode and anode materials, respectively, to enable construction of textile batteries has been published (Wang et al. 2005). Capacities of the order of $10 \, \text{mA h g}^{-1}$ over 30 cycles have been reported (Wang et al. 2008). Two woven, silver-coated polyamide yarns coated with PEDOT (a conducting polymer) and separated by a solid electrolyte have been proved to be sufficient to realize a very simple textile rechargeable battery (Bhattacharya et al. 2009). Although the performances are still poor, the concept might be useful to enable the implementation of better performing power sources, which can be seamlessly integrated into fabrics and garments.

7.6 Smart Textiles for Actuation

The term "actuation" refers to the action a system performs on the environment, usually the external one. Different forms of actuation are possible with reference to wearable systems. The main ones are heating and mechanical force generation. Textile-based devices and systems for actuation will be examined in the following.

7.6.1 Textile Heating Systems

Resistive fabrics were early developed for electromagnetic shielding and for body heating. Several commercial products are nowadays available on the market in the form of either yarns, fabrics or whole wearable systems (VDC 2007). Metal fibres embedded in the fabric, metal-coated textiles, textile-grade carbon fibres and polypirrole-coated fabrics have all been used, and not much research work appears to be needed in this area. Advances would be required in the development of energy sources of high specific capacity because of the very inefficient process of resistive heating.

7.6.2 Thermo and Electromechanical Actuation

Engineering actuative materials and embedding them in fabrics and garments may provide new design concepts to enhance functionalities of textiles for apparel and wearables.

Functionalities such as shape recovery, aesthetics, heat exchange and external motor control functions have been proposed and preliminarily examined. Not much work in this field is available in the literature and the technology has to be considered in its early infancy. The use of shape memory alloy fibres have been embedded into knitted fabrics and shape memory functions have been demonstrated for design and fashion exploiting thermal memory effects (Winchester and Stylios

2003). Smart textiles based on shape memory polymers have also been reported for thermally activating aesthetic functions (Chan Vili 2007).

A comprehensive overview of shape memory polymers and textiles is available (Hu 2007), where particular emphasis is given on clothes dynamically responding to changes in temperature and humidity, ensuring greater comfort to the wearer.

The use of dielectric elastomer (DE) electromechanical actuators has also been proposed to endow fabrics with motor functions in the field of rehabilitation (Carpi et al. 2009). At this time, integration of DE actuators into garments through a viable textile technology is simply not feasible. However, recent work aiming at producing DE actuators in form of fibres (Arora et al. 2007; Cameron et al. 2008) may lead in the future to electromechanically actuated fabrics.

7.7   Textile-Based Communication Devices

Input and output interfaces for communication play a very important role in wearable systems. Input–output textile-based communication devices have been proposed and their practical use already demonstrated.

7.7.1   Textile Keyboards

Since the early proposal of resistive textiles touch keyboards (Orth et al. 1998) much work has been done in this area and products are available in the market (VDC 2007). Textile keyboards have initially attracted considerable commercial interest because their flexibility and compliance make easy the incorporation into garments and apparels. It is not clear, however, what their impact could be in wearable systems since other input modalities might be preferable.

7.7.2   Photonic Fibres and Optical Displays

Communication apparel intended for social interaction needs interfaces for optical information restitution. In many applications, it is necessary to reproduce and display visual information generated by subsystems integrated into clothing. Textile-grade photonic fibres exploiting lasing, interference, lateral dispersion and other optical phenomena have been described (Balachandran et al. 1996; Gauvreau et al. 2008) as much as the possibility to weave them in the form of textile displays. Textile-based optical displays are nowadays available only in the form of woven plastic optical fibres

Fig. 7.5 Textile antenna

(Koncar 2005; Harlin et al. 2003) were pixels are addressed in a matrix and optoelectronic sources can be bulky and expensive.

Electroluminescent fibres, adapting the technology developed to realize flexible organic-light-emitting devices (OLEDs), have been realized (O'Connor et al. 2007) eventually enabling the fabrication of light-emitting fabrics. The integration of LEDs mounted on lightweight flexible substrates into clothing has been successfully performed and wearable displays are nowadays commercially available (www.lumalive.com).

7.7.3 Textile Antennas

The need for long range wireless communication for body area networks has promoted research in textile antennas design and fabrication (Salonen et al. 2004).

An ultrawide bandwidth (UWB) textile antenna has been designed, fabricated and characterized (Klemm et al. 2006). The very thin, flexible textile antenna has proved to possess excellent transient characteristics, although problem still needs to be solved at the manufacturing level to achieve proper quality and repeatable performances.

A textile antenna for wireless body area networks working in the 2.4–2.483 GHz band (industrial, scientific, and medical; ISM) has also been disclosed (Tronquo et al. 2006). This antenna has been integrated into the outer garment of firefighters to enable communication under field operations (Hertleer et al. 2009) (see Fig. 7.5).

7.8 Smart Fabrics and Interactive Textiles Platforms

Several attempts to design and build prototypes of wearable integrated platforms based on e-textiles have been undertaken in the last 10 years. Textronics has exploited this technology commercially (www.textronicsinc.com). Market studies project fast growth for this segment; however, its present size is marginal. As already mentioned in the introduction, all these systems are hybrid, exploiting various degrees of integration between silicon microelectronic components and clothes, and present technology should be considered transitional.

The first products, both out of market, have been the Philips/Levi ICD+ which incorporated a mobile phone and a music player (Farringdon et al. 1999) and the LifeShirt of Vivometrics (Grossman 2003), a wearable system for medical diagnostics and monitoring.

At the end of 1990s, Georgia Tech has proposed the wearable Motherboard (Gopalsamy et al. 1999) for monitoring "the golden hour" in military operations. It consisted of a shirt into which metal fibres and optical fibres were incorporated to be reversibly connected to off-the-shelf optoelectronic devices for monitoring wearer's vital signs and bullet penetration. A T-shirt consisted of a breathing-rate sensor, a shock/fall detector, a temperature sensor and electrodes for ECG recording was also developed in the course of the VTAM project (Weber et al. 2004). A sensorized baby pyjama, called Mamagoose, was developed by Verhaert in Belgium as a prevention tool for sudden infant death syndrome (www.verhaert.com). Heart rate captured by woven electrodes and respiratory activity recorded through capacitive sensors were processed and used to provide an alarm at risky situations. A sensorized shirt was developed by Smartex within the EC funded project Wealthy (Paradiso et al. 2005) aimed at providing solutions to monitoring cardiopulmonary conditions. The Wealthy platform has shown a very high level of integration between textile technology and a wearable electronic unit endowed with a wearable on-board processing and Bluetooth and GPS wireless transmission (see Fig. 7.6).

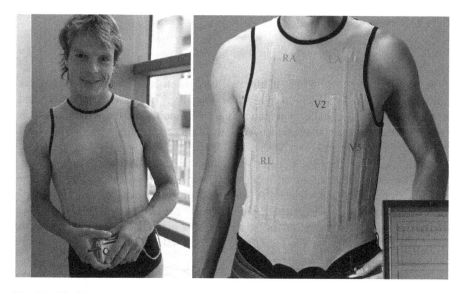

Fig. 7.6 Wealthy system

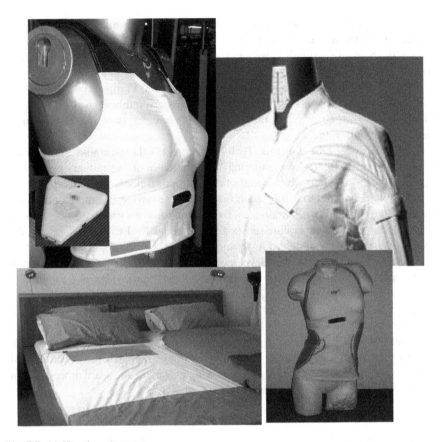

Fig. 7.7 MyHeart's prototypes

Several concepts of SFIT platforms have been designed and prototypes
realized in the EC funded project MyHeart (see Fig. 7.7) in which primary and
secondary prevention, acute intervention, chronic monitoring and rehabilitation in
cardiovascular disease management have been addressed through e-textile-based and
microelectronic solutions (FP6–2002-IST-507816).

Nowadays several research projects are under way to examine the appropriate-
ness of e-textile solutions in different areas, such as neuro and cerebrovascular
rehabilitation, safety and security. A very sophisticated platform, under development in the frame of the EC funded project Proetex (FP6–2004-IST-4–026987) is
shown in Fig. 7.8, it addresses the issue of endowing with smart functions both
inner and outer garments for firefighters and specialized personnel for disaster
management.

Fig. 7.8 Proetex's platform

References

Arora S, Ghosh T, Muth J (2007) Dielectric elastomer based prototype fiber actuators. Sens Actuators A 136:321–328

Balachandran R, Pacheco D, Lawandy N (1996) Photonic textile fibers. Appl Opt 35:91–94

Bhattacharya R, de Kok M, Zhou J (2009) Rechargeable electronic textile battery. Appl Phys Lett 95:223305

Bonfiglio A, De Rossi D, Kirstein T et al (2005) Organic field effect transistors for textile applications. IEEE Trans Inf Technol Biomed 9(3):319–324

Cameron C, Szabo J, Johnstone S et al (2008) Linear actuation in coextruded dielectric elastomer tubes. Sens Actuators A 147:286–291

Carpi F, De Rossi D (2005) Electroactive polymer-based devices for e-textiles in biomedicine. IEEE Trans Inf Technol Biomed 9(3):295–318

Carpi F, Mannini A, De Rossi D (2009) Dynamic splint-like hand orthosis for finger rehabilitation. In Carpi F, Smela E (eds) Biomedical applications of electroactive polymer actuators. John Wiley & Sons Ltd, New York, pp 443–461

Chan Vili Y (2007) Investigating smart textiles based on shape memory materials. Text Res J 77(5):290–300

Collins G, Buckley L (1996) Conductive polymer-coated fabrics for chemical sensing. Synth Metals 78:93–101

Coyle S, Wu Y, Lau K, De Rossi D et al (2007) Smart nanotextiles: a review of materials and applications. MRS Bulletin 32:1–9

Coyle S et al (2010) BIOTEX – Biosensing textiles for personalised healthcare management. IEEE Trans Inf Technol Biomed 14(2):364–370

Ctrysse M, Puers R, Hertleer C et al (2004) Towards the integration of textile sensors in a wireless monitoring suit. Sens Actuators A 114(2–3):302–311

De Rossi D (2007) A logical step. Nat Mater 6:328–329

De Rossi D, Della Santa A, Mazzoldi A (1999) Dressware: wearable hardware. Mat Sci Eng C 7(1):31–35

De Rossi D, Della Santa A, Mazzoldi A (1997) Dressware: wearable piezo-and thermoresistive fabrics for ergonomics and rehabilitation. In: Proceedings of 19th international conference of the IEEE-EMBS, Chicago, IL, 30 October to 2 November, vol. 5, pp 1880–1883

De Rossi D, Lorussi F, Mazzoldi A et al (2003) Active dressware: wearable kinesthetic system. In: Barth FG, Humphrey JAC, Secombe TW (eds), Sensors and sensing in biology and engineering. Springer Wien, New York, pp 379–392

Dhawan A, Muth JK, Ghosh T (2006) Optical nano-textile sensors based on the incorporation of semiconducting and metallic nanoparticles into optical fibers. Materials Research Society, Spring Symposium

Edmison J, Jones M, Nakad Z et al (2002) Using piezoelectric materials for wearable electronic textiles. In: Proceedings of the 6th IEEE international symposium on wearable computers, Washington, DC, pp 41–48

El-Sherif M (2000) Fiber optic sensors and smart fabrics. J Intell Mater Syst Struct 11(5):407–414

Fan X, Chu Z, Chen L et al (2008) Fibrous flexible solid-type dye-sensitized solar cells without transparent conducting oxide. Appl Phys Lett 92:113510

Farringdon J, Moore A, Tilbury N et al (1999) Wearable sensor badge and sensor jacket for context awareness. In: Proceedings of the 3rd symposium on wearable computers, San Francisco, CA, 18–19 October, pp 107–113

Gauvreau B, Guo N, Schiker K et al (2008) Color changing and colour-tunable photonic bandgap fiber textiles. Opt Express 16(20):15672–15693

Gibbs PT, Asada HH (2005) Wearable conductive fiber sensors for multi-axis human joint angle measurement. J Neuroeng Rehabil 2(7):1–7

Giorgino T, Tormene P, Lorussi F et al (2009) Sensor evaluation for wearable strain gauges in neurological rehabilitation. IEEE Trans Neural Syst Rehabil Eng 17(4):409–415

Gopalsamy C, Park S, Rajamanickam R et al (1999) The wearable motherboard: the first generation of adaptive and responsive textile structures (ARTS) for medical applications. Virtual Real 4:152–168

Gourmelon L, Langereis G (2006) Contactless sensors for surface electromyography. In: Proceedings of the 28th annual conference of the IEEE-EMBC, New York, NY, 30 August to 3 September, pp 2514–2517

Gregory R, Kimbrell W, Kuhn H (1989) Conductive textiles. Synth Met 28(1–2):823–835

Grossman P (2003) The lifeshirt: a multi-function ambulatory system that monitors health, disease, and medical intervention in the real world. In: Proceedings of the international workshop on new generation wearable system for e-health: toward revolution of citizens' health, life style management, Lucca, 11–14 December, pp 73–80

Hamedi M, Fochheimer R, Inganäs O (2007) Toward woven logic from organic electronic fibers. Nat Mater 6:357–362

Hamedi M, Herlogsson L, Crispin X et al (2009) Fiber-embedded electrolyte-gate field effect transistors for e-textiles. Adv Mater 21:573–577

Harlin A, Makinen M, Vuorivista A (2003) Development of polymeric optical fibre fabrics as illumination elements and textile displays. Autex Res J 3:1–8

Haynes A, Gouma P (2008) Electrospun conducting polymer-based sensors for advanced pathogen detection. IEEE Sens J 8(6):701–705

Hertleer C, Rogier H, Vallozzi L et al (2009) A textile antenna for off-body communication integrated into protective clothing for firefighters. IEEE Trans Antennas Propag 57(4):919–925

Hu J (2007) Shape memory polymers in textiles. Woodhead Textiles Series, No. 65, 360 pp

Ishijima M (1997) Cardiopulmonary monitoring by textile electrodes without subject awareness of being monitored. Med Biol Eng Comp 35(6):685–690

Kang T, Merritt C, Grant E et al (2008) Nonwoven fabric active electrodes for biopotential measurement during normal daily activity. IEEE Trans Biomed Eng 55(1):188–193

Klemm M, Troester G (2006) Textile UWB antennas for wireless body area networks. IEEE Trans Ant Propag 54(11):3192–3197, http://www.konarka.com

Koncar V (2005) Optical fiber fabric displays. Opt Photon News 16:40–44

Lanatà A, Sclingo E, De Rossi D (2010) A multi-modal transducer for cardiopulmonary activity monitoring in emergency. IEEE Trans Inf Technol Biomed

Lee J, Subramanian V (2005) Weave patterned organic transistors on fiber for e-textiles. IEEE Trans Electron Dev 52:269–275

Lorussi F, Galatolo S, De Rossi D (2009) Textile-based electrogoniometers for wearable posture and gesture capture systems. IEEE Sens J 9(9):1014–1024

Lorussi F, Rocchia W, Scilingo EP et al (2004) Wearable, redundant fabric-based sensor arrays for reconstruction of body segment posture. IEEE Sens J 4(6):807–818

Luid E, Jayaraman S, Park S et al (1997) A sensate liner for personnel monitoring applications. In: Proceedings of the first international symposium on wearable computers, Cambridge, MA, 13–14 October, pp 98–105

Maccioni M, Orgio E, Cosseddu P et al (2006) Towards the textile transistor: assembly and characterization of an organic field effect transistor with a cylindrical geometry. Appl Phys Lett 89:1–3

Marculescu D et al (2003) Electronic textiles: a platform for pervasive computing. Proc IEEE 91:1995–2018

Mattana G, Strickland A, Bonfiglio A et al (2010) Flexible, electrically conductive cotton yarns. Submitted to ACS Nano

Mattmann C, Clemens F, Troester G (2008) Sensor for measuring strain in textile. Sensors 8:3719–3732

Morris D, Coyle S, Wu Y et al (2009) Bio-sensing textile based patch with integrated optical detection system for sweat monitoring. Sens Actuators B139:231–236

Norris I, Mattes B (2007) Conducting polymer fiber production and application. In: Skotein T, Reynolds J (eds) Handbook of conducting polymers. CRC Press, Boca Raton, FL

O'Connor B, An K, Zhao Y et al (2007) Fiber shaped organic light emitting device. Adv Mater 19:3897–3900

O'Connor B, Pipe K, Shtein M (2008) Fiber based organic photovoltaic devices. Appl Phys Lett 92:193306

Orth M, Post R, Cooper E (1998) Fabric computing interfaces. CHI 98 Conference on Human Factors in Computing Systems, AMC Press, Los Angeles, pp 331–332

Pacelli M, Caldani L, Paradiso R (2006) Textile Piezoresistive Sensors for Biomechanical Monitoring, Proceeding of the 28th IEEE EMBS, New York, 30 August–3 September 2006 pp 5358–5361

Paradiso R, Loriga G, Taccini N (2005) A wearable health care system based on knitted integrated sensors. IEEE Trans Inf Technol Biomed 9(3):337–344

Paradiso R, De Rossi D (2006) Advances in Textile Technologies for Unobtrusive Monitoring of Vital Parameters and Movements, Proceeding of the 28th IEEE EMBS, New York, 30 August–3 September 2006, pp 392–395

Post E, Orth M (1997) Smart fabrics or "wearable clothing." In: Proceedings of the first international symposium on wearable computers, Pittsburgh, PA, pp 167–168

Post E, Orth M, Russo P et al (2000) E-broidery: design and fabrication of textile-based computing. IBM Syst J 19(384):840–860

Ramier J, Plummer C, Leterrier Y et al (2008) Mechanical integrity of dye-sensitized photovoltaic fibers. Renew Energ 33(2):314–319

Salonen P, Ramat-Samii Y, Kivitoski M (2004) Dual-band wearable textile antenna. In: Proceedings of the IEEE antennas and propagation society international symposium, vol 1, pp 463–466

Salvo P, Di Francesco F, Costanzo D et al (2010). A wearable sensor for measuring sweat rate. Sens J 10(10), October 2010, pp 1557–1558

Scilingo EP, Gemignani A, Paradiso R et al (2005) Performance evaluation of sensing fabrics for monitoring physiological and biomechanical variables. IEEE Trans Inf Technol Biomed 9(3):345–352

Sergio M, Manaresi N, Tartagni M et al (2002) A textile based capacitive pressure sensor. Proc IEEE Sens 2:1625–1630

Service R (2003) Electronic textiles charge ahead. Science 301:909–911

Shim, BS, Chen W, Doty C et al (2008) NanoLetters 8(12):4151–4157.

Tognetti A, Lorussi F, Bartalesi R et al (2005) Wearable kinesthetic system for capturing and classifying upper limb gesture in post-stroke rehabilitation. J Neuroeng Rehabil 2(8):1–16

Tronquo A, Rogier H, Hertleer C et al (2006) Robust planar textile antenna for wireless body LANs operating in 2.45 GHz ISM band. Electron Lett 42(3):142–143

VDC (2007) Smart fabrics and interactive textiles and related enabling technologies market opportunities and requirements analysis published. Venture Development Corporation (VDC)

Wang J, Too C, Wallace G (2005) A highly flexible polymer fiber battery. J Power Sources 150:223–228

Wang Z, Wang X, Song J et al (2008) Piezoelectric nanogenerators for self-powered nanodevices. IEEE Pervasive Comput 7(1):49–55

Weber AL, Blanc D, Dittmar A et al (2004) Telemonitoring of vital parameters with newly designed biomedical clothing VTAM. Stud Health Technol Inform 108:260–265

Wijesiriwardana R, Dias T, Mukhopadhyay S (2003) Resistive fiber-meshed transducers. In: Proceedings of the 7th IEEE international symposium on wearable computers, New York, 21–23 October, pp 200–209

Winchester R, Stylios G (2003) Designing knitted apparel by engineering, the attributes of shape memory alloys. Int J Cloth Sci Technol 15(5):359–366

Xu P, Zhang H, Tao X (2008) Textile-structured electrodes for electrocardiogram. Text Prog 40(4):183–213

Yadav A, Pipe K, Shtein M (2008) Fiber-based flexible thermoelectric power generator. J Power Sources 175(2):909–913

Part II
Applications

Chapter 8
A Survey of Commercial Wearable Systems for Sport Application

Sergio Guillén, Maria Teresa Arredondo, and Elena Castellano

8.1 Introduction

The aim of this chapter is to provide an overview on wearable systems used in sport and physical activities. Sportconcept covers many different types of activities and objectives of its practitioners. The use of ICT technologies in sport, and in particular of wearable systems, is therefore closely linked with the specificities of the sportive practice rules, conditions, environments and goals.

Sport is an activity that is characterized by the demand of a physical exertion. It is usually *institutionalized* (associations, federations and clubs), implies *competition* between individuals or between teams and is *governed* by a set of rules. Sport is also referred to as activities where the physical capacity, conditions and training are the fundamental circumstances that are primordial determinants of the outcome: win or loss. Nowadays, psychological conditions are of key importance, particularly in high competition where physical conditions are brought to the limit of the human being capacity. More broadly, the term also includes activities such as mind sports, card sports, board games or motor sports where mental acuity or equipment quality is major factors.

Sports that are subjectively judged are distinct from other judged activities such as beauty pageants and bodybuilding shows, because in the former the activity performed is the primary focus of evaluation, rather than the physical attributes of the contestant as in the latter (although "presentation" or "presence" may also be judged in both activities).

Sometimes, the concepts of sport and physical activity/fitness are mixed up. Physical fitness is considered a measure of the body's ability to function efficiently and effectively in work and leisure activities, to be healthy, to resist hypokinetic diseases and to meet emergency situations. Generally, a fitness practitioner must follow, what is named as *physical fitness program* which will probably focus on one or more specific skills, and on age or health-related needs, such as bone health.

S. Guillén (✉)
Instituto de Aplicaciones de las tecnologías de la información y de las comunicaciones avanzadas (ITACA), Universidad Politécnica de Valencia, 46022 Valencia, Spain
e-mail: sguillen@itaca.upv.es

A. Bonfiglio and D. De Rossi (eds.), *Wearable Monitoring Systems*,
DOI 10.1007/978-1-4419-7384-9_8, © Springer Science+Business Media, LLC 2011

Accordingly, a general-purpose physical fitness program must address the following essentials: cardiovascular fitness, flexibility training, strength training, muscular endurance, body composition and general skill training. (Wikipedia 2010) Physical fitness can also prevent or treat many chronic health conditions brought on by unhealthy lifestyle or aging.

Measurement is substantial to the practice of sport and fitness: measurement of the outcome of the practice, such as time, distance, length, weight, height and strength, as well as the results of the physical exertion such as oxygen consumption, heart rate (HR), temperature or biomechanical measurements. Thus, a big number of measurement systems have been developed and commercialized for use in almost all types of sport and fitness practice.

Wearable technologies, i.e. clothing and accessories incorporating computer and advanced electronics, data processing and wireless communication capabilities, suit very well the needs of having monitoring and real-time feedback for athletes. The decreasing cost of processing power and other components is encouraging wide-spread adoption and availability. ABI Research estimates that the market for wearable wireless sensors is set to grow to more than 400 million devices by 2014. Of course, health and fitness sensors are not the only use case for wearable sensor but they will likely dominate that market (Dolan 2009).

In the next sections, we will analyze the use of wearable technologies and systems for three types of measurements:

1. Physiological measurements, which are related to the response of the body to the demand of the physical exertion during training and competition.
2. Performance measurements, which are related to the outcome of the effort made by the sportsman/sportswoman, such as velocity, altitude, distance and many other parameters.
3. Biomechanical measurements, which are focused on the measure of movements of parts of the body, and are usually related to the study of the way to achieve optimal outcomes.

8.2 Wearable Systems for the Measurement of Physiological Parameters

This section will focus on wearable systems used for the measurement of vital signs from the sportsman/sportswoman's body. The parameters most extended and used for athletes are, in all the cases, the HR and the oxygen consumption.

8.2.1 Heart Rate

HR is determined by the number of heartbeats per unit of time, typically expressed as beats per minute (BPM). It can vary as the body's need for oxygen changes, such as during exercise or sleep. Athletes are interested in monitoring their HR to gain maximum efficiency from their training. HR is measured by finding the pulse of the body. This pulse rate can be measured at any point on the body where an artery's pulsation is transmitted to the surface. A precise method of determining pulse involves the use of an electrocardiograph, or ECG (also abbreviated EKG).

Commercial HR monitors are widely available, consisting basically of a chest strap with electrodes. The signal is transmitted to a wrist receiver for display. HR monitors allow accurate measurements to be taken continuously and can be used during exercise when manual measurement would be difficult or impossible (such as when the hands are being used).

In addition to the HR, current pulse meter usually incorporate the computation of many parameters associated with the main one, such as HR max, HR min and HR average, and other measurements associated with the performance, such as lap time, speed, pace and distance, or environmental information such as temperature and altitude. Indirect measurements are frequently added to the features of the most advanced devices: the maximum oxygen uptake (VO_2max), based on HR and HR variability at rest and personal information, and the number of kilocalories burned during an exercise session (Polar 2010)

Based on the same product components, *Polar* has launched a product to help monitoring training into collective sports: "The Team Pro system," which allows gaining a unique insight into the fitness capabilities of the team, includes base station, transmitter charger, ten transmitters and Team WearLink + straps, USB dongle and the software for PCs and PDAs. It adjusts and optimizes training intensities, records and studies fitness data (this is HR monitoring) in real time for 28 players and analyzes training intensity and recovery time using the Polar Training Load parameter.

8.2.2 Using Smart Clothes for Body Signal Measurements

Intelligent biomedical clothes are based on conductive and piezoresistive fabric developed to work as textile sensors, i.e. working as transducer of vital signs to electrical signals to be sent to a unit of microprocessors. (My Heart Project 2009)

From the textile point of view, different yarns and material are used to realize the textile interface: the ground yarn has to be elastic, light, comfortable, and breathable and antibacterial since the garment may be required to be worn during sporting activities. Another aspect is that yarn has to be conductive for the electrodes, such as silver-plated polyamide fibre and staple spun stainless steel yarns and stainless steel continuous filaments. On the other hand, to detect the body respiration movements, fabric sensors are knitted from piezoresistive yarns.

First commercial products are now offered in the market. Most of them are direct replacement of the chest strap by a sensitized garment, such as the NuMetrex® HR monitoring sports bra and men's T-shirt by *Textronics, Inc.* (Numetrex 2010).

Numetrex is a kit composed by a garment that incorporates the sensor for HR measurement, a data processor and transmitter and a watch for monitoring in real time.

Another example in this field is the *Addidas-Polar* (Adidas-Polar 2010.) product that consists of a T-shirt that incorporates a knitted strap with a sensor for HR measurement and a detachable electronic unit. The sensor part works with the Polar's wrist watch computer family.

8.2.3 Cardiopulmonary Response

Runners, cyclings, skaters and athletes can be interested in testing cardiopulmonary response. The ergospyrometry system allows the determination of a subject's metabolic response while exercising or working. One example of such a system is a wireless portable ergospyrometry system, *Oxycon Mobile* (Carefusion 2010.) by Sensormedics. The system has a light battery operated portable ergospyrometer that is mounted onto the subjects' body via a comfortable vest. The breath-by-breath data is collected through a facemask or mouthpiece and is sent to a host computer system via wireless transmission (telemetry), making the device suitable for use in a non-laboratory environment. Specific software allows the on-line and post-processing of data. Figure 8.1 shows the portable system.

8.3 Measuring Performance

In addition to vital signals measured from the athlete's body, there are plenty of devices in the market that help the user to control other different useful parameters during the sport practice, such as: time, altitude, speed, distance, cadence, stride, etc.

Different devices are used to assess and improve the *running* techniques and analyze individual performance. The basic device uses a 3D accelerometer to detect movement onto the three axes. These signals are further processed to detect activity, pace counting, speed and distance. For example and to mention a few, a complement that Polar commercializes for runners is a small device that can be mounted securely onto training shoe laces, which transmit data to the wristwatch computer. In conjunction with Apple Computers, Nike commercializes a similar device specially designed to fit in a mid-sole cavity in running shoes. The Nike sensor transmits data to an Apple's iPod mp3 player, which acts as user interface.

For *cyclists*, the same concept exists but in this case, the wireless sensor transmits current, average and maximum speed readings so that cyclist can get it right in training and ultimately, right on the big race day. Additional parameters that can be obtained are cadence, turn-by-turn directions, power data, speed, distance, time, calories burned, altitude, climb and descent. For extra-precise climb and descent data, a barometric altimeter is incorporated into pinpoint changes in elevation.

Fig. 8.1 Oxycon mobile system. Photography yielded by Sensormedics

For *golf sport*, Garmin offers a GPS packed with thousands of preloaded golf course maps. Approach uses a high-sensitivity GPS receiver to measure individual shot distances and show the exact yardage to fairways, hazards and greens. It measures the golfer position through GPS signal and the map shows an overview of the current hole that will be played. It can be mentioned, as well, that Suunto G6 is a watch worn on the wrist, which consists basically of three motion sensors that allow, after each swing, to check the main movements in golf: tempo, rhythm, backswing length and speed. This feedback helps identify the best shots, developing muscle memory to repeat only the good shots and analyze overall consistency.

The *measurement of calories* burned both during sport practice and during daily normal activity is a matter of interest of wearable systems manufacturers. In addition to the indirect calculator included as a feature in most of the pulse meter devices, there are special devices such as Bodybugg (24 Hour Fitness USA 2010), which is a calorie management system that combines four types of sensors in an arm band:

1. A *motion/steps counter*, which uses a three-axis accelerometer to measure the distinct patterns created by running and walking.
2. A *galvanic skin resistance*, which is sensitive to the sweating condition and detects the constriction or dilation of the vascular periphery.
3. A *skin temperature* that measures the surface temperature of the body.
4. *Heat flux* that measures the heat flowing from the body into the environment.

For *swimmers and triathletes* in training or racing, keeping a tab on laps is tedious, but it can also be easy to lose track at times, especially as the athlete gets tired. The British company Swimovate (Swimovate 2010) has a swimming computer, worn like a wristwatch, which takes the mental work out of keeping a lap count.

Other technique for measuring swimmers performance and technique is: using motion-sensor technology based on MEMS accelerometers to measure the acceleration of the wrist during the swim stroke. Advanced digital signal processing techniques and software algorithms have been developed that detect the arms moving through the water and analyze the data, giving read-outs for speed, distance and calories (Ohgi 2005).

The computer works for the four major strokes – freestyle, breaststroke, backstroke and butterfly – tumble-turns and straight out of the box – with no need for individual calibration.

In *underwater sports*, a diver must monitor several gauges to receive the crucial information about the dive: depth, time, tank pressure, decompression status and direction to name just a few. In this case, it is for security reasons. These parameters are crucial for making the diving security a reality. There are many manufacturers of dive computers, such as Suunto (Suunto 2010). Taking as example one of these products, Suunto D9 combines all dive-critical information in one instrument, conveniently mounted on the wrist. It incorporates a digital compass, which shows the general direction with a graphical compass rose, as well as the exact bearing with a numerical display. By using a wireless transmitter, the diver can also monitor tank pressure and air consumption data. An estimation of the remaining air-time is given throughout the dive along with the continuous depth monitoring, immersion time and decompression status.

8.3.1 Sensors over Sport Equipment

Another set of measurements is not taken from devices located on the body but on the equipment used for the sport practice, bike and even in the ambient in which sport is practiced, as in velodromes.

BAE Systems has installed a sophisticated performance monitoring system at the Manchester Velodrome to give British cyclists a further edge in training. (systems 2010.) The laser-timing technology, derived from a battle space identification system, represents an entirely new approach for monitoring performance in cycling, improving on previous break-beam systems, which are unable to

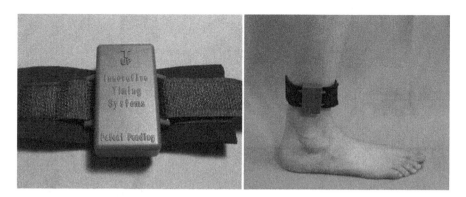

Fig. 8.2 Marathon chip. Photography yielded by Innovative Timing Systems

differentiate between individual athletes. Up to 30 cyclists will be able to train simultaneously with the new timing system, which uses a laser able to read a personalized code from a retro reflective tag attached on each bike. Installed at multiple points around the track, the system gives individual recordings for each cyclist with millisecond accuracy.

Another field using wearable systems is the sports racing, as marathons. A small chip is provided to each marathon participant. This chip can be placed anywhere on the athlete's body, e.g. in a shoe, inside the pocket, in a helmet. The chip (Fig. 8.2), which basically consists in a tag, is read by a series of antennas which are allocated overhead or of the site of the start, split or finish lines. Figure 8.3 shows the full systems, with the runners and their timing ships, antennas, station where the signals are transmitted and treated and a screen where a race classification can be seen.

As a marathon can be run from 50 to 500,000 participants, dimensions of the system are crucial. Antennas read each chip and send the signals to the base station where the signals are processed through specific software and make timing data accessible to all participants (Innovative 2007).

Mechanical (no ICT) wearable systems have been designed to support the users in realization of sport activity, for the improvement of both performance and physical safety during the practice. One example of this type of system is Ski-Mojo from Kinetic Innovations Limited (Innovations 2010). This is a mechanical system that is wore by ski practitioners to support in the correct posture, helping to reduce back pain and making the manoeuvres easier to perform for beginners. The device acts as a shock absorber, reducing the shock to the whole body caused by uneven terrain, improving stability and protecting the joints. In Fig. 8.4 Ski Mojo is showed.

Researchers of the Department of Mechanical Engineering, University of Padova, had been carried out an evaluation of muscular EMG activity variations wearing Ski-Mojo on a skiing treadmill (Petrone and De Bettio 2009). The aim of the investigation was the evaluation of the muscular activity patterns on a skier equipped with a Ski-Mojo device. Tests were performed on a skiing treadmill by an

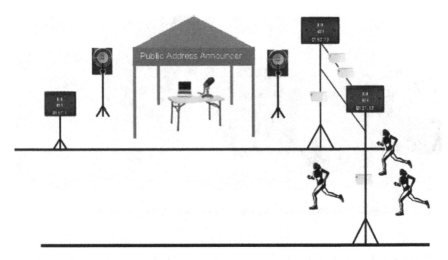

Fig. 8.3 Jaguar's marathon system. Graphic taken from Innovative Timing Systems' website

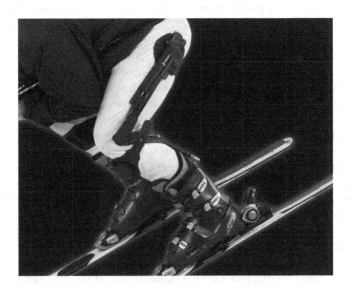

Fig. 8.4 Ski-Mojo. Photography taken from press pack on the website

experienced skier wearing the Ski-Mojo for the first time. He skies performing narrow slaloms turns (SLALOM) and wide slaloms turns (GIGANTE) with Ski-Mojo OFF and ON.

EMG signals were consistently lower when the Ski-Mojo was activated (ON). The average reduction of the EMG activation signal is reported for each muscle in the two conditions of Special Slalom (SS) and Giant Slalom (GS). Figure 8.5 shows the reduction per muscle considering Special Slalom (SS) in four muscles, rectus femoris, vastus medialis, tibialis anterior and biceps femoris. Red curve shows the

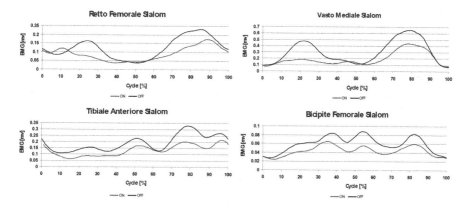

Fig. 8.5 EMG signals from rectus femoris, vastus medialis, tibialis anterior and biceps femoris muscles. RED curve with Ski Mojo ON and BLACK curve with Ski-Mojo OFF. Adapted from Petrone and De Bettio (2009)

signal with ski mojo activated and black curve with the ski mojo off. Reduction of muscle activation levels is ranging from -17% to -37%.

Another sport that is applying wearable computing systems is *taekwondo competition sparring*. Taekwondo is pushing energy into wearable systems which allow to ensure fairness in judging and to make the sport friendlier to spectators. The problem in achieving accurate scoring is the subjective judgement of what constitutes a valid scoring kick on the body. A valid scoring kick must be delivered "accurately and powerfully to the legal scoring of the body." This system is called "SensorHogu System" (Chi et al. 2009) and uses piezoelectric force sensors in body protectors to sense the amount of force delivered to a competitor's body protector, and wirelessly transmits the signal to a computer that scores and displays the point.

For security reasons, the signal is encrypted using a military grade to ensure no competitors can tamper with the wireless data packets. For robustness in terms of noise, the data stream is encoded using Frequency Agile Spread Spectrum technology.

In addition to the sensors body protector, SensorHogu contain three handsets, one for each of the judges who scores the match. There are two handsets for each judge, one for scoring the red player on the left hand, and the other for scoring the blue player on the right hand. Handset has two buttons, a trigger button scores a point for the body, and a side button scores two points for a head blow. According to scoring rules, at least two judges must press the same button on their handsets within a 1-s window for the point to score.

Concerning the architecture, SensorHogu uses a distributed architecture as it allows multiple signals to be processed in parallel at each device before the processed data is transmitted to a base station. The base station then forwards the data to the computer running the operator interface and score display. Base station collects the signals from the two SensorHogu and the three judges' handsets, and a monitoring laptop is used to collect, display and log the data. The time-stamped log tells how hard each player has been hit, and whether any of the judges' handset buttons is depressed. The laptop interprets these signals and updates the score board accordingly. The laptop is connected directly to a score display (an off-the-shelf LCD or CRT monitor).

This idea can be applied in many other fields in the future, for instance fitness innovations, such as "Dance Dance Revolution," games in which the players must have direct contact. Other applications could be for example: military use or police or self-defence training.

8.4 Biomechanical Measurements

Wearable systems are used in the study of the movements of different parts of the body. Usually, the applications are addressed to understand and to design models of spatial dynamics of the body.

One example is *BIOFOOT/IBV* (Instituto de Biomecánica 2002), a device dedicated to the research of the pressure patterns on the foot sole under different condition of activity. The basic system consists of the following components: A set of instrumented insoles of four different sizes, two signal amplifiers, transmitter, data receiver card and software license (Fig. 8.6).

Each insole has mounted 64 small size piezoelectric ceramic sensors, distributed selectively. The output signal is performed and sent to the amplifier, which is attached to the leg. The amplified signal is the input to the wireless transmitter. Wireless transmitter is located over a belt wearer in the waist. Signal is transmitted to the PC card wirelessly through a wide bandwidth channel (11Mb/s) and it is converted in data and graphics useful for interpretation using a proprietary software. BIOFOOT allows taking measures until 200 m of distance.

BIOFOOT has a large number of applications, such as:

- Ortho – prosthesis: design of footwear, orthosis and prosthesis.
- Surgery of the foot: pre-operative evaluation and functional post-operatory control.
- Sportive Medicine: study of the footwear and sports complements.

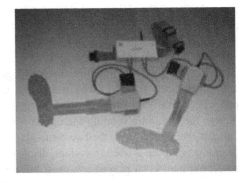

Fig. 8.6 BIOFOOT system. Photography taken from reference Instituto de Biomecánica (n.d.)

- Rehabilitation: follow-up and control of treatments.
- Dermatology: study of pressures on fabrics.
- Diagnosis of neuropathies and ideal design of insoles and worn for the prevention and the treatment of sores.

8.4.1 Rehabilitation

Determination of the exact foot pressures during the phase of walks is considered to be an essential component in the diagnostic evaluation and planning of the patients' treatment by location of pain, or by problems of insensibility in the feet, which can be caused by diverse types of diseases.

This application will let, during clinic practice: register in situ the distribution of pressures, to realize quantitative analyses of the human normal and pathological march. To evaluate the effect of prosthesis and orthosis of lower limb, to help the specialist in the detection and correlation among painful areas and areas of mechanical overcharge, to realize a detailed follow-up of the evolution of the patients' pathology, to help the specialist in the planning of the treatment most adapted for the patient, to document graphically the exploration and, in case of conservative treatments, to accompany the prescription on this information to attend the technician in the prosthesis confection.

8.4.2 Corporal Pain Detection

In this area, the system Biofoot/IBV allows to evaluate those pathologies, structural or functional, that reverberate in the support of the foot and in the distribution of pressures of contact during the different phases of the walk.

The dynamic evaluation of pressures supposes an advance opposite to the optical static traditional methods of exploration of the support planting. The deficits or alterations of support are established in graphical form and are correlated by the associate pathology.

8.4.3 Technical Orthopaedics

In the majority of the orthopaedic treatments, the design and construction of orthosis, including the selection of materials, is based on an iterative testing method "trial and error," using the static mold of the sympathetic foot as information. With these areas, the final adequacy of the orthosis to the patient depends, in very high degree, on the experience of the technician and on the capacity of the patient to transmit the problems of adequacy to the technician in charge of the confection.

This system will allow to analyze the effect of unload achieved with different materials, to study the effect of diverse insertions and complements, etc.

8.4.4   Sports Analysis

The analysis of the rapid gestures allows obtaining conclusions directly applicable to the training, to the evaluation of the sports technology (skill) and to the behaviour of the sports footwear and accessories.

The distribution of pressures is a determinant factor of the comfort of the footwear. A potential field of application of Biofoot/IBV is to evaluate the design of all kinds of footwear, which allows checking the adequacy in use, of the materials selected [of the sole and of the insole (staff)].

As an example in sports analysis, IBV has developed a methodology to analyze the influence of football boots design in the instep kick. A measuring system together with analysis software was designed and validated combining pressures and high speed video cameras able to quantify the pressure pattern on the foot dorsum and its relationship with ball behaviour (deformation and velocity) during the instep kick event (Ripoll et al. 2009a, b). Results demonstrated the influence of ball velocity and of the intention of precision in the impulse provided to the ball and the pressure pattern applied by the foot dorsum on the ball. These results are useful for kicking skill improvement techniques. Results also suggest that foot anatomy and footwear play an important role on the pressure patterns. This could have implications in injury prevention and soccer-specific footwear design (Ripoll et al. 2009a, b).

8.4.5   Sports Using an Implement: Bat, Club, Racquet

A similar idea has been developed to analyze grip forces in sports such as golf, tennis, cricket, baseball where the participant needs to use an equipment – bat, club, racket, etc. – to strike a ball or similar object. In such sports, the only point of contact that the participant has with the equipment is at the grip. Balance must be found between the force used to secure the object in the hand and wrist range of motion. Furthermore, grip and pinch strength vary based on wrist angle (Komi et al. 2008). In the cited study, the measure of grip force during a standard golf tee shot is done using two different measurement techniques. The first utilized a matrix type, thin film sensor applied to a golf grip. The second method involved 31 individual thin film force sensors strategically placed on two golf gloves (Fig. 8.7).

In this case, the system is built using pressure sensors by Tekscan using piezoresistors.

As a result, it was found that the forces produced by an individual golfer were very repeatable but varied considerably between golfers (force "signature"). Such

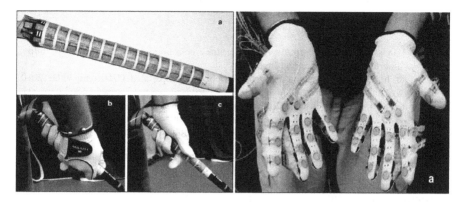

Fig. 8.7 (*Left*) Golf grip matrix sensor. (*Right*) Golfer gloves matrix sensor. Photographies are taken from reference Komi et al. (2008)

investigation of grip force can aid in future grip design, the creation of training aids, injury evaluation and prevention and answer existing questions about how elite players actually grip the club.

8.5   Conclusions

Technology used for monitoring vital signals from athlete while practicing sports is focused on the measurement of the HR using a sensor, which is located on the chest. The simplest case and the most used is a band that transmit sensored data signal to a wearable receiver, i.e. a wristwatch computer. Alternatives to this basic sensor system is now offered by smart sport clothes that are opening new opportunities for physiological parameter measurements by including other vital signs such as respiration, body temperature and ECG, and motion monitoring.

Measurement of performance is of similar interest for athletes, both during the training and during the competition. There are wearable instruments for this purpose for almost any kind of sports practice. In this article, we have identified some examples such as cycling, running, golf, swimming, underwater sports, or skiing, but it is far to be a completed list of instruments offered in the professional sport market that meet the practitioners' necessities.

Finally, the last groups of measures that can be taken are the biomechanical measurements which allow the sportsmen/sportswomen to detect wrong movements in their practice and help to correct them and improve their styles, avoiding possible injuries. Actually, this field is the most experimental one, and it is a subject of further research.

References

24 Hour Fitness USA, Inc. http://www.24hourfitness.com/training/bodybugg/. 2010. http://
  www.24hourfitness.com. Accessed June 2010
Adidas-Polar http://www.press.adidas.com/en/DesktopDefault.aspx/tabid-11/16_read-3002. 2010.
  http://www.adidas.com. Accessed June 2010
Carefusion   http://www.sensormedics.com/prod_serv/prodDetail.aspx?config=ps_prodDtl
  &prodID=3. 2010. http://www.sensormedics.com. Accessed June 2010
Chi H, Song J, Corbin G (2009) "Killer App" of wearable computing: wireless force sensing body
  protectors for martial arts
Dolan Brain (2009) Wireless health: State of the industry 2009 Year End Report. s.l. *Mobilhealth-
  news*
Innovative, Timing http://www.innovativetimingsystems.com/. 2007. http://www.innovativetimingsystems.
  com/. Accessed June 2010
Instituto de Biomecánica http://www.podocen.com/pdfs/dossier_ibv.pdf *IBV*. 2002. http://www.
  ibv.es. Accessed June 2010
Innovations,   Kinetic   Ltd.   http://www.skiallday.co.uk/sm.   http://www.skiallday.co.uk/sm.
  Accessed June 2010
Komi E R, Roberts JR, Rothberg J (2008) Measurement and analysis of grip force during a golf
  shot. (IMechE – Sports Engineering and Technology), 23–35
My Heart Project (2009) Deliverable 24.3. *On body electronics*. Project Deliverable, Aachen: My
  Heart Consortium
Numetrex http://www.numetrex.com/. 2010. http://www.numetrex.com. Accessed June 2010
Ohgi Y (2005) MEMS sensor application for the motion analysis in sports science. *COBEM 18th
  international congress of Mechanical Engineering*. ABCM 501–508
Petrone N, De Bettio G (2009) Evaluation of muscular emg activity variations wearing ski-mojo
  on a skiing treadmill. (DolomitiCert)
Polar http://www.polar.fi/en/products/maximize_performance/running_multisport/RS800CX.RS800CX.
  2010. http://www.polar.fi. Accessed June 2010
Ripoll EM, Nadal AP, Melis JO, Payá JG, García AU, Ferrer RF (2009) Foot dorsum pressure and
  ball impulse patterns in a soccer full instep kick with different requirements of ball speed and
  precision
Ripoll EM, Nadal AP, Melis JO, Payá JG, García AU, Ferrer RF, Gonza JC (2009) Puesta a punto
  de una metodología para el análisis de la influencia del calzado de fútbol en el golpeo del balón
Suunto http://www.suunto.com/es/Products/#filter=894. 2010. http://www.suunto.com. Accessed
  June 2010
Swimovate http://www.swimovate.com/. 2010. http://www.swimovate.com/. Accessed June 2010
systems, Bae. http://www.baesystems.com/Capabilities/Technologyinnovation/NewTechnologies/
  UKSport/index.htm. *BAE SYSTEMS uk sports*. 2010. http://www.baesystem.com. Accessed
  June 22, 2010
Wikipedia Foundation Inc. http://en.wikipedia.org/wiki/Physical_fitness Wikipedia. 2010. http://
  www.wikipedia.org . Accessed October 2010

Chapter 9
Wearable Electronic Systems: Applications to Medical Diagnostics/Monitoring

Eric McAdams, Asta Krupaviciute, Claudine Gehin, Andre Dittmar, Georges Delhomme, Paul Rubel, Jocelyne Fayn, and Jad McLaughlin

9.1 Introduction

The combination of an ageing population and the increase in chronic disease has greatly escalated health costs. It has been estimated that up to 75% of healthcare spending is on chronic disease management (mainly cardiovascular disease, cancer, diabetes and obesity) (World Health Organization 2010). It is now widely recognised that there is a need to radically change the present Healthcare systems, historically based on costly hospital-centred acute care, and make them more appropriate for the continuous home-based management of chronic diseases. The goals of the new approach are the improved management of the chronic disease through encouraging lifestyle changes and the effective early detection and treatment of any problem before it necessitates costly emergency intervention.

The most recent European technology and innovation funding programme initiative using such an approach is termed Ambient Assisted Living (AAL). *AAL aims to prolong the time people can live in a decent more independent way by increasing their autonomy and self-confidence...by improved monitoring and care of the elderly or ill person...while ultimately saving resources. The main objective is to develop a wearable light device able to measure specific vital signs of the elder or ill person, to detect falls and to communicate autonomously in real time with his/her caregiver in case of an emergency, wherever they are* (ICT for health 2007). There is, therefore, an urgent need of novel monitoring systems, which include new sensor technologies, mobile technologies, embedded systems, wearable systems, ambient intelligence, etc., which are capable of conveniently, discreetly and robustly monitoring patients in their homes and while performing their daily activities without interfering significantly with their comfort or lifestyle (McAdams et al. 2009; Fayn and Rubel 2010).

E. McAdams (✉)
Biomedical Sensors Group, Nanotechnologies Institute of Lyon, INSA de Lyon, Bât.
Léonard de Vinci, 20 avenue Albert Einstein, 69621 Villeurbanne Cedex, France
e-mail: eric.mcadams@insa-lyon.fr

A. Bonfiglio and D. De Rossi (eds.), *Wearable Monitoring Systems*,
DOI 10.1007/978-1-4419-7384-9_9, © Springer Science+Business Media, LLC 2011

To develop optimally designed, clinically viable monitoring systems, it is important to learn the lessons from healthcare innovations in the past, to be aware of the potential clinical applications of such technologies and to fully understand the technical constraints and possibilities which exist.

9.2   Historical Perspective

Those who cannot learn from history are doomed to repeat it. George Santayana

It can be said that "the history of diagnostic medicine is the history of its tools" to modify slightly a quote attributed to Lars Leksell. To fully appreciate the key factors influencing the successful development of wearable monitoring/diagnostic systems and the resulting changes, desirable or otherwise, to clinical practice and to the roles of the clinicians, hospitals and patients, it is important to learn lessons from the previous decisive periods in the history of medicine.

The history of diagnostic medicine can be loosely, but rather productively classified into periods characterised by their use of sensors vis-à-vis the use of human senses (Bynum and Porter 1993) see Table 9.1. Obviously, sensors are not the sole component in a monitoring system, but they do form the problematic first stage in biosignal measurement. The veracity of this comment is evidenced in the present *paucity* of successful clinical uptake of the many published wearable monitoring systems. In addition, given the interest of several of the authors in the area of Medical Sensors, this approach to the history of diagnostic medicine is admittedly appealing.

The "Non-Sense" period in Diagnostic Medicine (please excuse the deliberate pun): In this classification, the non-use of human senses is generally attributed to earliest man, at any rate before Hippocrates the so-called Father of modern medicine. However, rational use of human senses predates Hippocrates who along with Galan freely admitted that a large part of their medical knowledge came from earlier Egyptian works. *Since, then, Egypt and not Greece must be considered the original home of the medical art, we ought not to set up the Greek Aesculapius as the patron genius of medicine, but rather the physician whom the Egyptians gave this dignity, viz Imhotep* (Cumston 1936). As in the present day, rational "scientific" diagnosis based on astute observation has always co-existed with less rational

Table 9.1 The history of diagnostic medicine classified according to their use of sensors

Period in medical diagnostic history	Approximate historical dates
"Non-sense"	Earliest times – present
Human senses	"Hippocrates" – present
Augmented human senses	Nineteenth century – present
Replaced human senses	Mid-nineteenth century – present
Remote sensors	Late twentieth century – present
Wearable sensors	Twenty-first century

Fig. 9.1 Early physician using human senses

approaches. In the latter, the cause of disease was considered to exist outside the patient – an evil spirit or an angry god – and hence there was little need to study the patient's body in any great detail. A spiritual leader, such as a priest or shaman, was called upon to use one of a range of divination techniques (observing the stars, flights of birds, animal entrails, etc.) to diagnose the source of the problem and prescribe some cure or gift of appeasement to the offended deity.

The "Human Senses" Period in Diagnostic Medicine: As noted above, the belief that disease has a physical cause that can be diagnosed by careful observation of the patient, his/her diet, lifestyle, etc. is generally, but wrongly, attributed to Hippocrates and his followers. In this approach, a physician was encouraged to use all of his/her human senses (sight, hearing, smell, taste and touch) in making a prognosis/diagnosis (Fig. 9.1). He/she had little or no diagnostic technology and based his/her diagnosis largely on interview and a visual inspection of the clothed body. He/she (or more likely his/her unlucky assistant) would also smell and/or taste wounds, breathe, sweat and urine.

In crude engineering terms, human senses are based on a range of human sensor cells, which respond to differing inputs by generating an electrical signal which is sent to specialised areas of the brain for interpretation. It is important to grasp the view that the breakthroughs in diagnostic medicine are due to the augmentation and eventual replacement of these human senses, with their inherent limitations, by the development of less subjective and more accurate sensors. Each time a new generation of sensors has been developed, the role of the clinicians and their

Fig. 9.2 The introduction of the stethoscope

patients changed and the form of healthcare provision, for example the role of the hospital, changed dramatically – and will continue to do so.

The "Augmented Human Senses" Period in Diagnostic Medicine: This period could most probably be best exemplified by the invention of the stethoscope (Fig. 9.2) by Laënnec in 1819. Laënnec was able to describe, classify and correlate symptoms detected by his stethoscope with subsequent lesions detected at autopsy. As a result, he was able to present physicians with a complete diagnostic system for pulmonary and cardiac complaints, a key factor in the successful update of any new system. This period involved the developments of a range of other "scopes" including microscopes, ophthalmoscopes and endoscopes enabling the physician to "see into" the patient (before or after death) in ways he was never able to do before. The evolving fields of histology and cytology were made possible through the improvement of microscopes, started in the 1600s, and in the 1860s Pasteur discovered that bacteria cause disease. These advancements in optical techniques lead to an anatomical model of disease, with illness being attributed to defects in body architecture. Much of the advances in this period were made possible by the improvement in the status of hospitals at the time. With a concentration of patients with the same disease, it was possible to specialise, to follow the progression of a disease and to carry out post-mortems to locate disease loci.

As a result of these advances, visual and manual examination became more thorough and invasive and there was less emphasis placed on the patient's views. Physicians started to treat diseases not patients. The physician now felt the pulse, sounded the chest, measured blood pressure, peered into eyes, inspected the tongue and throat, etc. and, if necessary, laboratory tests were carried out. Not unsurprisingly, there was considerable resistance to this change from patients, the media and even physicians themselves.

Even though these novel sensing systems greatly extended the diagnostic possibilities for the physician, it must be noted that all these tools were still connected at one end with a particular sense of the physician. As such, these measurements remained qualitative, subjective and hard to communicate or teach to others.

The "Replaced Human Senses" Period in Diagnostic Medicine: During this period the physician's senses were progressively replaced by sensors, first by those that measured/recorded the same parameters in a more reliable fashion (such as temperature and pressure) and later by sensors that detected phenomena indiscernible to human senses [such as the electrocardiogram (ECG) and X-rays]. These innovations resulted in the presence of the physician no longer being needed during a diagnostic measurement and the subsequent creation of new healthcare professionals, such as clinical scientists and radiographers, whose role it was to carry out the ever-increasing number of diagnostic measurements. As diagnostic systems such as ECG and X-ray were cumbersome, expensive and difficult to operate, they were initially centred in large teaching hospitals with newly formed medical physics departments, further increasing the status of the hospitals and creating roles for medical physicists and biomedical engineers. A century earlier, hospitals were often places where patients who could not afford the services of a private physician were forced to go and conditions were often very primitive. Throughout the 1900s, the healthcare system's dependence on medical technology grew continuously and the modern hospital emerged as the centre of a sophisticated health care system, serviced by technologically sophisticated staff.

The "Replaced Human Senses" Period was characterised by the development of the study of human physiology and its application to the diagnosis of disease. Claude Bernard, who took nothing for granted, helped established the use of objective scientific measurements in physiology and medicine. Etienne-Jules Marey, a versatile and gifted engineer, was one of the pioneers of the graphic approach, laying much of the foundations of physiology and for Einthoven's later development of the ECG machine. Marey believed that *in the laboratory, as at the bedside of the patient, the skill of the individual, his practised tact, and the subtlety of his perceptive powers, played too large a part. To render accessible all the phenomena of life-movements which are so light and fleeting, changes of condition so slow or so rapid, that they escape the senses - an objective form must be given to them, and they must be fixed under the eye of the observer, in order that he may study them and compare them deliberately. Such is the object of the graphic method* (Reiser 2000).

In the physiological approach to disease diagnosis, illness is evidenced by changes in body function, which may not show up at post-mortem. The new physiological techniques were initially introduced into medical schools to help train physicians in the use of their senses. However, eventually the new equipment was wheeled out onto the ward to carry out the measurements/recordings directly. The physician's senses had been replaced by the new sensing systems. Objective data could now be recorded, analysed and compared with those of the past or future thus facilitated training and communication *Every EKG when and where it may have been recorded, is immediately comparable with every other EKG* (Bynum and Porter 1993). The successful uptake of the new technology was due to the tenacity of a few key researchers, such as Willem Einthoven, who laboured to improve the devices and establish the clinical usefulness of the measurement techniques.

The "Remote Sensors" period in Diagnostic Medicine: With the widespread introduction of telemetry, it has become common place for a physician to observe and interview a patient located at a distance from the clinic. Such exchanges are often termed "Telemedicine." This term has been defined by the World Health Organisation as the *practice of medical care using interactive audio visual and data communications. This includes the delivery of medical care, diagnosis, consultation and treatment, as well as health education and the transfer of medical data* (Mandil 1996).

Of more interest to the present discussion is Telehealth monitoring (or Tele-monitoring), which involves *the remote exchange of physiological data between a patient at home and medical staff at hospital to assist in diagnosis and monitoring.... It includes (amongst other things) a home unit to measure and monitor temperature, blood pressure and other vital signs for clinical review at a remote location (for example, a hospital site) using phone lines or wireless technology* (Curry et al. 2003). Many examples of simple home-based, telemonitoring sensor technologies now exist. For example, basic blood pressure cuffs, glucose meters, pulse oximeters and heart monitors are available and patients with heart disease or diabetes, for example, can transmit their "vital signs" from the comfort of their homes to their health care professional and get feedback or other follow-up, when appropriate. These basic home-based "Telemonitoring" systems are already enabling leading medical centres around the world to more effectively keep patients healthy and out of the hospital – and to decrease costs.

The Present/Future "Wearable Sensors" period in Diagnostic Medicine: For certain medical conditions, it is necessary to monitor a patient "continuously" or "on demand" while going about their everyday business. In such cases, the periodic use of basic home-based "Telemonitoring" systems is not sufficient and in most cases body-worn sensing systems are required.

It is interesting to note that, in contrast with much of the past, the present evolution is due more to governmental-pull than to (solely) technological-push. Although it is sometimes claimed that the technologies required for this clinical revolution already exist and that it is other aspects such as device interoperability, financial reimbursement and clinical uptake which are holding back progress, the authors, as sensor specialists, firmly believe that the key sensor-related technologies are not as advanced as widely believed. This misconception is probably due to the positive descriptions in the literature of prototype monitoring systems, which are clinically untested. It is one thing to wear a prototype device with all its attached leads and sensors for a few minutes to harvest a suitable, short, artefact-free trace for an attractive publication, it is another for the system to be accepted by a patient and to work robustly for an extended period as they go about their daily lives. This problem is partly due to the difficulty in organising and carrying out clinical trials.

As a result, scientists often tend to focus on the use of models, phantoms, patient simulators, etc., standard tools of their trade but which ignore the key component in the problem to be solved – the patient, particularly the patient– device interface. The authors firmly believe that when the other aspects are in place, enabling the rigorous real-live assessment of "wearable" systems, many of those published will be found wanting and the apparent complacency concerning the level of the technological readiness, in particular that of the sensors, will evaporate.

If truth be said, academics tend to apply for funding so that they can continue their fundamental research, with often little real understanding of the clinical need they are supposed to be addressing. As a consequence, they tend to continue their on-going fundamental research and successfully publish but rarely reach the stage of clinical transfer of relevant solutions to clinical problems, as hoped for by funding bodies. This is not solely due to the academics' alternative focus or lack of clinical awareness. There is very often a lack of adequate clinical representation in funding body panels and as reviewers. This is probably due to a combination of factors, not least of which is the clinician's workload and lack of motivation at being involved in such "scientific" programmes. Projects are therefore most often assessed by other engineers and scientists who may intuitively think that they are great ideas and are impressed by the quality of academic research and publications. However, they are not best suited at assessing the clinical relevance and feasibility of the work.

The comments made above do not only apply to academics, similar shortcomings exist within industry, especially within small and medium enterprises (SMEs). It has been claimed that 70–80% of new products that fail do so, not for lack of sufficiently advanced technology, but because of a failure to understand real users' needs (Von Hippel 2006). There is therefore a great need for a "one-stop-shop" to arrange and manage clinical trials for companies, especially SMEs, and for academics. However, developers need to do more than just bring new technologies to end-users to ask them what they think. Additionally, clinicians and end-users need to be involved at the very start of the innovation process to suggest, assess and give rapid feedback on the clinical need for a product and the relevance/robustness/etc. of the proposed device. The involvement of clinicians (and patients) in the innovation process will sensitise them to the technological possibilities and motivate them to be actively involved in the innovation process rather than being the passive recipients of the latest technological "gadget."

9.3 Present and Possible Clinical Applications

It is often vital to monitor in an ambulatory situation the following parameters: Heart activity (ECG, heart rate, blood pressure, pulse); Lung activity (respiration rate, respiration depth, tidal volume, oxygen saturation); Brain activity (EEG, vigilance, relaxation); Digestion (gastric emptying); Emotions and stress levels; Body characteristics (temperature, posture, position, activity), etc. Some of these parameters are already catered for to some degree, e.g. Holter systems for Cardiac activity have been widely used for many years and, as a result, have greatly increased understanding of heart disease and have led to significant improvements in patient care. Other important physiological functions have not benefited to the same degree due to the difficulties in obtaining accurate measurements outside of a laboratory, for example for assessing breathing or stress. *One apparent area for application of this new technology is the monitoring of breathing patterns over extended periods in the study of respiratory disorders such as chronic obstructive pulmonary disease, pulmonary emphysema, restrictive lung disease, or asthma* (Wilhelm et al. 2003). Hitherto unexplored parameters should therefore not be ignored by scientists; however, the situation is something of a "catch 22." If the sensor systems do not as yet exist, there is often little clinical demand for them until they do and their relevance established by a pioneer.

The design and relevance of wearable sensing systems will depend on, and will determine, the ambulatory monitoring applications.

9.3.1 *"Holter-Type" Monitoring*

In many clinical areas, it is important to record measurements on patients over extended periods (generally for a maximum of a few days) as the patient goes about their every-day-life to accurately diagnose their condition. Again, the ECG Holter monitor would be the best known example. Often symptoms do not present themselves "on cue" in the doctor's surgery and hence the need for ambulatory monitoring, for example to detect periodic arrhythmias, to detect the location of an epileptic focus, or to study breathing difficulties. Such continuous recording furnishes clinicians with a much clearer picture than the occasional "snapshot" collected during a patient visit. Multi-parametric continuous ambulatory monitoring helps quantify the number of events, differentiate between several possible causes and helps identify any contributing factors such as stress, sleep, food, medications, activity, etc. For example, emotions and stress can affect a wide range of conditions and hence the concurrent monitoring of the former helps assess their contributions much more effectively than the usual patient diaries or retrospective reports. [An additional advantage of unobtrusive ambulatory monitoring systems is the avoidance of the well known "white coat" syndrome in which patients exhibit elevated blood pressures in clinical settings due to their increased anxiety in a clinic environment.] Multi-parametric monitoring which includes measures of posture and activity also help identify and possibly compensate for the effects of movement-induced artefacts on the other parametric traces.

In "Holter-type" diagnostic recording applications, it is not always necessary to transmit the monitored data continuously to a remote monitoring station. Often the recorded data is accessed at the end of the recording period and/or transmitted periodically.

9.3.2 "Post-Intervention" Monitoring

In post-intervention monitoring, such as that of a heart attack victim recovering in a coronary care unit, there is a role for "Holter-type" monitoring systems to progressively "un-tether" the patient from their bed and to encourage them to take part in some closely monitored movement and recreation for their physical and mental well-being. In such applications, the sensors will be accurately applied by clinical staff and the parameters monitored continuously. Such monitors are vital in assessing the efficacy of on-going treatment and in the planning of subsequent medication/treatment.

As patients heal, there is a continued need for their monitoring as they gradually re-integrate into normal everyday life. As they do, the design of the monitoring systems, the sensors and positions used and the handling of the data will alter. As part of their rehabilitation, cardiac patients, for example, are encouraged to exercise while still in hospital or later on an out-patient basis. Patients are generally wary of doing so due to the fear of having a further heart attack. Apart from the use of stationary treadmills and exercise bikes, many exercise regimes do not lend themselves to the continuous monitoring of patients and there is therefore a need of ambulatory monitoring systems, if only to reassure the patient. The design of such systems would more optimally be in the form of a suitable wearable "smart garment" or "smart patch" (see next section). It would be preferable, given the vulnerability of the patient, that the measured data be monitored continuously; either remotely by a member of clinical staff or through onboard data processing coupled with the capability of alerting supervising staff.

9.3.3 "On-Demand" Monitoring

As patients recover and return home, they may still require medical assessment of their condition from time to time. The same applies to those diagnosed with chronic diseases or perhaps to the "worried well" (Continua Health Alliance 2009) with a significant family history of a serious disease. In some cases, "continuous" measurement may be required, for example in the continuous measurement and control of glucose concentrations in diabetics, enabling the provision of better adjustment of insulin dosage. In this case, however, a clinician does not need to be involved, or at least only on an intermittent basis. Regular checks can be arranged with a patient's general practitioner or, more conveniently, can be carried out remotely using home-based monitoring systems. In the next section, for example, a cancer-detecting bra is presented which incorporates temperature sensors to detect cancer-related changes in breast tissue. The question arises, however, does the patient really need to have such a system incorporated into an item of clothing or would it not be more appropriate to simply have some form of monitoring system housed, for example, in the patient's bathroom?

There exists a significant need for "on-demand" (rather than continuous) "clinical support" for patients on the move. For example, a recovering heart attack victim may feel the occasional chest twinge and desire urgent medical feedback. "Smart garments" are generally not appropriate for such applications as the patient cannot be expected to wear such a system everyday if the need is only very occasional. However, a small, portable monitoring system, a sort of "professional-in-my-pocket" would be more suitable and patient-compliant, especially if it is integrated into a standard accessory such as a wallet, watch or mobile phone. Depending on the condition, such an "on-demand" system could give direct feedback to the patient and/or send the information to a remote monitoring station. Some portable devices already exist, for example the Personal ECG monitor (PEM) from the European IST EPI-MEDICS project includes a reduced easy to place lead set that allows reconstruction of the standard 12-lead ECG and embeds intelligence that decides when and where to send the ECGs, when needed (Rubel et al. 2005). A further example is SHL Telemedicine's CardioSen'C™ personal cellular-digital 12-lead ECG monitor capable of transmitting to their Telemedicine monitoring centre via any type of phone connection, including regular fixed lines and cellular phones. The healthcare team at the SHL monitoring centre evaluates the transmitted data and provides immediate feedback and reassurance to the patient/subscriber. When necessary, they will instruct the patient on what action to take and/or contact emergency medical services, providing them with all the available medical data, thus saving critical time and ensuring rapid diagnosis and treatment.

The success of such an approach obviously depends on the ability of the patient to apply the sensors in the correct anatomical positions. This will in turn depend on the design of the system and on the circumstances of the patient. From a design point of view, the accurate location of the sensors for the recording of many "vital sign" parameters is not trivial and warrants much research as misplaced sensors can give rise to misdiagnoses.

Medical conditions which require the relatively straightforward application of sensors can enable the direct feedback to patients to reassure them or to encourage lifestyle changes. Such direct feedback to patients can be used to empower and motivate them, improve their awareness and potentially allow them to better control their condition.

9.3.4 *"Emergency/Disaster" Monitoring Systems*

There also exists a significant market for "victim patches" (Bonfiglio et al. 2007) to be applied by "first responders," such as ambulance staff, fire-fighters, and by a ship's or aircraft's crew, to victims at the scene of an accident, disaster, fire or in a remote location. The parameters monitored by the "victim patch" can be viewed by the staff on site, if appropriate, or by qualified clinicians at a remote monitoring station or a local casualty department with the aim of optimising their survival management. Potentially suitable, disposable and/or reusable "Smart Glove" systems are marketed by Ineedmd and Commwell containing embedded pre-positioned electrodes, miniaturised electronics and transmitters. Alternatively, Tapuz Medical markets a flexible ECG chest electrode belt, which hooks under the arms of the victim, and it is claimed that it correctly locates the electrodes irrespective of victim's chest size.

Several "victim patches" are being developed as part of the EU project Proetex, including one based on Intelesens' vital signs patch. The monitoring system is a small, disposable, chest-worn, adhesive sensor patch and a small potentially reusable transmitter module which is clipped on to the patch. The advantage of such a system is that the devices can be rapidly applied by "first responders" to many victims without the need to undress them or to hold the systems in place. By linking the systems to a local hospital, the devices could enable seamless admission of the victims to the causality department and enable triage before, during and following admission.

9.4 Sensing Constraints and Possibilities

The essential characteristic of a biosignal is that of change as a function of time or space. Biosignals can be classified in many ways, for example in terms of their medical application, the transduction mechanism used (e.g. for Temperature Sensors: Thermistors, thermocouples...), etc. Biosignals are associated with various forms of energy and can be thus divided into the six important groups listed in Table 9.2. Some of these biosignals are intrinsic to the body (for example, biopotentials emanating from the heart or brain), whereas others are modulated when external energy sources are applied to the body (for example, stress levels detected via changes in the skin's electrical impedance).

In a biomedical sensor, a physical quantity that has been found to correspond with a physiological phenomenon of interest to the clinician is detected and transduced into another form of energy, generally an electrical signal that can be transmitted, processed and/or recorded.

Table 9.2 Classification of biosignal according to associated form of energy

Form of energy	Parameters	Examples of biosignals
Electrical	Voltage, current, resistance, capacitance, inductance, etc.	ECG, EEG, EMG, EOG, ENG
Mechanical	Displacement, velocity, acceleration, force, pressure, flow, etc.	Blood pressure, pulse wave velocity
Thermal	Temperature, heat flow, conduction, etc.	Body core temperature, skin temperature
Radiant	Visible light, infra-red radio waves, etc.	SpO_2, photoplethysmography
Magnetic	Magnetic flux, field strength, etc.	Magnetoencephalography, flow meters
Chemical	Chemical composition, pH (derived from several forms of energy), etc.	Glucose, cholesterol, creatine kinase

The clinical usefulness of such biosignals lies in the historic observation that a given signal measured, for example on the skin's surface reflects an inaccessible organ's behaviour, for example, and that certain changes observed in the biosignal indicate a dysfunction in the biological processes involved. It must be remembered that the choice of such signals, the sensors (transduction mechanism) and positions used are not "set in stone" but are often "accidents of history." For example, ECG electrodes are placed on the arms and legs because these were the only body parts that could realistically fit into Einthoven's large saline-filled bucket electrodes, required to decrease the contact impedance as the input impedance of his galvanometer was so low. It was only when the amplifier was improved that the area of electrodes could be decreased and the buckets replaced by limb plates, still in use today in some circles. Further improvements in amplifier design in the 1920s and 1930s enabled the minia-turisation and "transportability" of the device and the further decrease in electrode area. The latter point enabled their attachment to the patient's chest and, eventually once the clinical usefulness of such additional measurements was established, the precordial leads V1–V6 were internationally accepted and standardised. *The monitoring device and amplifier determined the electrode size, design, and location of the electrodes, which in turn determined the clinical application and the presentation of the physiological data* (McAdams 2006).

The clinician will therefore have to choose from the evolving possibilities which signal or set of signals is the most appropriate for the physiological phenomenon he/she wishes to study. As a further example, in the monitoring of glucose in diabetes there are a wide range of possible sensing systems including those based on amperometric, potentiometric and impedimetric measurments on blood, sweat and interstitial fluid, to mention only some of the possibilities vying for clinical and commercial success. Often several signals are necessary to assess the various aspects of, say, an organ's function/dysfunction and thus gain a more complete picture. For example, in assessing the activity of the heart it is possible to measure signals related to bioelectricity, flow, motion, volume, pressure, and/or biochemistry. Each signal will necessarily describe a different aspect of cardiac activity.

Part of the evolving nature of biomedical engineering research and development is the discovery of new sensor transduction mechanisms and associated instruments (Lovell 2007). To meet the challenge and to fully harness the potential of wearable health monitoring, it is important to develop a new generation of sensor-driven technologies (Vodjdani 2008).

The sensor platforms required for approaches such as "wearable" monitoring (Fig. 9.3) can be divided into the following four groups for the purposes of this review:

- "Holter-type" systems with standard sensor designs and locations
- Body-worn sensor patches and bands
- Body-worn bands and harnesses
- "Smart garments" for long-term applications

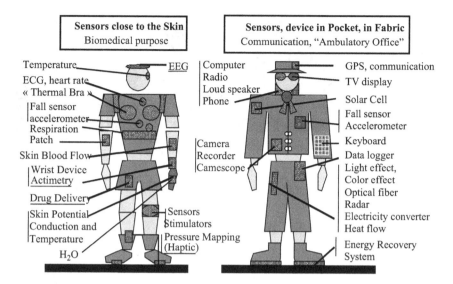

Fig. 9.3 Some sensing possibilities (Dittmar et al. 2005)

9.4.1 *"Holter-Type" Systems*

Many of the multi-biosignal systems presently envisaged by a range of organisations resemble standard Holter monitoring systems (Fig. 9.4a). They involve a recording/transmitting device which is attached to the patient's belt, a necklace or located in some form of waistcoat. The device is connected to standard sensors placed in standard locations and thus involves unwieldy leads. Such systems (Fig. 9.4b–d) are best suited for monitoring applications similar to those involving Holter monitoring (for research purposes and/or on concerned and thus motivated patients for a few days to help diagnose their arrhythmias or other health concern) but are generally too obtrusive and cumbersome for elderly patients, especially those with physical disability. The multiple wires connected to the sensors spread across the body will limit the patient's activity and level of comfort to the detriment of patient compliance. In addition, the long connecting wires and traditional sensor designs can give rise to large amounts of motion-induced biosignal artefacts (McAdams et al. 2009).

9.4.2 *Sensor Patches and Bands*

Adhesive patches (Fig. 9.5) with totally integrated sensors and a mounted miniaturised telemetry device can be worn by patients for short- to mid-term monitoring

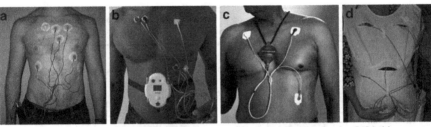

Traditional Holter Monitor   NASA's Lifeguard   Cardionet's monitor   Lifebelt's system
                                                      Photo provided
                                                      by CardioNet, Inc

Fig. 9.4 Holter-type systems

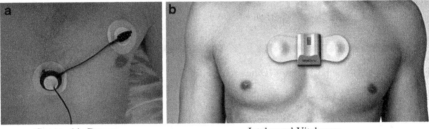

    Camntech's Dream   Intelesens' Vitalsense
       Director

Fig. 9.5 Sensor patches

applications (up to one week or intermittently over an extended period). To develop such monitoring patches for a given patient's condition, one must be able to monitor, with sufficient accuracy and with a minimum of artefact, the key bio-signals and parameters of interest from within the small patch "footprint" located on a discrete, comfortable and convenient site on the body. This will generally require a major change in the transduction mechanisms, the sensor designs and locations traditionally used. One must move away from (literally in this case) the standard anatomical locations and sensor types (McAdams et al. 2009).

Intelesens, for example have developed a range of Wireless Vital Signs body-worn patch systems (see basic system on Fig. 9.5b). The systems include a miniaturised short range body-worn wireless monitor with on board intelligence to monitor for and trigger on medical events, e.g. cardiac arrhythmias; a matching belt-worn device using cellular links to send data immediately to the clinician and an easy to apply, disposable sensor patch for high quality collection of the vital signs which include ECG, respiration, temperature, accelerometers and, it is antici-pated, blood oxygen.

There are only a limited number of body sites that lend themselves to the comfortable detection of most key biosignals using an adhesive sensor patch. Suitable sites will most likely be on or near the torso and located on sites, which will not experience significant twisting or stretching of the skin, otherwise the patch

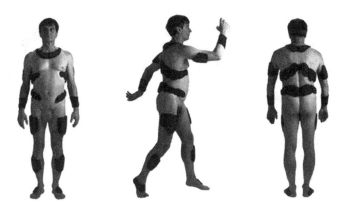

Fig. 9.6 Body sites for comfortable "wearability" (© 1998 IEEE)

will cause shearing of the skin layers. Gemperle et al. indentified spaces on the human body where solid and flexible forms could be comfortably and unobtrusively located without interfering significantly with human movement (Fig. 9.6) (Gemperle et al. 1998). Although they were interested in wearable computing, their guidelines are also relevant to the positioning of sensor patches and bands.

9.4.3   Body-Worn Bands and Harnesses

Gemperle et al.'s sites for comfortable "wearability" can and have been used to locate devices with sensors integrated into bands strapped around the torso or a limb, thus avoiding, to some extent, the problems of skin irritation associated with adhesive patches. Gloves (not shown on Fig. 9.6), wrist/forearm bands, arm bands, torso belts and harnesses and head bands/hats (not shown) are all possible areas for sensor attachment.

A "Smart Glove" can readily be envisaged for use in periodic "Home-based" monitoring, in which case the patient simply puts on an appropriately designed sensor-embedded glove to their hand once or twice a day for the measurement and transmission of their vital signs. Sensors can be built not only into the inside of the glove, e.g. to measure skin blood flow, but can also be placed on the outside of the fabric and then held by the patient against their skin in the desired location. In the simplest case, an ECG monitoring electrode can be located, for example on the inside of the right-handed glove, in contact with the skin, and another electrode be located on the outside of the glove, for example on the palm of the glove. By simply bring the hands together, a one lead (Lead I) ECG can be obtained. If further ECG (and other information) is required, additional sensors can be integrated into the exterior surface of glove and the glove pressed against the chest in the appropriate locations. Several years ago, Ineedmd Inc. patented a glove platform (Fig. 9.7a) in which were embedded miniaturised electronics, sensors and

| An INEEDMD Glove | INSA's Emosense | AMON | Exmovere's Telepath |

Fig. 9.7 Examples of wrist-worn devices

transmitters (Ineedmd Homepage 2009). Wires and sensors were built into a glove that, once fitted on the right hand, was then positioned on the left chest for registering vital signs data. Commwell has introduced a similar concept called the "PhysioGlove" which, it is claimed, *assures the performance of diagnostically accurate and reproducible 12 Lead ECG recordings by the patients themselves or by minimally trained personnel, within less than a minute* (Commwell Homepage 2005). Such "Smart Gloves", however, are less than optimal for "continuous"-monitoring applications but could be used for short-term monitoring of motivated patients or in professions where the wearing of gloves is acceptable, e.g. military, firemen, pilots, racing car drivers, cyclists.

The wrist is a promising location for monitoring systems and tends to benefit from enhanced patient compliance. The Biomedical Sensors Group of the Nanotechnologies Institute of Lyon at INSA Lyon, France initially developed a prototype "smart glove" for neuro-physiological investigations. This has since evolved into a wrist-worn device called Emosense (Fig. 9.7b). It is an ambulatory monitoring and recording system comprising sensors, amplification and wireless data transmission. The device includes a range of integrated sensors for the measurement of skin blood flow (the Hematron sensor), skin temperature, skin conductance, skin potential and heart rate. These parameters have enabled the monitoring and study of autonomic nervous system activity, providing information on emotional and sensorial reactivity, vigilance and mental state. It has been used to assess the reactions and abilities to cope of car drivers, the elderly, athletes and the visually impaired under a range of conditions.

A similar system is marketed by Exmovere (formerly Exmocare). The bluetooth-enabled biosensor wristwatch service, apparently now called the Telepath (Fig. 9.7d), monitors its wearer's pulse, heart rate variability, skin conductance and activity level and can send a report regarding the wearer's emotional and physiological state to a loved one or caretaker, via email, SMS or instant messaging (Exmovere Holding Homepage 2010).

Although there has been much research throughout the world aimed at developing an "ambulatory blood pressure" monitoring system for hypertension and similar studies, the basic approach of measuring oscillometric blood pressure with an inflatable arm cuff has hardly changed. The most promising "wearable" monitor to date is the Advanced care and alert portable telemedical MONitor (AMON) system (Fig. 9.7c), which effectively incorporates a standard automatic inflation

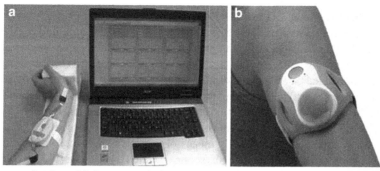

Intelesens' Pulse wave velocity sensor   BodyMedia's Bodybugg armband

Fig. 9.8   Examples of arm-worn devices

wrist-based blood pressure meter (Anliker et al. 2004). (It also measures pulse rate, oxygen saturation, body temperature and an ECG.)

However, an alternative (or preferably, an additional), more convenient continuous measurement is required to assess hypertension in the home or mobile environment. Pulse wave velocity (PWV) is a likely candidate and researchers around the world are developing systems based on this biosignal. Arterial stiffness is a major cause of cardiovascular disease and PWV is a well-established technique for obtaining a measure of arterial stiffness between two locations in the arterial tree. Intelesens has been researching a wireless, forearm-mounted PVDF sensor system (Fig. 9.8a) for the measurement of PWV enabling the reliable and inexpensive measurement of arterial stiffness associated with not only hypertension but also serving as a useful index in assessing atherosclerosis – now regarded as an early warning for cardiac dysfunction and diabetes (Intelesens Homepage 2010). It is envisaged that wrist-worn systems can be extended further up the forearm to enable discrete, comfortable monitoring of this vital parameter. Not only is the forearm suitable from a wearability point of view, it is also potentially suitable as a monitoring site for skin temperature, galvanic skin response, accelerometers, etc.

BodyMedia has developed the BodyMedia FIT® armband monitoring system (Fig. 9.8b) for the health and fitness market. The armband platform includes sensors that measure heat flux, galvanic skin response, skin temperature, near body temperature, and contains a three axis-accelerometer to enable users to gauge the intensity of their workouts and estimate their energy expenditure. Although the present system is designed for the fitness and weight loss market, BodyMedia also develops a professional product line, called SenseWear®, that may be used in research or medical applications.

Simple bands/straps around the chest with embedded sensors are already commonly used in sports monitoring and generally involve heart rate sensing from a basic 1 lead ECG from non-standard locations. Further sensors can be and have been added to chest/limb bands.

Electrode arrays built into straps can be used for electrical impedance tomography (EIT), the imaging of the distribution of the electrical properties throughout the encircled body segment. EIT has been used to study the heart, lungs, gastric emptying, breast cancer, etc. (Jossinet 2005). Similar arrays of electrodes built into chest bands can be used for impedance plethysmographic studies of, for example, blood volume, cardiac output, lung water content and even body composition. The use of several electrode-embedded bands in parallel around the thorax or a limb can be used for more advanced impedance plethysmographic studies. The extension or deformation of the band can be used to furnish information on respiration. VivoMetrics' LifeShirt System uses inductive plethysmography to measure changes in the cross-sectional area of the rib cage and abdomen over time by means of two parallel, "sinusoidal" arrays of insulated wires woven into the garment (see next section). The data is used to calculate the amount of air inhaled/exhaled during respiration.

VivoMetrics recently created a mini, harness-version of their traditional full garment "LifeShirt" called "VivoResponder" (Fig. 9.9a) to meet the needs of first responders, fire-fighters, etc. It is a lightweight chest harness with a range of embedded sensors that continuously monitor breathing rate, heart rate, activity, posture and skin temperature. VivoMetrics have developed a similar system called "VivoChampion Trainer" to monitor athletes. The harness makes it possible to ensure firm contact with the torso over a range of promising skin sites. Depending on the designs of the sensors used, the system would appear potentially very well suited to the monitoring of a wide range of parameters over extended periods of time. A similar, bra-like system for women would also be very promising and enable the positioning of many sensors and sensor types in firm contact with the skin. The "cross-your-heart" design of bra would appear be the most appropriate, increasing the contact area with the thorax. A heat-sensing bra was pioneered a decade or so ago by Hugh Simpson, M.D. (Simpson and Griffiths 1989). Chronobra (Fig. 9.9b), as it was dubbed, measured the deep temperature of the breasts using an array of built-in heat sensors. It was believed that such measurements could detect the early signs of breast cancer and that the bra could become an accurate breast cancer screening technique that women could use at home. A more recent example is the so-called Cancer bra being developed by scientists at the

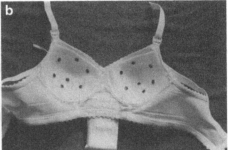

VivoMetrics'VivoResponder   Simpsons' Cronobra (Savage 2008)

Fig. 9.9 Examples of chest harnesses

University of Bolton (Bolton University News 2007). This smart bra concept will incorporate a microwave antennae system woven into the fabric, which will detect temperature changes in breast tissue. It is claimed that the bra will not only detect cancers before tumours develop, but will also be able to assess the effectiveness of ongoing cancer treatment.

With the appropriate design of sensors, hats and head-bands could be used for a range of measurements including EEG, EMG, EIT, brain core temperature and perhaps even ECG if necessary.

There are a wide range of applications within the EEG monitoring area for an ambulatory system that includes a suitable hat or cap incorporating electrodes located in Standard "10–20" electrode placement, for example for the study of epilepsy over prolonged periods and during everyday life. Even for routine clinical use, such an electrode system would be highly desirable. Many efforts have been made to develop an electrode "cap," for example *Neuroscan's Quik-Cap* (Fig. 9.10a); however, the correct application of a large number of electrodes on a hairy scalp is exceedingly difficult and still awaits an optimal solution. Nonetheless, there are possible EEG applications for systems involving a reduced number of electrodes, possibly positioned on more convenient sites, and accessible under a simple head band or hair band. These include the study of relaxation, vigilance, sleep patterns and the use of EEG in biofeedback and brain–computer/machine interfaces.

Due to the location of the hypothalamus, the thermoregulatory control centre in the brain, cerebral temperature is one of the most important markers of fever, circadian rhythms and of physical mental activities (Benzinger and Taylor 1972). Unfortunately, it is difficult to measure. The Biomedical Sensors Group at INSA Lyon has developed a brain core thermometer (BCT) sensor (Fig. 9.10b), for non-invasive ambulatory and non-ambulatory applications, based on the Zero heat flow principle (Fox and Solman 1971; Dittmar et al. 2006).

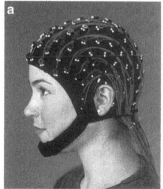

Neuroscan's Quik-Cap   INSA's BCT Device

Fig. 9.10 Examples of head bands and caps

9.4.4 Smart Garments

For short monitoring applications, the "Holter-type" systems can be used by highly motivated patients, for example those with a health problem seeking diagnosis and treatment. Longer term monitoring, up to one week of continuous monitoring can be accommodated with an adhesive patch system, obviously with the development of the appropriate sensors. However, patches worn for over a week, even for less with many subjects, will cause skin irritation problems due to the adhesives used. For longer term monitoring applications, suitable wearable "smart garments" are therefore required.

At first glance, "smart garments" appear very promising. The device and leads could be easily integrated into seams and pockets in the clothes. As the body's surface area is relatively large, around 1.5 m^2, and approximately 90% of it is covered with clothing, the clothing could conceivably contain sensors in close proximity with the skin and thus over the key organs, etc. one may wish to monitor (Dittmar et al. 2004). In theory, one could locate sensors almost anywhere on the clothed body, even over body sites traditionally used in standard monitoring, e.g. the sites for standard 12-lead ECG monitoring (McAdams et al. 2009). However, the relative movement between loose fitting clothing and the skin would give rise to problems associated with quality of contact, motion artefacts and patient comfort and hence only a very limited proportion of the body surface is suitable for the application of sensors via tight-fitting clothing or elasticated bands/sections in otherwise "normal" clothing. As a result, it is generally not possible to make firm sensor contact with many traditional monitoring sites and, much like traditional Holter monitoring, sensors have to be repositioned in non-standard sites to enable firm, comfortable contact with the skin which do not give rise to artefacts due to, for example, excessive body hair, muscle noise (i.e. EMG), and body flab (i.e. motion artefacts in ECGs). Novel sensing technologies are therefore required which will enable the monitoring of vital signs from novel locations if the full potential of such wearable garments is to be realised. The "smart garments" available commercially or reported in the literature can be loosely grouped according to the level of integration of the sensors, leads, etc. into the textiles (Van de Velde 2010).

- *Garment level*: "Late-stage integration"
- *Fabric level*: "Integrated sensors"
- *Fibre level*: "Ubiquitous sensors"

However, the categorisation of smart garments is not that clear cut. As companies seek to develop multi-sensor garments with completely integrated sensors at a fibre level, inevitably some of their sensors are more integrated than others and their garments tend to straddle the above categories.

In "late-stage" integration, existing sensors and associated hardware are "retro-fitted" into "regular" clothes at the garment level. Pockets are formed to hold the various parts of the hardware and thus the garment often resembles a multi-pocketed waistcoat. Standard sensors or their connectors are anchored on the inside of the garment so that they are held in the appropriate position against the skin once the garment is applied. In many respects, these systems resemble the "Holter-like" systems reviewed above, with the same limitations involving patient comfort and compliance. Most "smart garment" projects tend to start with this approach, possibly reflecting the researchers' primary interest in their monitoring and telemetry hardware rather than on the sensor–patient interface. Unfortunately, few projects progress past this stage as it involves solving the thorny patient compliance and signal artefact problems, problems that are not readily solved by the typical electronic engineer and which do not yield a high publication-effort ratio. It could be argued that many of today's inventions fail due to the lack of motivation in the inventors to see the process through to completion. Academics, once they have their publication and an attractive photo or two of the concept for presentations to come, have little inducement to continue developing the system over many years until it is clinically viable.

Early versions of VivoMetrics LifeShirt resembled the "waistcoat" design described above. However, VivoMetrics is one of the few companies that have been in the area for a considerable period of time, over 10 years and the design of their garment and sensors has evolved. At present, the core of the system involves an array of sensors embedded in a lightweight, washable, sleeveless undergarment (Fig. 9.11a) made of highly stretchable material. Inductance plethysmographic sensors for pulmonary monitoring are woven into the fabric and the system has electrodes for a single channel ECG to measure heart rate and a dual-axis accelerometer to record patient posture and physical activity. Additional peripheral devices can be attached to the shirt to enable the measurement of blood pressure, blood oxygen saturation, core body temperature, skin temperature, etc.

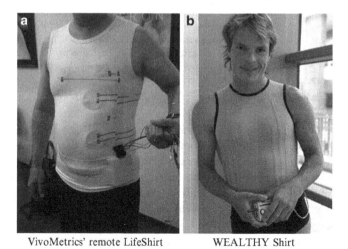

VivoMetrics' remote LifeShirt   WEALTHY Shirt

Fig. 9.11 Examples of "smart garments"

Sadly, VivoMetrics ceased operations in July 2009 and filed for bankruptcy protection in October 2009. The LifeShirt® is currently not being marketed, though their technology portfolio is apparently being offered for sale. Vivometrics was working on a new, remote monitoring version of their system, capable of monitoring the wearer's skin temperature, respiration, activity/posture, and heart rate (ECG) and wirelessly report these parameters to a remote monitoring site via the Internet or directly via GSM wireless. The system consists of ECG and Respirometry (magnetic coils) sensors distributed throughout a shirt that communicate with a central electronics pod that attaches to the garment. Within the pod itself, there are actigraphy sensors (3-axis accelerometers) and a temperature sensor, as well as the amplifier, etc. Development of the new shirt, sensors and communications network, and system architecture was done by Tronics MedTech, Inc. (Tronics MedTech Homepage 2009).

Smartex has also been around for over 10 years and is pioneer in the area of electronic textiles; fabrics where sensor function, electronics and interconnections are woven into them using circular and flat bed knitting technologies. They have been involved in numerous EU projects including "WEALTHY," a wearable fully integrated system (Fig. 9.11b), able to simultaneously monitor ECG, respiration, posture, temperature and a movement index. The stretch garments have integrated fabric ECG electrodes (leads I, II, III, V2 and V5), four fabric impedance electrodes (impedance pneumography), four piezoresistive fabric sensors (thoracic and abdominal respiration and movement) and embedded temperature sensors. The connecting conductive fibres are woven with stretchable yarns.

Other systems include RBI's Visuresp, Biodevices' Vital Jacket, Sensatex' Smartshirt, and SmartLife's HealthVest

> *The recent and continuous trend toward home-based and ambulatory monitoring for personalized healthcare, although exciting and potentially leading to a revolution in healthcare provision, necessitates even more demanding performance..... Electrodes must, therefore, (1) require no prepping, (2) be located in the correct location once the smart garment is put on, (3) make good electrical contact with the skin, (4) not give rise to motion artifact problems, (5) not cause discomfort or skin irritation problems, and (6) be reusable and machine-washable. Although much work has been carried out in this novel area, it is not surprising given the above list of required performance criteria that the electrodes/sensors tend to form the bottleneck in the success of the overall monitoring systems* (McAdams 2006).

While truly long-term monitoring awaits the successful development of totally integrated "smart garments," one potential solution is the use of implanted devices similar to an implantable event (Holter) recorder (also known as an implantable loop recorder). The small monitoring device and associated sensors are placed under the skin during a minor operation and are commonly used to monitor heart rhythms for up to a year or more (Medtronic Homepage 2010). Such devices are increasingly embedding additional sensors such as blood pressure and/or accelerometers to improve the monitoring of heart failure and their use for other applications could be envisaged.

9.5 Discussion and Conclusion

History has shown that if a new monitoring system is to be successful, it must be pioneered by a product champion who tenaciously presses on beyond the crude prototype stage, at which most would-be innovators stop, and develop a clinically viable product and firmly establish its clinical relevance through clinical trials and publications in the relevant clinical journals and conferences. The innovator must therefore have knowledge of the clinical needs, environment and procedures which will affect the success of the new system; preferably being a scientifically competent clinician or a scientist working closely with clinicians. In this regard, it would be of great benefit if governments would act to simplify the procedures required to organise and manage the vitally required clinical trials of promising and much needed healthcare products. The creation of "living labs" should also be continued so as to encourage the involvement of clinicians and patients in the earliest phases of the innovation process.

Many of the systems which do not get passed the prototype stage fail because the inventors have not considered the clinical requirements nor those of the patient. It has been pointed out that most new products that fail do so because of a failure to understand real users' needs, not for lack of sufficiently advanced technology. As Abraham Maslow pointed out "to the man who only has a hammer in the toolkit, every problem looks like a nail." Very often, the scientists and engineers involved in the "wearable" monitoring arena, concentrate on the sophistication of their hardware with little attention paid to the less-glamorous, problematic sensor–patient interface, with the "simple" sensors being added on at the end of the project. The problem is compounded by the fact that scientists tend to use bench tests, models, patient simulators, etc., which all ignore the key component in the problem to be solved. Not surprisingly, the devices appear to work well until they are attached to real patients under real-life conditions. It must be added that these sensor-related problems will not necessarily be miraculously solved by making sensors smaller, as appears to be widely believed.

The technology must accommodate the needs of the patient, not the other way round. It is therefore important to develop new generations of sensor-driven technologies (Vodjdani 2008). One should therefore start with the (potential) clinical need, identify key biosignals related to the physiological processes of interest and seek to develop a platform of novel sensing technologies capable of monitoring the physiological processes from convenient locations using novel transduction mechanisms. The appropriate hardware should then be developed around these constraints/requirements. This alternative approach should lead to systems which actually work under real conditions and should lead to novel, patentable innovations.

There are many possible monitoring scenarios, some of them reviewed in this chapter, and it is not possible to optimally address them all with the same "wearable" technologies. It is important to recognise the (present) limitations and advantages of the various approaches. Basic "Holter-type" devices are still often the best compromise for the short-term monitoring of a motivated patient to diagnose his/ her illness. Adhesive patches appear well suited for "victim patches" and many monitoring applications for up to one week. Sensor belts and harnesses have applications in longer term monitoring situations while truly long-term monitoring awaits the successful development of totally integrated "smart garments" or implantable systems. Portable systems exist which are suitable for certain long-term "on-demand" applications.

Acknowledgements This research work has been supported by a Marie Curie Early Stage Research Training Fellowship of the European Community Sixth Framework Programme under the contract number MEST-CT-2005–021024 within the project Wide Area Research Training in Health Engineering (WARTHE).

References

Cumston CG (1936) An introduction to the history of medicine. Kegan Paul, Trench, Trubner & Co. Ltd., London

Mandil SH (1996) The potential requirements of COPINE health sector users. World Health Organization, Geneva (WHO/AOI/96.14)

Reiser SJ (2000) The technologies of time measurement: implications at the bedside and the bench. Ann Intern Med 132(1):31–36

Commwell Homepage (2005–2010) http://www.commwellmedical.com/. Accessed 3 Mar 2010

Jossinet J (2005) Bioimpedance and p-Health. Stud Health Technol Inform 117:35–42

McAdams E (2006) Bioelectrodes. In: Webster JG (ed) Encyclopedia of medical devices and instrumentation, 2nd edn. Wiley InterScience, New York

Von Hippel E (2006) Democratizing innovation. The MIT Press, Cambridge, MA

Bolton University News (2007) University develops smart bra for breast cancer screening. http://www.bolton.ac.uk/News/News-Articles/2007/Oct2007–1.aspx. Accessed 3 Mar 2010

ICT for Health (2007) Three new projects have been launched. http://www.ehealthnews.eu/research/410-ict-for-health-three-new-projects-have-been-launched. Accessed 3 Mar 2010

Lovell N (2007) www.bsl.unsw.edu.au/docs/2007/PhysMeas_NL_v11.pdf. Accessed 3 Mar 2010

Vodjdani N (2008) Ambient Assisted Living Joint Programme objectives and participation rules, 1st Global Village Workgroup, February 7th, Paris, France. http://handicom.it-sudparis.eu/gvi/presentation/NV%20Global%20village07022008.pdf. Accessed 11 Oct 2010

Continua Health Alliance (2009) The next generation of personal telehealth is here. http://www.continuaalliance.org/static/cms_workspace/Continua_Overview_Presentation_v14_2.pdf. Accessed 3 Mar 2010

Ineedmd Homepage (2009) http://www.ineedmd.com/. Accessed 3 Mar 2010

Exmovere Holding Homepage (2010) http://www.exmovere.com/healthcare.html. Accessed 3 Mar 2010

Tronics MedTech Homepage (2009) www.tronicsmedtech.com. Accessed 3 Mar 2010

Intelesens Homepage (2010) Extracts from recent publications on ST+D pulse wave velocity system. http://www.intelesens.com/pdf/pulsewaveresearch.pdf. Accessed 3 Mar 2010

Medtronic Homepage (2010) http://www.medtronic.com/physician/reveal/. Accessed 3 Mar 2010

Van de Velde W (2010) i-wear intelligent clothing. http://www.docstoc.com/docs/24800152/I-wear-Intelligent-Clothing. Accessed 3 Mar 2010

World Health Organization (2010) Ageing and life course. http://www.who.int/topics/ageing/en/. Accessed 3 Mar 2010

Fayn J, Rubel P (2010) Towards a personal health society in cardiology. IEEE Trans Inf Technol Biomed 14(2):401–409

Simpson HW, Griffiths K (1989) The diagnosis of breast pre-cancer by the chronobra. Chronobiol Int 6(4):355–369

Bynum WF, Porter R (1993) Medicine and the five senses. Cambridge University Press, New York

Fayn J, Rubel P (2010) Towards a personal health society in cardiology. IEEE Trans Inf Technol Biomed (in press). doi: 10.1109/TITB.2009.2037616

Fox RH, Solman AJ (1971) A new technique for monitoring the deep body temperature in man from the intact skin surface J Physiol 212(2):8P–10P

Benzinger TH, Taylor GW (1972) Cranial measurement of internal temperature in man. In: Hardy JD (ed) Temperature: its measurement and control in science and industry. Reinhold, New York

Gemperle F, Kasabach C, Stivoric J et al (1998) Design for wearability. In: Proceedings of the second IEEE international symposium on wearable computers, Los Alamitos, California, pp 116–122

Curry RG, Trejo Tinoco M, Wardle D (2003) Telecare: using information and communication technology to support independent living by older, disabled and vulnerable people. www.icesdoh.org/downloads/ICT-Older-People-July-2003.pdf. Accessed 11 Oct 2010

Wilhelm FH, Roth WT, Sackner MA (2003) The LifeShirt: an advanced system for ambulatory measurement of respiratory and cardiac function. Behav Modif 27(5):671–691

Anliker U, Ward JA, Lukowicz P et al (2004) AMON: a wearable multiparameter medical monitoring and alert system, IEEE Trans Inf Technol Biomed 8(4):415–427

Dittmar A, Axisa F, Delhomme G, Gehin C (2004) New concepts and technologies in home care and ambulatory monitoring. Stud Health Technol Inform 108:9–35

Dittmar A, Meffre R, De Oliveira F et al (2005) Wearable medical devices using textile and flexible technologies for ambulatory monitoring. In: Proceedings of the 27th annual international conference of the IEEE EMBS, Shanghai

Rubel P, Fayn J, Nollo G et al (2005) Toward personal eHealth in cardiology. Results from the EPI-MEDICS telemedicine project. J Electrocardiol 38(4 Suppl):100–106

Dittmar A, Gehin C, Delhomme G et al (2006) A non invasive wearable sensor for the measurement of brain temperature. Proc IEEE Eng Med Biol Soc 1:900–902

Bonfiglio A, Carbonaro N, Chuzel C et al (2007) Managing catastrophic events by wearable mobile systems. Lect Notes Comput Sci 4458:95–105

McAdams E, Nugent CD, McLaughlin J et al (2009) Biomedical sensors for Ambient Assisted Living. In: Chandra Mukhopadhyay S, Lay-Ekuakille A (eds) Advances in biomedical sensing, measurements, instrumentation and systems, Springer, New York

Savage L (2008) What happened to the cancer-detecting bra? J Natl Cancer Inst 100(1):13

Chapter 10
Emergency and Work

Annalisa Bonfiglio, Davide Curone, Emanuele Lindo Secco, Giovanni Magenes, and Alessandro Tognetti

10.1 Introduction

In most recent years, technological advances have brought in consumer electronics many portable applications that have become part of our daily life. Miniaturized headphones, mp3 players are only an example of this trend. Leveraging on the low cost and versatility of these devices, some companies have launched new products combining the portability of these systems with the possibility of using these devices as support for some common human activities. In parallel with this technological and market evolution, awareness raised among public opinion about the need of contrasting accidents occurring to those people who work in harsh conditions and need to increase safety and possibly efficiency of intervention. This need is particularly enhanced for professional categories such as fire-fighters and Civil Protection rescuers. In scenarios such as large fires, earthquakes, floods, terrorist incidents or large industrial accidents, professional rescuers must intervene maximizing efficacy whilst minimizing their own risks; presently, they are only endowed with protection devices that act as purely passive shields or, as for emergency radios, necessarily need to be activated by the person in danger. Unfortunately, this is not always possible, and it is clear that a system that allows the automatic detection of potential dangers and reaction of the single operators would be highly desirable. Therefore, times are now mature for the emerging branch of wearable electronics (Lukowicz 2005; Lymberis 2004; Marculescu et al. 2003), together with advances in information processing technology and telecommunication (Jovanov 2008) could become part of a system for improving the safety and the efficiency of emergency interventions. A necessary requirement for such a system is that the person and the whole operational procedure employed in emergency scenarios should not change significantly their normal situation and, in particular, for the persons involved, the system should not divert their attention

A. Bonfiglio (✉)
Department of Electrical and Electronic Engineering, University of Cagliari, Cagliari, Italy
e-mail: annalisa@diee.unica.it

A. Bonfiglio and D. De Rossi (eds.), *Wearable Monitoring Systems*,
DOI 10.1007/978-1-4419-7384-9_10, © Springer Science+Business Media, LLC 2011

from the potentially dangerous scenario in which they operate. For these reasons, the idea of realizing a wearable system with these characteristics is particularly attracting. Thanks to the aforementioned technological advances, it is possible to develop information infrastructures fully integrated in garments that collect, process, store and transmit information about the wearer (Binkley 2003; Park et al. 2003) and about the surrounding environment collected by sensors embedded in the garments too. In this way, the garment, so far a passive protective device, becomes an active equipment, able to detect in a short time possible dangers for the wearer and to automatically activate an assistance procedure. Rescue for rescuers, but not only: also the efficiency of the operation may be improved thanks to the fact that the single operator becomes part of a network continuously controlled by a central station which, at any time, is able to know location and any other useful information concerning the single nodes of the network and may therefore intervene to adapt the procedure to the needs of the occurring operational scenario. These concepts that are being developed for a special category of workers, namely the professional rescuers, may be easily extended to the much wider field of technical garments for workers, as there is a number of professional figures that potentially can gain not only in safety but also in productivity from being equipped with wearable information/monitoring systems. This kind of approach (Lukowicz et al. 2007; Klann 2007) has been chosen in one of the first European projects dedicated to the application of wearable systems to work. Wearit@work (and its continuation Profitex) (http://www.weariatwork.com; http://www.project-profitex.com) is a project dedicated to the development of wearable computing applications for special categories of workers. The main concept behind these projects is the idea of endowing the wearer (being a fire-fighter, or an airplane technician, or a healthcare or production management staff member) with a wearable device able to provide him/her with the essential information at any stage of the operation. Although not specifically addressing all technological issues concerning the wearable system in itself (for instance, ergonomics, power autonomy, etc.), this project underlined the importance of some crucial requirements that these architectures must possess: in fact, to be really useful, they must provide only the necessary information (nor more, nor less) in the easiest and most useful way, without interfering with the capability of the worker to pay attention to the external, often potentially dangerous, events. All these issues have been investigated directly with the end-users to define a precise role for the wearable devices into the normal operational procedures followed by these workers. Only following this approach the wearable device can really become an enabling technology that improves the safety and efficiency of the operator and can be easily accepted in the normal working practice.

Some examples of wearable devices have also been recently developed in other projects, mainly in Europe. Some of them are already being produced as, for instance, the Viking Fire Jacket (http://www.vicking-life.com). In this garment, a simple sensor-alarm system has been inserted: a heat sensor is placed on the jacket shoulder (with temperature sensors in the outer and inner layers) and a series of LEDs placed in well visible sites in the garment are activated when potentially dangerous heat levels are detected. In this way, both the wearer and the other team members around him/her can be warned about a possible anomalous increasing temperature, before it results in injury for the fire-fighter.

Safe@Sea (http://www.safeatsea-project.eu/index.html) is a recently funded European Project that aims at producing a smart uniform for fishers and people who work aboard. In addition to the investigation of the best materials for endowing this uniform with the finest characteristics in terms of thermal insulation, resistance to tears, buoyancy, also ICT systems for localization and overboard detection and communication are among the objectives of this project.

For the first time, a European Project named PROETEX (http://www.proetex. org) (started in 2006) has aimed at building a whole wearable monitoring system for emergency operators. The system includes sensors, actuators, energy storage and scavenge equipment, telecommunication and signal processing units. The project output is a set of progressively more advanced versions of the future, smart, rescuer's uniform. The global system consists of three components: an inner garment (IG) aimed at measuring cardiac activity, breathing rate, temperature and dehydration; an outer garment (OG), i.e. a jacket that hosts systems that provide measurement of the external temperature, of the heat flux through the jacket, toxic gases' detection, GPS coordinates determination, wearer's activity monitoring; a pair of boots that host a toxic gas detector and an activity detection system wirelessly connected to a central electronic terminal, also hosted in the OG, that collects all the data and transmit them to a remote monitoring station.

10.2 Designing a Wearable Systems for Emergency and Work: Main Problems and Constraints

There are several constraints that must be taken into account for designing such a complex system: on the one hand, those related to the user needs, meant as comfort, easy maintenance of the systems, adaptation to the different scenarios of intervention; on the other hand, those related to each of the different and heterogeneous technologies involved in the system; finally, those related to both, for instance, power autonomy that is for sure a constraint imposed by the technology, but also a requirement related to the kind of intervention in which the system has to be used. For all these reasons, design is a crucial activity in this kind of projects, and must be done not only at the engineering level; it unavoidably needs the intervention of many other professionals: first of all, the end-users of the systems, the representatives of medical staff specialized in emergency interventions, and experts in bio-ergonomic design that take into account both the comfort needs and the best location of the monitoring systems for physiological measurements. Only the combination of these different professionals and skills ensures achieving the goal of obtaining a really useful system, where the sensors systems guarantee the complete monitoring on the person and of the external parameters without being consciously perceived by the users.

10.3   Components of the Wearable System

10.3.1   Sensors

Many different sensors can be presently embedded in garments. A panoramic view
of the main sensors embedded in garments may be found in Chap. 1. They can be
classified according to several criteria. For instance, the measured parameter, the
sensor location and the possibility to be directly integrated in the textiles. The first
point is the starting base for the whole design. Depending on the application, there
may be many different, possibly interesting, parameters. It is important to define
them with the help of the end-users and to establish a list of features that the
appropriate sensors should have: sensitivity, time of response, repeatability of
measurement, duration of the measurement (for single and periodic measures),
energy requirements, bio-compatibility. Considering the measured variable, sensors can be divided into two classes: those that measure the operator's physiological
parameters, and those that detect variables related to the surrounding environment.
Based on this classification, the decision about where locating the different sensors
can be more or less straightforward, i.e. sensors for external variables generally
need to be put in the external layers of garments, while those for body monitoring
normally require to be put in the inner layers. The issues of comfort and protection
must be taken as a primary priority. In any case, embedding sensors within a
garment must not compromise the integrity of protective layers nor adding a
possible source of danger for the wearer. A typical example is the fire-fighter
jacket: as it is the main fire-protecting shield, no holes or any other heat conductive
paths should be generated due to the insertion of sensors in the garment.

For those parameters that are measured by sensors directly in contact with the
skin, the right location results from a proper compromise between comfort and
sensitivity of the measurement to the natural status of the body. For instance, some
sensors are very sensitive to body motion, and for this reason the sensor output may
be affected by motion artifacts. It is obvious that artifacts should be minimized but
this cannot be simply done by reducing the human motion (for instance, by
increasing the pressure between the sensor and the body) because this could
compromise the whole comfort. Bio-ergonomic design gives a relevant contribu-
tion for solving this kind of problems: interesting applicative examples can be
found in Chap. 12 where the astronauts' future suit is presented.

Finally, it is important to underline that fusion of data coming from different
sensors may provide additional useful information related to the particular application (Curone et al. 2010). For instance, the evaluation of the person's activity,
useful in emergency scenarios to prevent dangerous situations, can be typically
done by accelerometers. Fusion of accelerometers and physiological data may
enable the feasibility of an activity classifier able to generate an automatic identification of conditions that may imply potential dangers for the monitored subject
over extended periods of time. In addition to easily detectable conditions directly
provided by the accelerometer signals (i.e. "resting motionless lying down"), a data

fusion algorithm identifies conditions such as "resting with high heart rate (HR)" (which is potentially abnormal if not preceded by any physical activity or if kept for too long after an intense physical activity), which requires a simultaneous analysis of both signals.

10.3.2 Energy

Power autonomy is one of the main constraints to consider in designing a wearable system. This can be achieved by combined consideration of different aspects: energy storage, energy generation and energy management. All these aspects contribute to solve the problem of giving the system the necessary energy to perform at the best level for the longest time, compatible with the requirements of the application. For storage, many different kinds of efficient batteries are commercially available. The only criterion for the choice is to cope with the need of power autonomy and maintenance of the system. Batteries should last for a reasonable amount of time and if there are many different systems embedded in the textile with a dedicated battery, it is not feasible to recharge each of them every time they are used. Automatic recharging systems should be used, as for instance, inductive charging: this kind of system may be "hidden", for instance, in the coat rack that holds the garment when it is not worn. Another issue is the battery weight. Recently, some interesting example of flexible battery has been developed (Bhattacharya et al. 2009). The great advantage is the weight reduction and the possibility to obtain a flexible device that can be directly embedded within the garment.

Energy generation is another piece of the puzzle: ideally, a large amount of energy could be derived from the human body and from its interaction with the surrounding environment. Practically speaking, so far any attempt to derive reasonable amounts of energy from human motion (with piezoelectric elements) has not been really satisfying. To increase the efficiency of piezoelectric generation, ceramic materials should be used and these are not really compatible with wearability. Thermogeneration is another possible source of energy and so far it gave the most promising results. A complete description of the application of thermogenerators in wearable systems is available in Chap. 2.

Another potentially interesting technology is photovoltaic cells. Clearly, it puts a constraint as the cells must be positioned in the external layer of the garment. In addition, as the amount of energy that can be derived by these systems is proportional to the surface of the solar cell, this must be as large as possible. This poses two problems, concerning the application to garments. First, cells must be flexible to cope with the drapability of the garment, and second, depending on the application and on the employed materials, a large cell module could compromise the transpirability of the garment. Recently, relevant progresses have been made in Organic Solar Cells (Hoppe et al. 2008), whose main characteristics is flexibility and low cost. So, potentially, they are the best candidates for obtaining the desired features for a power autonomous wearable system. Unfortunately, the main issue to solve before ensuring a wide applicability of this technology is the need of a robust but still flexible encapsulation that preserves organic semiconductors (that constitute the active layer of the cell) from the contact with oxygen and humidity.

The last but not least issue to consider in designing a power system for wearable devices is *energy management*. Normally, the highest power consume is due to transmission system. Therefore, it is essential to limit as much as possible the transmission frequency and the amount of time during which the transmission system is active. This must be done without forgetting to take into consideration the application constraints, both in terms of the sampling frequency of the signals to transmit and of the total time of autonomy that the system has in certain applications.

10.3.3   Communications

Communication systems in a wearable application normally provide two types of connections: (1) a Personal Area Network (BAN), whose peripheral nodes provide information from the different sensors and systems embedded in the garment to a central node and (2) a long range communication system (LRS) whose task is to assure a reliable link between the person and a remote Network Coordinator Node (NCN). For emergency scenarios, the appropriate distance between the operator and the Node generally varies between 1.5 and 2 km. An intermediate Router at a shorter distance should be carried by an operator and installed near the border of the disaster area, possibly in line of sight with the NCN: the router should have an interface for long range communications towards the remote NCN and another for medium range towards the operator node. A directional antenna and a power amplifier are needed to face problems arising from the environmental propagation.

Concerning the communication protocols, the choice is driven by different possible criteria. For emergency applications, power consumption and adaptability of the protocol to the particular application are the main ones. For instance, in the PROETEX project, described in the following chapters, Tetra, WiMAX and WiFi were taken into consideration. At the time of the project, Tetra was extensively used in emergency scenarios, but the bandwidth it can offer is quite narrow, so it has not been considered further for the particular application. On the other hand, WiMAX has been designed to provide broad band access to a wide area with mobility support. At the time of the project, most of the available products were mainly thought for fixed access while small devices were not optimal under the point of view of power consumption. Therefore, finally WiFi was chosen as it represented the best compromise between link capacity and power consumption, being usually employed for notebooks, PDA and cellular phones.

10.3.4   Electronics and Data Processing

In a standard wearable monitoring system, all data coming from different sensors must be interfaced with an electronic box, which provides communications with the single systems (including sensors, actuators and communication modules). A bus connection (for instance, RS-485) allows reducing the number of wires and can also support the extension in the number of modules that can be connected to the electronic box by linking them in parallel in any point of the bus cable.

Each sensor/actuator module is based on a microcontroller that provides A/D conversion and signal storage. It also implements a reduced set of the bus commands, with fully functional communication and error control.

To reduce the communication traffic on the bus, it is wise to implement on board signal pre-processing routines to send on the bus only small amounts of data extracted from the raw signals recorded by the sensors.

10.4   The Proetex Wearable System

The full set of PROETEX prototypes consists of three uniforms addressing Civil Protection', urban and forest fire-fighters' requirements, integrating sensors and electronics inside a T-shirt or *Inner Garment* (IG, see Fig. 10.1), a jacket or *Outer Garment* (OG, see Fig. 10.2) and a pair of *Boots* (see Fig. 10.3). Data collected from sensors of the IG and OG (Body Area Network) are transmitted real-time to a self-powered *Professional Electronic Box* (PEB, hosted in the OG) through a serial RS485 protocol. In the meanwhile, data coming from the sensors in the boots are transmitted wireless to the PEB by means of a Zigbee protocol. Finally, all data are sent out to the NCN of the operations, far away from the disaster area (Operation Area Network). This communication is performed with a *Long Range Communication System* working with a Wi-Fi protocol.

All equipments are identified by a unique device number allowing to manage data flow between each operator and the NCN. A *Monitoring Software*

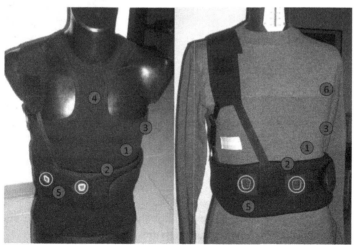

1-Textile electrodes (heart rate sensor); 2-Piezoelectric cable (breathing rate sensor); 3-Body temperature sensor (under the left armpit); 4-blood oxigen saturation sensor (on the CP version only); 5-Vital signs board (IG electronics); 6-dehydration sensor.

Fig. 10.1 The inner garment. On the *left* the prototype developed for Civil Protection rescuers, on the *right* the one for fire-fighters

1-Visual alarm module (1x in the FF OG, 2x in the CP OG); 2-acustic alarm module; 3-external temperature sensor; 4-Wi-Fi communication module; 5-Professional Electronic Box (PEB); 6-triaxial accelerometers (2x); 7-GPS module; 8-Flexible antenna (2x); 9-Flexible battery; 10-CO sensor; 11-Heat flux sensor (only FF OG)

Fig. 10.2 The outer garment (OG) with a list of all embedded systems

Fig. 10.3 The fire-fighter boot that included a gas sensor in the side pocket and an activity sensor under the internal sole

visualizes real-time all sensors' data to the NCN, automatically activating alarms when dangerous contexts are detected. Three, progressively optimized, versions of the prototypes have been produced and tested during the 4-year project.

10.4.1 Inner Garment

The IG is the subsystem devoted to monitor the health status of the emergency operators without interfering with their activities.

Textile electrodes and fabric piezoresistive sensors have been integrated, in a one step process, in a comfortable shirt to reveal cardiopulmonary parameters (namely,

Heart Rate and Breathing Rate), whereas non-textile sensors have been embedded in the shirt for Body Temperature and Oxygen Saturation monitoring (see Fig. 10.1).

The textile electrodes (as well as the piezoresistive sensor) have been realized using SMARTEX technology (Paradiso 2005): a stainless steel based yarn has been knitted together to a ground yarn by using a tubular intarsia technique to get a double face, whereas the external part is not conductive to insulate the electrodes from the environment. For the piezoresistive sensor realization, a Belltron® 9R1 yarn from Kanebo has been used. As ground yarn of the first IG prototype, a cotton yarn has been used; for the prototype release, the ground yarn was realized using a fire resistant yarn based on meta-aramid fibres. This solution was adopted for complying with the new trend of the market of the underwear for emergency operators.

The architecture of the 2nd IG prototype is based on a T-shirt having two main areas devoted to specific tasks:

- an elastic region including all the sensors;
- a region containing a detachable on-board electronics (Vital Signs Board – see Fig. 10.1);

The textile sensors and electrodes are connected to the electronics modules through textile conductive cables integrated in a one-step process in the shirt. Apposite textile compatible connectors (developed by Ohmatex, Denmark, subcontractor of the project) are encapsulated in the fabric and allow the physical connection between the cables and the electronic components.

In the final IG prototype, the Vital Signs Board has been connected to the PEB (placed in the OG) by means of a wireless connection.

Heart rate sensor – HR is detected by acquiring an ECG lead through two textile electrodes integrated in the IG. An algorithm, implemented in the microprocessor of the Vital Signs Board, extracts HR value from the raw ECG signal. Hydrogel membranes have been chosen in the first IG prototype to improve the signal quality whereas in the following prototypes specific textile solutions have been adopted to improve the contact electrodes-skin without hydrogel membranes.

Breathing rate sensor – The first IG prototype has been realized in two versions, which are characterized by a different approach to monitor the respiratory activity (Lanatà et al. 2010):

- Version 1 includes a fabric piezoresistive sensor that changes its electrical resistance when stretched. This sensor is sensitive to the thoracic circumference variations that occur during respiration.
- Version 2 is based on four textile electrodes able to carry out an impedance pneumography. This methodology consists of injecting a high frequency (50 kHz) and low amplitude current by means of the outer electrodes and measuring the impedance changes on the thorax by the inner ones: a relationship between the air flow through lungs and the impedance variation allows the respiratory cycles monitoring.

Two different versions of Breathing Rate sensor have been realized for the second IG prototype: a new piezoelectric sensor in wire form has been implemented on one shirt (version 1). This passive sensor shows a high signal-to-noise ratio together with low sensitivity to motion artifacts. This device replaces the impedance pneumography. A specific module has been realized for the piezoelectric sensor: the front-end converts the charge variation of the piezoelectric sensor through a charge amplifier and the microcontroller performs the necessary operations to extract the breathing rate from the raw signal.

The other version includes, as for the first prototype, a fabric piezoresistive sensor (version 2).

Body temperature sensor – An LM92 temperature sensor has been sealed in a Polyamide Foil and then it has been embedded in the shirt in a proper pocket at the left armpit level. Insulation textile layers have been foreseen to shield the body temperature measurement from the effects of the environmental temperature.

SpO$_2$ sensor – An optical transducer based on a proprietary technology, made of several couples of optical emitters and receivers, has been integrated in a unit at breastbone level. A built-in processor triggers the best located transmitters to dynamically select the highest signal levels; a processor selects samples, stores values in memory and sends them to PEB through the RS485 bus.

10.4.2   Outer Garment

The jacket or OG is the subsystem devoted to protect the rescuer, monitoring his/her activity and surrounding environment.

The OG has been produced in three different configurations, depending on the application: specifically, the prototype for Civil Protection is based on the official uniform of the Italian Civil Protection operators. Urban and forest fire-fighters prototypes use fire-proof jackets adapted from the uniform for French fire-fighters.

All OG configurations include two tri-axial accelerometers and an external temperature sensor: furthermore, a newly developed textile motion sensor and a commercial carbon monoxide monitoring sensor have been added. Forest fire-fighters and Civil Protection operators have also an integrated GPS module, whereas urban fire-fighters do not, since they operate inside buildings where reliable GPS signals are rarely available. Moreover, a Heat Flux Sensor has been included in both urban and forest fire-fighters OG, to prevent operators from skin burn when facing flames.

Regarding data transmission, all OGs incorporate a LRS. The Civil Protection system has also an Alarm Module, which is a subsystem launching visual and acoustic warnings when one or more sensors detect operator dangers beforehand.

Even if all sensors are powered by an embedded Lithium–Ion Polymer (LiPo) battery inside the PEB, a new flexible battery has been experimented in urban fire-fighter OG to increase functionality and ergonomics of garment. The battery allows a working time over 2 h with a nominal voltage of 3.8 V.

External temperature and heat flux sensors – The External Temperature Sensor is placed under the OG external coating at shoulder level; this set-up avoids environmental disturbances and optimizes higher temperature detection as requested by the French fire-fighters (partners of the project). The initial requirement of measuring up to 1,200°C was proven to be over-dimensioned, therefore, a 5.9 × 2.1 × 0.9 mm ATEXIS (http://www.atexis.fr) platinum sensor with a −70°C to +500°C range and platinum-coated nickel wires has been finally chosen. The Heat Flux sensor is placed in the proximity of the platinum sensor, inside the third comfort layer of the OG at shoulder level as well. The sensor has a 50 mm diameter, 420-μm thickness and a sensivity of 75 μV/W/m^2. The presence of the Heat Flux sensor is necessary because the very efficient protection provided by the OG textile equipment makes difficult the feeling of thermal environment (Oliveira et al. 2009).

GPS module – An ANN-MS-0-005-0 Active GPS Antenna from U-Blox (http://www.u-blox.com) was chosen. It provides an accurate measure of the absolute position of the user, when in open space (outdoor and far from buildings or high obstacles). Unfortunately, sensor's performance drastically decreases when approaching or entering into buildings. In any case, this information is considered as really important by rescuers working in large operative areas, where they cannot be directly visually monitored.

Alarm module – The alarm module consists of a power red led driven by a microcontroller, which makes the led flashing at different frequencies depending on the type of alarm. It includes also an audio alarm board with a buzzer control. Two self-powered alarm modules have been integrated in the Civil Protection OG on the front side of the trunk and in the back face at shoulder level. Redundancy guarantees higher visibility in case of structure collapse or flooding. An alphabet of coded blinking can be defined to distinguish the seriousness of danger as detected by the system.

Accelerometers – Motion sensors have been designed to detect posture, accidental falls to the ground and immobility of the operator (Anania et al. 2008). They are realized by means of two accelerometer modules, one placed in the jacket collar (for monitoring trunk movements) and the other in the right sleeve (to achieve more accuracy in activity detection since an operator can move his arm while not moving trunk). Each module is based on a tri-axial accelerometer (ADXL330 by Analog Devices (http://www.analog.com)) and a low power microprocessor (Texas Instruments MSP430F149) for A/D conversion, real-time signal processing and transmission of the extracted information (activity and fall detection flags) to the PEB.

Textile motion sensor – It is applied to the external part of the OG insulation layer, and it is used to detect immobility. The transducer is made of conductive elastomeric strip (200 × 20-mm wide) printed on an elastic fabric and integrated in the sleeve's elbow region. Since elastomer shows piezoresistive properties, movements can be detected by analyzing sensor signal. The processing algorithm is implemented in the microcontroller (Texas Instruments MSP430F149) of the sensor's module (Tognetti et al. 2007) that is used to send the information (an activity flag) through the RS485 bus.

CO sensor – Carbon monoxide is an extremely toxic gas with density comparable to air; therefore, a CO sensor is placed in the OG lapel near the user's mouth and nose. The sensor is integrated in the outer shell layer of the OG, while electronics is protected from the heat by fixing it under the insulated layer. A waterproof and gas permeable coating protects the sensor. The device (CO-D4 by Alphasense (http://www.alphasense.com) electrochemical) is selective and sensitive for CO range between 0 and 1,000 ppm; its output is fed through a transimpedance amplifier to convert current to voltage so that it can be read by the analog port of a microprocessor (Texas Instruments MSP430F149). The microcontroller performs data conversion, extracts CO concentration and implements the protocol routines required to send the information (CO concentration in ppm, a warning flag if concentration is more than 50 ppm and an alarm flag if more than 100 ppm) through the RS485 bus.

10.4.3 Shoes

Boots prototypes have been developed and tested both for Civil Protection and fire-fighter brigades. From a textile and structural point of view, the prototypes already respect UE standards and are already arranged for integrating sensors and energy harvesting elements in the sole. The heel zone is wide to guarantee the stability of the users, the upper part is made with leather and materials resisting to fire and heat, elastic flex points have been inserted in the upper part to allow high flexibility and a new lacing system has been realized to improve functionality. Furthermore, to avoid abrasion and increase comfort, breathable and waterproof lining have been added inside.

CO_2 sensor – Since CO_2 is heavier than air and starts to accumulate at ground level, the Boots have been equipped with a CO_2 transducer, which enables faster detection of toxic gas before it reaches the respiratory tract. A specific housing has been created in the upper part of the boot, capable of maintaining the sensor in contact with air. Due to this positioning, the sensor must be compact, robust, low power consuming and with electronics insulated from extreme condition at floor level. A CO_2 D1 sensor by Alphasense has been chosen and tested. Power supply is furnished by a rechargeable battery whereas communication with PEB is realized by means of a Zigbee module connected with the sensor: finally, the whole system – sensor, electronics, battery and Zigbee module – is hosted in a 58 × 48 × 16-mm box inserted in the boot housing.

Activity sensor – Two organic semiconductor transistors printed on top of a thin plastic layer have been put under the shoe sole and act as pressure sensor (Manunza et al. 2007). Their signal is sent to an electronic module put in a side pocket of the boot. Power supply is furnished by a rechargeable battery whereas communication with PEB is realized by means of a Zigbee module connected with the sensor.

10.4.4 Communications and Electronics

Besides wearable technologies for sensing and data processing, as mentioned before, a remote communication infrastructure (between the rescuers and the local coordinator) has been developed within the project. This is realized by means of a software interface and a LRS (Magenes et al. 2009).

Monitoring software – It manages data received by PEB; data detected by sensors and collected by PEB are asynchronously wireless transmitted to the remote software. Data exchange is based on "query-answer" protocol. Queries, generated by the software, request data from the device: then PEB generates answer strings and sends them to the PC, where data are displayed in text and graphical form and saved in files for further off-line analysis; in the meantime, an on-line processing generates automatic alarms when the variables overcome previously set-up thresholds.

Long range communication system – Communication between the PEB and the PC is set up by means of a Wi-Fi-based communication system, allowing the communication of data up to 1 km far from the operator. Wi-Fi protocol was selected as a good compromise between portability, lightness and performance. In essence, the LRS is made of three units or nodes: the NCN, placed near the PC hosting the monitoring software (node 1); the Router node, next to the border of the disaster area (2), and the operator's equipment (3). To have a high modularity, the key unit of the network is a unique single board computer placed in all the three nodes: this is an ALIX3c2 board (http://www.pcengines.ch) embedded in a $113 \times 163 \times 30$-mm aluminium box; the board runs LINUX operating system and is powered by a high performance rechargeable LiPo battery integrated in the box (7.4 V, 1,600 mAh). Each board has one mini-PCI slot with an installed Wi-Fi LAN card: card repeats queries and answers between PC and PEB. Specifically, on node 3, the module links with PEB by using the RS485 bus and wireless transmit data using two 52×48-mm wide planar textile antennas (placed in the front and back sides of the jacket) (Hertleer et al. 2009). The bandwidth of the antennas is more than 280 MHz, whereas return loss and influence of humidity have been analyzed; the interaction between electromagnetic field radiation and operator body was investigated too, and it was found below SAR limits as requested by EU regulations. On node 1, the board interacts with the PC and the monitoring software through the Wi-Fi card of the PC. Node 2 behaves like a relay station in between nodes 1 and 3: the embedded computer is placed on a tripod in the middle of the operation area, collecting data from all three nodes of rescuers and sending data to node 1. To cover a communication range of more than 1 km, a 13.5 dBi sector antenna with a 3 dB pattern equal to $90° \times 15°$ (http://www.stelladoradus.com) is adopted between the farthest nodes (1 and 2).

10.5 Conclusions

In summary, this chapter deals with the application of wearable electronics systems to work in emergency. These systems may provide operators with a powerful mean for increasing safety and efficiency.

This kind of systems requires a huge design effort not only in terms of single components of the system, but also for their integration in a system that must have, as main features, comfort, robustness, easy maintenance and power autonomy. Integration is a concept that refers not only to the hardware design, but also to the signals generated by the whole system: signals constitute information in itself, but the fusion of multiple, heterogeneous signals provide a new information referred to context recognition. For the first time, the wearable system allows to have a simultaneous, remotely controllable, wide range (physiological, external) information without interfering neither with the human operation nor with the human attention (on the contrary, enriching context awareness). New investigations of automatic detection of dangerous conditions are uniquely enabled by this innovative technology.

These concepts that are being developed for a special category of workers, namely professional rescuers, in a European Project named PROETEX (EU VI FP), may be easily extended to the much wider field of technical garments for workers, as there is a number of professional figures that potentially can gain not only in safety but also in productivity from being equipped with wearable information/monitoring systems.

References

Anania G, Tognetti A, Carbonaro N et al (2008) Development of a novel algorithm for human fall detection using wearable sensors. 2008 IEEE Sensors Annual Conference, pp. 1336–1339

Bhattacharya R, de Kok MM, Zhou J (2009) Rechargeable electronic textile battery. Appl Phys Lett 95:223305–223308

Binkley PF (2003) Predicting the potential of wearable technology. IEEE Eng Med Biol Mag 22:23–27

Curone D, Tognetti A, Secco EL et al (2010) Heart rate and accelerometer data fusion for activity assessment of rescuers during emergency interventions. IEEE Trans Inf Technol Biomed 14(3):702–710

Hertleer C, Rogier H, Vallozzi L et al (2009) A textile antenna for off-body communication integrated into protective clothing for firefighters. IEEE Trans Antennas Propag 57:919–925

Hoppe H, Saricifci S (2008) Polymer solar cells. In: Marder SR, Lee KS (eds) Photoresponsive polymers II, advances in polymer science. Springer, Berlin–Heidelberg

Jovanov E (2008) A survey of power efficient technologies for wireless body area networks. In: Proceedings of the 30th annual international conference of the IEEE engineering in medicine and biology society, p. 3628

Klann M (2007) Playing with fire: user-centered design of wearable computing for emergency response. Mobile response 2007: mobile information technology for emergency response, 116–125

Lanatà A, Scilingo EP, Nardini E et al (2010) Comparative evaluation of susceptibility to motion artifact in different wearable systems for monitoring respiratory rate. IEEE Trans Inf Technol Biomed 14(2):378–386

Lymberis A (2004) Research and development of smart wearable health applications: the challenge ahead. In: Wearable e-Health Systems for Personalised Health Management: State of the Art and Future Challenges. IOS Press, Amsterdam, The Netherlands, pp. 155–161

Lukowicz P (2005) Human computer interaction in context aware wearable systems. In: Artificial Intelligence in Medicine (ed). Berlin, Germany: Springer-Verlag pp. 7–10

Lukowicz P, Timm-Giel A, Lawo M et al (2007) WearIT@work: toward real-world industrial wearable computing. IEEE Pervasive Comput 6:8–13

Magenes G, Curone D, Lanati M et al (2009) Long distance monitoring of physiological and environmental parameters for emergency operators. In: Proceedings of the 31st annual international conference of the IEEE engineering in medicine and biology society, pp 5159–5162

Manunza I, Sulis A, Bonfiglio A (2007) Pressure and strain sensing using a completely flexible organic transistor. Biosens Bioelectron 22:2775–2779

Marculescu D, Marculescu R, Zamora NH et al (2003) Electronic textiles: a platform for pervasive computing. Proc IEEE 91:1995–2018

Oliveira A, Gehin C, Delhomme G et al (2009) Thermal parameters measurement on fire fighter during intense fire exposition. In: Proceedings of the 31st annual international IEEE EMBS conference, pp. 4128–4131

Park S, Jayaraman S (2003) Enhancing the quality of life through wearable technology. IEEE Eng Med Biol Mag 22:41–48

Paradiso R (2005) Knitted textile for the monitoring of vital signals. WO2005053532, Patent Pending

Tognetti A, Bartalesi R, Lorussi F et al (2007) Body segment position reconstruction and posture classification by smart textiles. Trans Inst Measure Control 29:215–253

Chapter 11
Augmenting Exploration: Aerospace, Earth and Self

Diana Young and Dava Newman

11.1 Introduction: Exploration and Discovery

Advances in robotics technology, computer technology, and materials science are enabling the development of hybrid human–machine system designs that allow humans to perform at higher levels, to function in extreme environments, and to better recover from or compensate for injury. The design goals of these human–machine systems are presented, with particular emphasis on wearability and reliability. Accompanying challenges of sensing, actuation, and control are also discussed. We also explore the innovative and interdisciplinary design process, which includes modeling, experimentation, and prototyping.

As an example of the potential impact that human–machine systems may have in the areas of assistive mobility devices, athletic performance, and the exploration of space and other extreme environments, the BioSuit™ System is discussed as a case study. Using mechanical counter pressure (MCP) to keep astronauts alive, the BioSuit™ presents an attractive alternative to conventional gas-pressurized spacesuits. The BioSuit's soft, compliant architecture and actuation may enable significantly greater mobility for astronauts during future exploration missions to the moon or Mars. We believe the principles of design used in the BioSuit™ System, based on the concept of soft exoskeletons, may offer new solutions for rehabilitation here on Earth. A new generation of soft architecture exoskeleton suits, which will include embedded sensing and actuation, is proposed as promising hybrid human–machine medical devices for the treatment and rehabilitation of gait disorders.

Exploration in outer space has enabled the creation of wearable technology concepts for extreme environments. For astronaut explorers in either the cold or the hot extremes of space ($\pm150°C$), wearable technology concepts not only provide enhanced information to aid in missions, but also are essential for safety and survival. The spacesuit is essentially the world's smallest spacecraft, and is designed to keep the

D. Young (✉)
Media Laboratory, Massachusetts Institute of Technology, 77 Massachusetts Avenue,
Cambridge, MA 02139, USA
e-mail: young@media.mit.edu

A. Bonfiglio and D. De Rossi (eds.), *Wearable Monitoring Systems*,
DOI 10.1007/978-1-4419-7384-9_11, © Springer Science+Business Media, LLC 2011

Fig. 11.1 Astronaut Buzz
Aldrin walks on the surface of
the moon near the leg of the
lunar module (LM) "Eagle"
during the Apollo 11
extravehicular activity
(EVA), photo taken by
Astronaut Neil Armstrong
using a 70-mm lunar surface
camera (photo courtesy of
NASA)

astronaut alive by providing pressure, atmospheric life support that includes oxygen supply and carbon dioxide removal, maintaining thermal comfort and humidity control, and providing environmental protection and mobility. All elements necessary to survive the harsh environment of space are provided by the spacesuit (including power and propulsion), thus, allowing astronaut explorers to perform extravehicular activities (EVAs) outside of their spacecraft or planetary base.

The first human to experience spaceflight was Yuri Gagarin, a Russian cosmonaut, who made a single orbit of the Earth on April 12, 1961 in the Vostok 1 spacecraft, and the first woman in space was cosmonaut Valentina Tereshkova, on June 16, 1963 in the Vostok 6. On May 5, 1961 the United States became the second nation to put a human in space, when Alan Shepard performed a suborbital flight in the Freedom 7, as part of the Project Mercury. Due to the requirement to maintain an adequate balance of thrust from rocket boosters and spacecraft mass, the early spacecrafts used in Project Mercury were quite cramped, containing a habitable volume of only 1.6 cubic meters (Nasa 1959). On July 20, 1969, the Apollo 11 mission landed the first humans on the Moon. This historic event was immortalized in astronaut Neil Armstrong's famous photo of Buzz Aldrin walking on the Moon surface (shown in Fig. 11.1). Later, with the creation of Skylab (1973–1979), the United States' first space station, habitability, and human factors for longer missions would become a greater priority (Shayler 2001).

11.2   A Brief History of Wearable Technology for Space

A spacesuit contains many functions required for sustaining life for several hours during EVA, or space walks (Newman and Barratt 1997). The breathable air supply is maintained at about 95% oxygen, with the ability to expel carbon dioxide through a lithium hydroxide filter and water (NASA 1998). In preparation for an excursion into space, astronauts spend 1–4 h in a high oxygen environment within a spacecraft or space station to wash out, or remove, dissolved inert gas (nitrogen) from the blood stream, to prevent decompression sickness (the "bends") when donning (putting on) their spacesuits. The air supply is compensated by a pressure control system, designed to maintain a nominal pressure of 30 kPa, or the equivalent of being 8.4 km (27,480 ft) above sea level on Earth. This is a very high altitude, but since the total oxygen concentration and partial pressure are adjusted, astronauts can perform useful work in spacesuits operating at this nominal pressure. If suits operate at higher pressures, in the range of 56 kPa, then pre-breathe times can be reduced to 30 min, and even higher pressures would offer a "zero prebreathe" option, but currently, the limited mobility in gas-pressure suits keeps the operating pressure around 30 kPa. For the pressure system, redundant bladder layers are used to provide a flexible system that can prevent catastrophic or critical failure.

Despite the extreme cold temperatures of space (approximately 3 K), cooling the interior of the spacesuit is a critical requirement, as the heat generated by the astronaut inside the suit cannot be easily removed. In early missions, such as the American Gemini program, the challenges imposed by thermal regulation were too great and resulted in the abrupt termination of missions (Nunneley 1970). In a spacesuit, heat may only be removed by means of thermal radiation and conduction (as convection is not possible in a vacuum), but since the suit is thermally insulated to manage extreme cold and hot temperatures of sunlight and shadow (approximately $\pm 150°C$), respectively, temperature regulation must be achieved through cooling. The Apollo missions pioneered the use of the liquid cooling and ventilation garment (LCVG) that circulates cool water throughout a tubular network close to the skin (NASA 1971). In addition, current spacesuits also implement technology to heat the gloves: each finger has a small heating element to maintain thermal regulation in the hands (Grahne et al. 1995).

The first level of environmental protection for the astronaut is shielding against ultraviolet radiation. This is most important in the visor, through which light in the visible spectrum passes. To address this need, the Apollo spacesuit visor used an ultraviolet-stabilized polycarbonate shield (NASA 1971). Micrometeoroids (particles of meteoroids), which may have mass up to 5 g and travel up to 1 km/s (generating impact forces equal to or greater than those generated by a bullet on Earth), are also an important concern. Therefore, current suits include a hard upper torso composed of fiberglass, metal and fabric, as well as soft components comprising up to 14 layers of materials such as aluminized Mylar®, Gor-Tex®, Kevlar® and Nomex®, as shown in Fig. 11.2 (ILC Dover, Inc 1994). (These advanced spacesuit materials have enabled numerous protective clothing and outerwear products on Earth today, as discussed later in this chapter.)

Spacesuits also incorporate electronic and computational systems to manage and control the following functions (NASA 1971; Skoog 1994):

- System monitoring of oxygen, carbon dioxide, and water content.
- Biomedical monitoring of crewmember vital signs, such as heart rate, body temperature, breathing frequency, and radiation exposure.
- Information display of gas levels and power supply to astronaut.

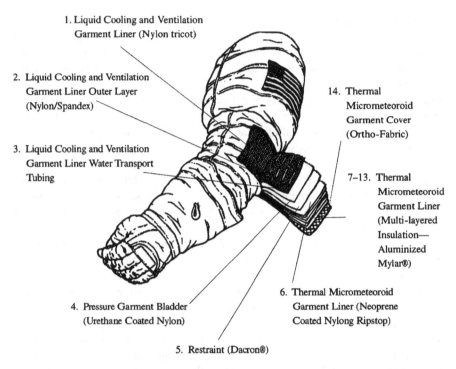

1. Liquid Cooling and Ventilation
   Garment Liner (Nylon tricot)

2. Liquid Cooling and Ventilation
   Garment Liner Outer Layer
   (Nylon/Spandex)

14. Thermal
    Micrometeoroid
    Garment Cover
    (Ortho-Fabric)

3. Liquid Cooling and Ventilation
   Garment Liner Water Transport
   Tubing

7–13. Thermal
    Micrometeoroid
    Garment Liner
    (Multi-layered
    Insulation—
    Aluminized
    Mylar®)

6. Thermal Micrometeoroid
   Garment Liner (Neoprene
   Coated Nylong Ripstop)

4. Pressure Garment Bladder
   (Urethane Coated Nylon)

5. Restraint (Dacron®)

Fig. 11.2 A cut-away of the extravehicular mobility unit (EMU) spacesuit arm depicting the numerous layers and functions (photo courtesy of NASA)

- Fault detection, isolation, and recovery. (In many spacesuits, no single system can lead to a catastrophic or critical state. This means power supplies, pressure regulators, and data communication systems typically have redundancy.)
- Power supply. A spacesuit is designed to operate for seven continuous hours, with emergency backup capacity of 30 min. The average metabolic capacity of the entire system is 300 W, supplied by a silver–zinc battery system. (This represents an energy capacity of about 7.5 MJ, or almost 30 times the capacity required by a standard notebook computer on Earth.)
- Wireless communication system, equipped with an audio (including microphone and earphone links) and data link (transmitting the astronaut and suit diagnostics above).

Included in the life support system within a spacesuit, approximately 0.6 L of water is available to the astronaut, and hygienic slips are present to contain human waste. The combined suit has a mass of approximately 125 kg (about 275 lbs on Earth), and takes about 15 min to put on or take off with two people (Skoog 1994; Newman and Barratt 1997).

A spacesuit is not only a wearable, protective life support system for the extreme environment of space, but may also even be viewed as the world's smallest spacecraft. Current spacesuits contain a propulsion backpack, or Simplified Aid

Fig. 11.3 US Astronaut
Bruce McCandless
demonstrates the manned
maneuvering unit for
propulsion in the weightless
environment of space in 1984
(photo courtesy of NASA)

for EVA Rescue (SAFER), for use only during emergencies. Onboard space
propulsion systems historically provide compressed nitrogen as a propellant, and
a control mechanism in the hands providing six degrees of freedom control. The
initial system, the Manned Maneuvering Unit (shown in Fig. 11.3), was retired in
1984 (NASA 1998). Despite the vision of spacesuits as spacecraft, in practice,
EVA, or spacewalks should be appreciated for the high risk exploration activity
they represent – most space missions tasks today are accomplished with an astronaut achieving mobility by means of a large robotic arm, as exists on the Space
Shuttle and International Space Station (ISS) (Hoffman 2004).

11.3 Recent Technological Advances in Technology for Space Exploration

11.3.1 Navigation Systems

Today, human spacesuit research benefits from innovations being developed on
Earth for enhanced navigation, wearable kinematics, and assistive locomotion
technologies. In consideration of future planetary missions, key challenges for
mission success and safety include: greatly enhanced mobility (Newman et al.
2007); real-time navigation aids (Johnson et al. 2009, 2010); and maintenance of
spatial orientation on the lunar surface for surface-based astronauts (Oman 2007).
The reduced gravity on the moon (one-sixth that on the Earth) and the lack of visual
cues and landmarks make judging distances difficult. NASA and researchers at MIT
and the Ohio State University are developing mission planning and support tools.
The Surface Exploration Traverse Analysis and Navigation Tool (SEXTANT)

developed at MIT used in conjunction with the Individual Mobile Agents System (iMAS) developed at NASA Ames, provides a system to assist with pre-mission planning, exploration scenario simulation, real-time navigation, and contingency planning during planetary EVAs. Data sources include digital elevation models, metabolic cost prediction, traverse waypoints, landmarks, and sun ephemeris data. Constraints, such as maximum slope to traverse, maximum metabolic rate, or walk-back distance, are implemented. Users select output metrics to be displayed, including metabolic cost, maps of surface slope, accessibility or visibility, a low-energy-cost direction of travel heuristic, sun-relative angles, and lunar sun illumination to better sense surface features and assess distance as determined by surface contrast differences due to the sun angle. The goal of the combined SEXTANT/iMAS system is twofold: to allow for realistic simulations of traverses to assist with hardware design, and to give astronauts an aid that will allow for more autonomy in EVA planning and re-planning. The user can specify traverses on an elevation model generated from topography data from the Lunar Orbiter Laser Altimeter (LOLA) instrument aboard the Lunar Reconnaissance Orbiter. SEXTANT calculates the sun position with respect to points along the traverse and the time the explorer arrives at each of them. A pre-built horizon elevation database is used to determine whether the explorer is in shadow or sunlight. This data is then used to compute the thermal load on suited astronauts, or the solar power generation capacity of rovers for the entire EVA. Furthermore, the display of shadows on SEXTANT's 3D mapping interface helps to increase situational awareness by giving astronauts or rover operators a view of the terrain as it actually appears. The sun position calculation in SEXTANT also directly influences the computation of the most efficient traverse path. The combined SEXTANT/iMAS system produces auditory alerts when certain traverse parameters have been exceeded or when the explorer deviates sufficiently from the planned traverse (Fig. 11.4).

The Lunar Astronaut Spatial Orientation and Information System (LASOIS) developed at the Ohio State University aims to enhance the spatial orientation of astronauts during exploration missions (Li et al. 2009). Since there is no satellite network around the moon providing absolute coordinates as we have on Earth with the Global Positioning System (GPS), it is important to enable absolute orientation in the lunar environment through a limited set of satellite coverage, lunar markers,

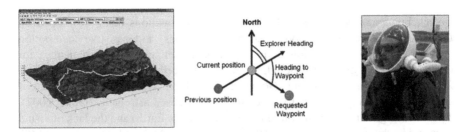

Fig. 11.4 SEXTANT 3D mapping interface (*left*). Determination of relative heading between explorer and waypoint (*middle*). A geologist using iMAS during simulated EVA (*right*)

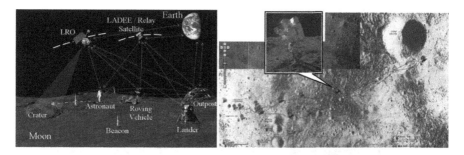

Fig. 11.5 On the *left*: Integrated sensor network concept for LASOIS (photo credit: NASA/OSU). An astronaut that is part of this network would use an IMU and other sensors to orient themselves on the lunar surface. On the *right*: a simulated astronaut path (photo credit: NASA/OSU/Google)

and inertial measurements. Future suits will be equipped with MEMS IMUS, lightweight stereo cameras, step sensors, and displays within the spacesuit. Spatial localization is to be accomplished by fusion of these multiple sensors using and complemented by an imaging system capable of tracking terrain targets. The system will be simulated in a lunar-like desert environment on Earth using satellite and airborne imagery coupled with the local sensors on the spacesuit (Fig. 11.5) (Li et al. 2008).

Human–machine navigation systems, predictive parametric analysis tools, and real-time decision aids for astronaut explorers should be implemented using wearable computers, and perhaps synthetic vision capability in the EVA spacesuit helmet, into advanced spacesuit and planetary mobile rover designs (Carr et al. 2003; Marquez et al. 2005).

11.3.2 BioSuit™ Development: A Wearable Second Skin

One of the key requirements of human planetary surface exploration is a spacesuit that enables astronaut locomotion. Unlike microgravity EVA, which involves limited translation that is performed almost entirely with the hands and arms, future planetary surface EVA requires locomotion, long traverses, climbing, and extensive bending (Fig. 11.6). These activities place new demands on spacesuit mobility and dexterity that can only be attained through implementing designs that facilitate natural locomotion and minimize energetic expenditures. The mobility permitted by future planetary spacesuits must therefore greatly surpass the mobility provided by current and previous spacesuits. High joint torques in the Apollo lunar suits forced astronauts on the Moon to walk with a considerably altered "bunny hop" gait to compensate for reduced knee and hip mobility. The current NASA EMU and Russian Orlan spacesuits (Abramov 2002) are designed for work in weightlessness, where lower body rigidity is actually beneficial in reducing the metabolic cost of producing counter-torques for body stabilization given the nature of the microgravity environment; however, this lower body rigidity and the substantial mass (approximately 125 kg) of the EMU or

Fig. 11.6 The BioSuit™ system offers the potential to be significantly lighter and much less bulky than current gas-filled spacesuits while facilitating joint mobility, which permits astronauts to perform unprecedented extreme exploration EVA

similar gas pressurized spacesuits make these systems inadequate for planetary surface locomotion and exploration mission tasks.

Mechanical counter pressure (MCP) has been proposed by Annis and Webb (1971) and Clapp (1983) and others to greatly enhance astronaut mobility and dexterity as well as offer improved safety and other advantages for planetary EVA. In an MCP suit, the body is pressurized using elastic tension in a skin-tight garment rather than the gas in a traditional spacesuit. The conceptual elegance of this approach has the potential not only to increase mobility, as will be discussed in greater detail, but also to simplify life-support system design, to reduce the astronaut's energy expenditure in using the suit for exploration, and to diminish the risk of depressurization and other EVA hazards.

The first MCP design, the Space Activity Suit (SAS) proposed by Webb (Webb 1968; Annis and Webb 1971), was a very creative idea before its time. The SAS used seven layers of highly elastic material to squeeze the wearer while a bubble helmet and chest bladder provided adequate breathing pressure. While the SAS initiated the MCP concept and demonstrated advantages of mobility, low energy costs, and a simplified life support system, the difficultly in donning/doffing the SAS was the largest limitation. NASA did not pursue MCP suit development for flight operations. A few additional MCP investigations such as those by Clapp (1983), Korona and Akin (2002), Tourbier et al. (2001), and Waldie et al. (2002) have advanced the state of the art of MCP glove design. These studies have investigated elastic materials or low-modulus MCP. The work of Tanaka et al. (2003a, b) investigated important physiological issues, such as blood flow and fluid distribution, as well as pressure distribution, and has provided preliminary evidence that MCP can be realized with no adverse physiological effects.

Given the performance characteristics of our biological skin, we envision a "second skin" suit capable of augmenting our biological skin to withstand the absence of a pressurized environment. If designed properly, Webb showed that an MCP suit could expose regions of skin no larger than 1 mm^2 to vacuum. There are two major advantages suggested by this result, namely, improved safety and thermal cooling. Especially on planetary surfaces, where the astronaut will be exposed to highly abrasive environments and activities, tears become an issue of increased concern. In a gas-pressurized suit, a small tear means not only the loss of pressure, but also the loss of breathable oxygen. In an MCP suit, however, this would not be the case, as Webb's result suggests that should a small hole appear in an MCP suit, the user would be unharmed. There would be no loss of breathable oxygen, and the skin would not suffer any damage. Should the hole be larger than 1 mm^2, the wearer would still have sufficient time to return to a pressurized environment due to the fact that the effects of the reduced pressure would be highly localized. Thermal cooling is the second advantage suggested by Webb's research. In an MCP suit, an air-permeable fabric enables normal thermal cooling, including evaporation of sweat. The MIT BioSuit™ System leverages new concepts and technologies to overcome obstacles encountered in previous MCP designs. Designs are conceived to allow the explorer the same speed and comfort of donning as clothes by separating the don/doff and pressure production processes: the suit would gradually shrink around the user in a controlled manner utilizing active materials only after it is donned. We envision the BioSuit™ to eventually provide a "second skin" capability incorporating biomechanical and cybernetic augmentation for human performance. By working at the intersection of engineering, design and physiology new emergent capabilities and interrelationships are sought. Radiation exposure and shielding as well as micrometeorite protection deserve significant consideration and novel solutions to realize future exploration missions.

In attempting to maximize the flexibility of the BioSuit™ System design to cope with the inevitable requirements and design changes inherent in all complex systems, we investigated the evolution of requirements for the extravehicular mobility unit (EMU) as a case (Saleh et al. 2004). We explored a fundamental environmental change, namely, using the Shuttle EMU aboard the ISS, and tracking the resulting EMU requirements and design changes. The EMU, like most complex systems, has faced considerable uncertainty during its service life due to changes in the technical, political, and economic environment. These have resulted in requirements changes, which in turn have necessitated design modifications or upgrades. Results suggest that flexibility is a key attribute that needs to be embedded in the design of long-lived systems to enable them to efficiently meet the inevitability of changing requirements after they have been fielded (Jordan et al. 2005a, b). Table 11.1 summarizes implemented design requirement changes for the EMU stemming from the political decision to use the Shuttle EMU aboard the ISS (Saleh et al. 2004). Table 11.2 lists the requirements that should serve as the guiding principles in the design of an advanced locomotion MCP spacesuit (Bethke et al. 2004; Newman et al. 2004).

Table 11.1 Summary of extravehicular mobility unit (EMU) design changes

Environment change – use shuttle EMU aboard ISS	
Requirement change or procedure change	Design or procedure change
Make EMU sizable on-orbit	Adjustable cam sizing in softgoods; Sizing rings in arms and legs; HUT replaceable on-orbit; HUT redesign from pivoted to planar
Increased EMU life	Recertification of EMU components; Change in static seal material; Noise muffler redesign; Flow filter redesign; Coolant water bladder material change

Environment change – physical environment of ISS	
Requirement change	Design or procedure change
Min. metabolic rate lowered	LCVG bypass designed; Heated gloves redesigned
PNP < 0.995 over 10 years	Track orbital debris; Define Allowable Penetrations
Different radiation exposure	Carefully plan all EVAs
Risk of propellant exposure	1-h bake-out procedure; Lengthen SCU

Environment change – technical environment advances	
Requirement change	Design or procedure change
Advance in joint technology	Joint patterning and materials changed; Bearing design and materials changed
Delicate assembly tasks	Glove design and materials changed
Increased EMU life	Battery redesign; CCC upgrade to regenerable canister; Carbon dioxide sensor upgraded
Secondary system in case of crewmember separation	SAFER

Table 11.2 Requirements for an advanced locomotion mechanical counter pressure (MCP) suit

Spacesuit function	Requirement
Pressure production	Continuously maintain fabric tension to apply at least 23 kPa (170 mmHg) of pressure at body surface
Pressure production	Locally expose no more than 1 mm^2 surface area of the skin (Webb 1968)
Pressure distribution	Distribute pressure evenly, with no more than 2.67 kPa (20 mmHg) spatial variation in pressure (Carr and Newman 2005)
Mobility	Require no more than 2 N-m of extra work (joint torque) against the suit to flex the knee to 90° (current EMU requires 3.74 N-m to bend the knee to 72° (Schmidt et al. 2001))
Mobility	Allow full unsuited range of lower body joint rotations
Operational	Feasibility don and doff times of less than 10 min
Operational	Feasibility don and doff by an individual wearer

11.4 Modeling the Human Body in Motion

The BioSuit™ System and future, wearable soft exoskeletons for children with cerebral palsy (CP) are both envisioned to provide a second skin capability. Therefore, it is critical to understand the detailed stretching and deformation of human skin as well as the changes in volume, surface area, and shape that occur during body movement and locomotion. Human body skin strain-field maps can be produced from laser scan imaging and precision motion analysis, then eigenvector analysis based on the principle of the skin structure exhibiting lines of nonextension (LoNE), or a natural 3D geometrical pattern, is revealed during limb motion, or during different flexion/extension static postures (Bethke et al. 2005; Sim et al. 2005; Wolfrum et al. 2006; Newman et al. 2007). Originally proposed by Iberall (1958, 1964, 1970), the LoNE concept represents skin contours, or topography, that neither elongate nor shrink during body motions. Simply stated our skin strain-fields reflect the tension, compression, and associated strains of skin as the body moves. Identification and modeling of skin strain-fields and nonextending 3D patterns of the skin during motion allows us to consider a true second skin design for assistive actuation technology, i.e., a "smart" soft exoskeleton that moves with the body. The set of minimum normal strain directions suggests the orientation, or "weave" direction, of the tensile fibers for a wearable second skin suit design (Bethke et al. 2004, 2005; Wolfrum et al. 2006). Newman determines the strain-fields by placing passive markers on the skin. Then, the distance between the markers can be determined during limb motion. The current 3D digital design capabilities include MATLAB (Natick, MA) and a graphical user interface (GUI) designed to take laser scan point cloud data from a Cyberware WB4 scanner or precision video-based motion analysis data as inputs and provide the LoNE as output (Fig. 11.7). Once the directions of zero stretch are calculated, they can be connected with other directions of zero stretch to form the LoNE for the human body. Cases with minimal compression or extension can also be connected when appropriate. This technique, while developed for optical strain-field calculations, can be modified for use with strain gauges and has applications to the design of second skin suits for Earth applications as well as space. For instance, a soft exoskeleton design to enable assistive locomotion for children with CP is currently being developed. Figure 11.8 illustrates an implementation of the LoNE in Newman's work on developing advanced spacesuits for NASA astronauts, and Fig. 11.9 displays the current state-of-the-art BioSuit™ mockup spacesuits based on this second skin design.

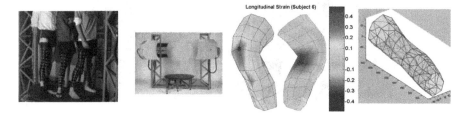

Fig. 11.7 Digital design procedure for second skin design: 3D laser scans, skin-strain field map showing up to 40% (0.4) positive and negative strains at the knee and behind the knee, and resulting pattern of suggested weave

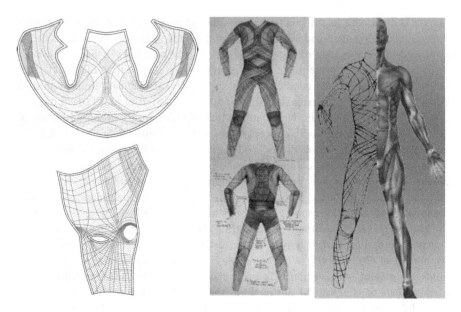

Fig. 11.8 An implementation of second skin lines of nonextension (LoNE) advanced spacesuit design. Thorax pattern (*top left*), thigh pattern (*bottom left*) (credit: Dainese), BioSuit™ concept, LoNE superimposed on musculature

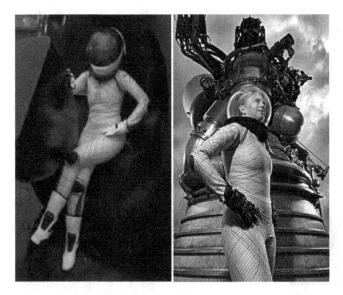

Fig. 11.9 MIT BioSuit™ mockup spacesuits implementing mechanical counter pressure (MCP) and patterned maximum mobility designs (invention and engineering by Prof. Dava Newman; design by Trotti and Associates, Inc.; fabrication by Dainese; photo credit: Donna Coveny and Doug Sonders)

11.5 BioSuit™: Inspired Technology Roadmap for Space and Earth Applications

A Technology Roadmap, as well as recommendations for investment in 10 technological areas for future EVA planetary systems and Earth applications of wearable systems and second skin designs, is suggested (Newman 2005). The roadmap suggests further research and investment in the categories of design, pressure production, and wearable technologies. To realize advanced concepts, four design areas are highlighted: 3D laser scanning, 3D and conductive textiles, electrospinlacing, and biomimetics (or nature-inspired design). 3D laser scanning is currently available, but further implementation of the described skin strain-field mapping is recommended. Currently, there exist state-of-the art 3D knitting machines, but none incorporates active materials. The capability to weave in 3D conductive textiles and wearable technologies is recommended in the near term. Electrospinning, or electrospinlacing, is currently demonstrated in research labs, but further investment is recommended to realize stable spun 3D nanofibers with anisotropic properties (Krogman et al. 2009). Nature is often a powerful teacher – for example, one promising research direction is to study how the cardiovascular and musculoskeletal systems of the giraffe provide sufficient counter pressure to maintain proper supply of oxygenated blood to the head throughout its large range of movement (4–5 m, from ground level to full height). This capability, which essentially relies on an "anti-gravity" capability incorporated into the giraffe's neck musculature, is a marvel of nature's design.

Pressure production is the next category for technology investment, and promising technologies include smart materials (shape-changing polymers), ferromagnetic shape memory alloys (FSMAs), and smart gels or fluid-filled bladders. Smart materials offer a suit design solution to decouple the don/doff and pressure production processes. The wearer could activate the suit to stretch into a loose-fitting suit for donning, and once donned the material would be deactivated to shrink around the user, thus creating MCP; or vice versa, i.e., the suit is originally pressurized to about two-third the desired MCP and then smart materials are activated to further shrink the suit around the body to attain the specified pressurization. Of the numerous smart materials considered to date, shape memory polymers (SMPs) appear to be a promising candidate for MCP garments for the former application because of their large maximum strain (typically greater than 100%), which could facilitate easy donning and doffing; whereas FSMAs might be the most promising active materials for the latter application, where once activated the FSMAs can provide additional constant pressure around the body and are the best candidate to provide human force and torque levels, hence their nickname of "muscle wires."

There exists potential for the BioSuit™ technology to be used as a countermeasure for astronaut deconditioning and for musculoskeletal rehabilitation on Earth for pathology (i.e., multiple sclerosis (MS), CP, and osteoporosis). An intriguing consequence of using electroactive materials may be the ability to send biomechanical signals to the body's tissues. This ability could allow a second skin design to provide a countermeasure for the degenerative effects of microgravity or certain pathologies on the musculoskeletal system. Research in Rubin (2001) indicates that bone cell deposition in the leg may be stimulated by longitudinal vibration at a much higher frequency (30 Hz) and much lower amplitude (5 microstrain) than experienced during normal walking on Earth (1 Hz and 3,000 microstrain). If high frequency and low amplitude signals can be generated by active polymer materials, then these wearable materials could be incorporated into the boots or legs of a second skin suit, and an electrical forcing function could drive them to vibrate the leg and mechanically stimulate bone growth. Low-strain mechanical inputs may compensate for the absence of gravity on bone, which suggests using waves to focus electromechanical signals of a desired frequency and amplitude longitudinally throughout the skeleton and/or at specific bone sites.

The final category for technology investment is in wearable technologies, specifically, the development of wearable bioinstrumentation and actuation based on a network of integrated fabric sensors and advanced signal processing techniques, on a textile platform. Three design goals include: to provide a comfortable and reliable health monitoring system, to measure multiparameter data, and to improve safety. Biosensors embedded in textiles should measure electrocardiography (ECG), pulse oximetry, temperature, and other biomedical signals; biochemical sensors that monitor body fluids; and finally, dosimeters used to measure the local radiation environment. Technology development to enable the envisioned wearable network includes: ultra low-power concepts, data processing, and display modules for reliable diagnosis suitable for extreme environments, and intelligent, real-time decision support and navigation systems (Johnson et al. 2009, 2010). The BioSuit™ wearable sensor network is shown in Fig. 11.10 (Canina et al. 2006).

11.6 Transitioning Spacesuit Technology for Earth Applications

Although spacesuits are designed to enhance the physical capabilities of humans during space exploration, the wearable technologies developed for this goal have led to innovative garment designs to enable exploration and to improvement of our quality of life here on Earth. Selected examples are discussed below.

11.6.1 Current Applications

Insulating and protecting astronauts from the extreme temperature conditions encountered during space flight is a primary requirement in spacesuit technology. The importance of this requirement was underscored on January 27, 1967, when, in

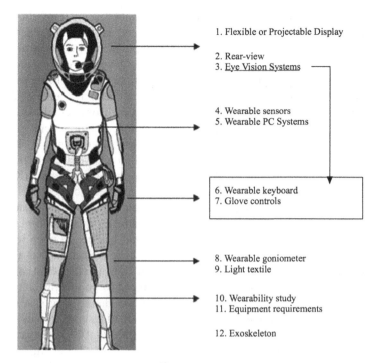

1. Flexible or Projectable Display

2. Rear-view
3. Eye Vision Systems

4. Wearable sensors
5. Wearable PC Systems

6. Wearable keyboard
7. Glove controls

8. Wearable goniometer
9. Light textile

10. Wearability study
11. Equipment requirements

12. Exoskeleton

Fig. 11.10   BioSuit™ system wearable sensor network

the first trials of the Apollo space program, a flash fire occurred during a launch pad test that resulted in the death of three astronauts (Virgil "Gus" I. Grissom, a veteran of Mercury and Gemini missions; Lieutenant Colonel Edward H. White II, who performed the first U.S. space walks (Gemini); and Roger B. Chaffee) (NASA 2009). This tragic event spearheaded extensive research and development into fire-resistant textiles for spacesuits and vehicles. In particular, NASA began collaboration with the U.S. Air Force, which in the late 1950s began synthesizing polybenzimidazole (PBI), a high-temperature stable polymer. Together, they spent considerable effort developing nonflammable and thermally stable textiles using PBI, which are now used extensively in astronaut flight suit and clothing. In addition, since becoming commercially available in the early 1980s, PBI fibers have been integrated into specialized garments for performance in extreme temperature scenarios here on Earth, such as those encountered by firefighting and emergency response crews (Raheel 1994).

In addition to the development of thermal insulation materials, technology first developed for space exploration has been extended to create wearable cooling systems for Earth (Flouris and Cheung 2006). For instance, a specialized cooling suit, which includes 50 m of 2-mm wide tubing used to circulate cool liquid within the garment, has been developed to help maintain comfortable working temperatures for the mechanics of the McLaren Formula 1 racing team (Raitt 2008). This method of cooling, used in the LCVG (discussed earlier in the chapter), has also been extended for the treatment of MS. MS, a chronic disease that affects both cognitive and physical ability, is caused by the loss of myelin, a coating around the nerves that enables signal propagation, which therefore inhibits proper communication within the brain. By cooling the core body temperature, nerve conduction can be temporarily restored to demyelinated nerves. However, direct cooling methods, such as immersion in a pool, are inappropriate for severely disabled patients, and homeostatic mechanisms, such as vasoconstriction and shivering, may prevent the lowering of core body temperature. Developed from LCVG technology, cooling garments, such as a head cap and/or torso vest (Ku et al. 2000), have been developed that overcome the homeostatic mechanisms, and alleviate MS symptoms (Schwid et al. 2003). By circulating a 45°F water-based fluid through the tubing integrated within the garment, efficient heat transfer enables lowering of the core body temperature, and has been shown to reduce MS symptoms for up to 3 h after a 45-min cooling session (Beenakker et al. 2001).

Much of the research conducted to help astronauts adapt to the effects of gravity in space has given rise to many applications designed to improve health here on Earth. To help prevent bone loss and muscle atrophy during long space flights, NASA astronauts exercise with added resistance during treadmill activity in space (NASA 2009). In preparation for the weightless conditions of EVA in space, while still on Earth NASA makes use of extensive underwater (pool) training for space-suited astronauts (Pollock and Fitzpatrick 2002). The athletic and rehabilitative benefits of temporarily reducing the ground reaction load (or body weight) have been demonstrated in water training (Gehlsen et al. 1984; Newman et al. 1994; Broman et al. 2006), as well as treadmill training using suspension harnesses (Newman et al. 1994; Jackson and Newman 2000; Colombo et al. 2000; Carr and Newman 2007) and pneumatic bladders (Macias et al. 2007).

Innovations in spacesuit design to help counteract the long-term musculoskeletal physiological deconditioning effects of weightlessness while in space (Kozlovs-kaya et al. 1995) have also contributed to the development of new therapeutic garments for motor control disorders. In particular, a modification of the Russian Pingvin (or Penguin) Suit, known as the Adeli Suit, has been developed. This garment, which relies on internal elastic bands and pulleys to exert force against the body, holds promise as a tool for rehabilitation in the treatment of those who suffer from CP (Bar-Haim et al. 2006; Turner 2006). Another suit design, the Gravity Loading and Countermeasure Suit™ (GLCS), recently developed at MIT, features high-tension elastic vertical fibers, to simulate the load of Earth's gravity on the body, as well as low-tension elastic circumferential fibers to enable easy donning and doffing (Waldie and Newman 2010). In addition to its potential contributions to future space missions, the GLCS may also find important application to counteract gravitational forces to improve postural development and voluntary movement of CP patients on Earth. Also, given its lightweight and comfortable design, it may be of particular benefit to the treatment of children. Figure 11.11 shows the Adeli Suit and the GLCS systems.

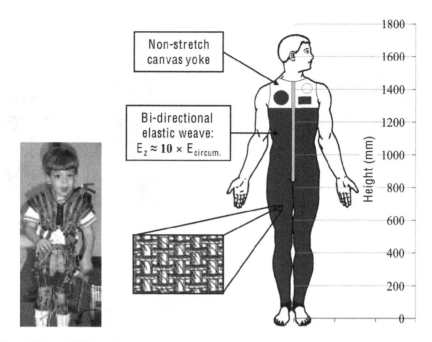

Fig. 11.11 Adeli Suit (*left*) and Gravity Loading Countermeasure Suit™ (GLCS) (*right*) systems

11.6.2 Wearable Future Second Skin Suits

The development of suit technologies for exploration and rehabilitation suggest promising research opportunities to explore the development of orthotic suits that incorporate both soft architectures and active elements. Such orthotic suits, or second skins, may be especially beneficial for the treatment and rehabilitation of decreased motor function in children, such as those suffering from CP, for whom lower limb orthotic devices are beneficial to improving gait (Radtka et al. 2005). To realize the appropriate time-dependent mechanical control for these active orthotic suits, sophisticated actuation and sensing are required.

11.6.2.1 Actuation

Future active orthotic suits will utilize actuation to improve biological muscle function. Currently, many active orthoses use high torque, brushless DC motors (Blaya et al. 2003; Blaya and Herr 2004) (see Fig. 11.12) or pneumatic bladders (Davis et al. 2003; Ferris et al. 2005, 2006). By using DC motors, high strain and

Fig. 11.12 A variable
impedance ankle-foot
orthosis to assist for drop-foot
gait. An electronically
controllable spring controls
the joint angle using
capacitive force transducers
and a rotary potentiometer to
measure ankle angle (Blaya
et al. 2003; Blaya and Herr
2004)

efficiency may be achieved, but obtaining high power density, high-energy storage capacity, and small size (compared to normal mammalian skeletal muscles) are remaining challenges. Pneumatic bladders address many of these aforementioned issues, but at the expense of slow, low-bandwidth responses (Davis et al. 2003). Much effort has been invested to explore the potential of other technologies for the design of artificial muscle in active orthoses, but many of these have well understood shortcomings: shape memory alloys, such as SMPs and FSMAs (poor energy conversion efficiency); piezoelectric materials (low strain percentage); and magnetostrictive materials (low strain percentage, low power density, and elastic energy density).

Several new classes of actuators are currently under investigation for implementation of artificial muscle. The first class comprises various types of electroactive polymers (Pelrine et al. 2000). Within the dielectric elastomer family, these polymers exhibit characteristics of strain, energy density, and response time that compare favorably to those found in mammalian skeletal muscles. This technology holds promise for the design of large artificial muscles (Herr and Kornbluh 2004), such as those required for the development of prosthetic arms (Carpi and Smela 2009), and testing of its application for small muscles (Senders et al. 2010) has already begun.

One additional class of materials, carbon nanotubes (CNTs), shows potential for application to artificial muscles. CNT actuators may yield order of magnitude improvements over commercial actuators if a bulk material can be made that achieves performance comparable to individual nanotubes (Baughman et al. 2002). To date, no bulk material has achieved a strain over 0.2% (Baughman et al. 1999), a value that is over three orders of magnitude lower than mammalian muscle. Nevertheless, with so much innovation occurring in the research areas of actuation and artificial muscle, over the coming years it is expected that some of these technologies will enable a variety of new actuation materials that will be applicable to the development of second skin orthoses.

11.6.2.2 Sensing

There are several primary sensing schemes that may be used in conjunction with appropriate control and actuation elements in second skin orthotic devices. These include sensing systems that measure neural activity in the brain, muscle activation, and "intrinsic" sensing, i.e., sensing that is confined to the device (with no direct connection to the user).

Direct cortical control, in which neural activity is used to control artificial devices to replace biological function lost through disease or injury, is perhaps the most ambitious strategy for control of future orthotic devices. It has been shown that accurate predictions of arm movement can be generated from the recorded activity of a set of cortical neurons (Georgopoulos et al. 1986; Schwartz and Moran 1999). Recent work demonstrates that accurate real-time control of neuroprosthetic devices using such cortical signals is possible, and when the subjects (rhesus macaque monkeys) are given visual feedback of their movement trajectories (Taylor et al. 2002), accuracy may be improved with daily practice. Also, by using microelectrodes chronically embedded in the cerebral cortex and extraction algorithms, promising results have been achieved in the control of prosthetic arms (Schwartz 2004; Hochberg et al. 2006; Jackson et al. 2006). For humans with tetraplegia, correlations between spiking activity and intended movement have been observed in the primary motor cortex, indicating a promising future for the control of devices when movement is imagined by the patient (Truccolo et al. 2008).

Control of active assistive devices may also be achieved through direct measurement of muscle activation using electromyography (EMG), a technique for measuring the electrical activity of skeletal muscle as it is stimulated, using temporary electrodes mounted on the surface of the skin or wire electrodes installed in the tissue through surgical means. Using EMG, the control of a powered orthosis may be directly linked to existing biological control of muscle. Exploration of this approach for application to active ankle-foot prostheses (Au et al. 2005) and orthoses (Ferris et al. 2006) has already yielded promising results.

Although the above sensing techniques are extremely promising, as they provide means for directly connecting the operation of an orthotic device to the intended movement of its user, given their necessary physical connections to the human body, they pose great practical challenges to the design of mobile, wearable devices. Therefore, in many cases intrinsic sensing is desirable. In particular, sensing relevant force and/or strain profiles, as well as the kinematics of a user's motion, is useful for many assistive mobility applications. In the following discussion, the current state of the art in force/strain measurement and wearable kinematics systems is briefly reviewed, with a focus on capturing lower limb movement.

Force and Strain Measurement

In the design of second skin orthoses for the rehabilitation of gait disorders, embedded force and strain sensors may be used to detect relevant parameters of movement, such as ground reaction force and center of pressure, as well as measuring strain-fields within the second skins themselves (with strain information of sufficiently high resolution, it may be possible to infer kinematics of motion). Although there are many ways to capture this information, only methods that may be easily implemented and customized within the construct of the suit itself are of interest here.

Force and/or strain can be measured by means of capacitive sensors (force applied to the plates of a capacitor changes the distance between the plates, thereby changing the capacitance), as implemented for the detection of ground reaction force (created by the foot contact with the ground) (Blaya et al. 2003; Blaya and Herr 2004). This method has great applicability, but does require more complex signal conditioning than those described below – since a capacitor cannot pass a current at 0 Hz (DC), the signal must be modulated at an intermediate frequency, usually less than 1 kHz, and therefore has relatively low bandwidth.

There are several technologies used to detect changes in force and/or strain with high bandwidth, in particular, polymer thick film (PTF), piezoresistive materials, and conductive rubbers can be used to make force sensitive resistors (FSRs). In addition, foil strain gauges are commonly used for force and strain sensing. These devices, which are commercially available for installation on a variety of substrates, have large bandwidth (they are essentially resistors with very small capacitance and inductance), negligible hysteresis, and small form factor. These features are highly desired, especially for measurement of lower limb movement, as human locomotion exhibits rapid changes in force that must be accurately recorded. (Optical strain gauges are interesting alternatives, but are not yet available in small form factors.) The use of foil strain gauges to measure tension of artificial pneumatic muscle has also been demonstrated (Ferris et al. 2005).

However, the most promising strain gauge technologies for application to second skin orthoses are those that are fabric-based, such as those described in Scilingo et al. (2003), Lorussi et al. (2004), Wu et al. (2005), and De Rossi et al. (2005). These fabric implementations utilize polymer coatings that demonstrate a change in resistance in response to strain. (For detailed discussion on strain-sensing fabrics, see Chap. 7.)

Wearable Kinematics Systems

With the advent of small inertial measurement units (IMUs), composed of accelerometers and/or gyroscopes, local sensing may be implemented in small wearable devices to measure human motion, without the encumbrances of wires and heavy electronics. For many applications, such as counting steps or measuring stride length, detecting changes in acceleration (Schneider et al. 2003) or angular velocity (Miyazaki 1997; Aminian et al. 2002), respectively, may be sufficient indicators of motion.

For many applications, more quantitative information describing human motion, such as orientation, velocity, and position, is required. Over short periods of time, relative orientation may be obtained through direct integration of the gyroscope measurements, and relative velocity/position may be obtained by integrating the accelerometer measurements. However, estimation with respect to an absolute reference frame is often much more useful than relative measurements. Furthermore, over longer periods of time, the drift inherent in these commercial sensors presents a significant obstacle to this direct calculation of orientation and velocity and position. Fortunately, information from other sensors, such as magnetometers and GPS, and appropriate sensor fusion algorithms may be used to combine measurements and obtain these parameters in absolute units, relative to the absolute reference frame of the Earth. These methods offer tremendous opportunities to study the biomechanics of human motion outside of laboratory and clinical settings, such as those required when using state-of-the-art optical motion capture systems (Brodie et al. 2008; Lapinski et al. 2009).

In particular, tilt and orientation may be accurately estimated using gyroscopes, accelerometers, and complementary filtering, as has been achieved for implementation in assistive devices to improve balance (Weinberg et al. 2006). To address the accumulating error produced by integrating the gyroscope measurements to obtain angle, and/or integrating the accelerometer measurements to obtain velocity and position information, additional sensing may be employed. For example, Schepers et al. (2007) demonstrated absolute estimation of ambulatory foot orientation and position using accelerometers and gyroscopes, in conjunction with force sensors that detect foot-ground contact and force. By detecting when the foot is on the ground, the bounds of integration can be appropriately reset (when the foot makes contact with the ground, the height (normal to the ground) of the sensor is approximately 0 m, and the velocity is 0 m/s). Similarly, the bounds of integration may be reset using measurements from the accelerometers, for cyclic walking (Dejnabadi et al. 2006; Moreno et al. 2006). The accuracy of integration may be further improved with fusion algorithms that use quaternion-based representation of orientation. Such algorithms allow for efficient real-time operation while effectively preventing "gimbal lock," a problem seen when Euler angles are used (Favre et al. 2006). In particular, this technique has been shown to be useful for estimating knee joint angles during walking (Favre et al. 2008).

Other related work includes the use of accelerometers and gyroscopes to measure the kinematics, in particular the angles and angular velocities of the thorax, pelvis, and upper leg, during the activity of rising to standing from a seated position in a chair (Boonstra et al. 2006). A similar method investigated the efficacy of using these inertial sensors to track the orientation and position of the torso during several locomotion tasks, including transitioning from standing to sitting and the initiation of gait (one step) (Giansanti 2005). It is important to note that these demonstrations focus on well-defined movements that produce relatively low accelerations and angular rates. To extend estimation of kinematics to a broader set of human motions, including those that feature high accelerations and angular velocities, more complex sensor fusion algorithms are required.

Nonlinear Kalman filters, such as the extended Kalman filter (EKF) (Brown and Hwang 1997) and the unscented Kalman filter (UKF) (Julier and Uhlmann 1997), represent a class of sensor fusion algorithms that can correct for the drift exhibited by commercial inertial sensors, while providing absolute unit estimation. The EKF/ UKF uses a nonlinear model of the dynamics (including noise) of the state to be estimated. The EKF/UKF evolves the state according to this model and then updates the state based on observations from the appropriate sensors to minimize the error of the estimated state. For example, using gyroscopes and accelerometers, the EKF may reliably reduce the error generated by integrating the angular velocities by using observations of the acceleration measurements (include contributions from gravity and inertial movement), enabling robust estimation of 2D orientation with the lowest error, relative to the absolute reference frame of the Earth.

Recent work has demonstrated the effectiveness of this technique for tracking orientation of the torso (during crate-lifting tasks) (Luinge and Veltink 2005) and orientation of the hand (Sabatini 2006). Similarly, this method may be extended to estimate joint angle, as in Luinge et al. (2007), in which data from two IMUs, one mounted on the upper arm and the other on the lower arm, were input to an EKF to obtain elbow joint angle. A common variation of this method includes magnetometers, in addition to accelerometers and gyroscopes, to obtain 3D orientation. (Because rotation about the normal to the surface of the Earth is invariant in the gravitational field, accelerometer measurements alone are not adequate to correct the angle about this normal.) Recently, Schepers et al. (2010) demonstrated the use of local magnetic fields (generated by the wearable electronics system) to improve the reliability of magnetometers in the EKF estimation of hip and shoulder joint angles.

In addition to the above applications, the EKF may also be applied to the study of human gait, for which accurate measurement of lower body kinematics is essential.

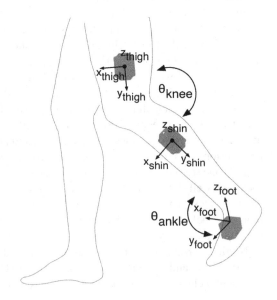

Fig. 11.13 One inertial measurement unit (IMU), which captures 3D acceleration and 3D angular velocity, is placed on each lower limb segment. These data are input to the extended Kalman filter (EKF) to determine the 2D orientation of each limb segment and the resulting knee and ankle joint angles in the sagittal and coronal planes (Young et al. 2010)

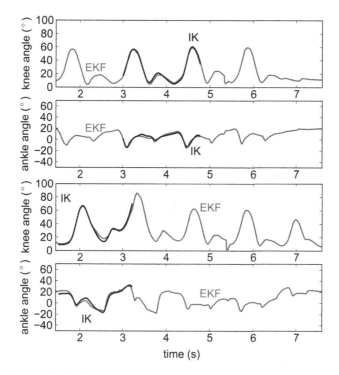

Fig. 11.14 Knee and ankle joint angles (in the sagittal plane) estimated using the IMUs and EKF to those obtained using optical motion capture and inverse kinematics (IK) for a level ground trial (*top*) and a stair ascent walking trial (*bottom*) (Young et al. 2010)

In a manner similar to that shown in Luinge et al. (2007), knee and ankle joint angles may be obtained using three IMUs: one IMU (measuring 3D acceleration and 3D angular velocity) mounted on each of the three limb segments of the leg, as shown in Fig. 11.13. Using the inertial data from these devices as inputs to the EKF, the 2D orientation of each limb segment is estimated and these results are then combined to obtain the joint angles in the sagittal and coronal planes (Young et al. 2009, 2010). Figure 11.14 shows an example comparison between the joint angle results obtained by this approach to those obtained by the traditional approach using optical motion capture and inverse kinematics (IK) software (which includes a model of the human body).

It is important to note that human locomotion poses significant challenges to the estimation of kinematic parameters using inertial sensing and fusion algorithms, as the swinging movement of the legs even at moderate walking speeds may produce high accelerations and angular velocities. Also, the ground reaction forces generated may produce large mechanical noise, and any extraneous movement of the IMUs may easily further degrade the performance of the estimation. Therefore, the practical issue of sensor mounting is of extreme importance. Fortunately, inertial sensors are now available in sufficiently small form factors to enable a great deal of design flexibility and remain promising for integration within soft architecture designs, such as the second skin suits proposed here.

11.7   Concluding Remarks

Exploration for space as well as for extreme conditions on Earth inspires ambitious wearable electronics systems for human activities. The research undertaken to develop space systems to support and maintain human life gives rise to many innovations to augment human capability on Earth. In particular, the design of future astronaut spacesuits now finds application in the area of assistive biomedical devices for rehabilitation. These therapeutic garments, which will have sensing and actuation elements embedded within soft architectures, are expected to be instrumental in the rehabilitation of gait pathology and may be particularly applicable for the treatment of children.

Acknowledgements The authors thank Professor Hugh Herr for his collaboration and the MIT Portugal Program and NSF Grant CNS-0932015 for research support of the authors.

References

Abramov IP, Glazov GM, Svertshek VI (2002) Long-term operation of "Orlan" space suits in the "Mit" orbiting station: Experience obtained and its application. Acta Astronaut 51(1–9):133–143

Aminian K, Najafi B, Büla C, Leyvraz PF, Robert P (2002) Spatio-temporal parameters of gait measured by an ambulatory system using miniature gyroscopes. J Biomech 35:689–699

Annis JF, Webb P (1971) Development of a space activity suit. NASA contractor report CR1892, NASA Lewis Research Center, Cleveland, OH

Au SK, Bonato P, Herr H (2005) An EMG-position controlled system for an active ankle-foot prosthesis: an initial experimental study. In: Proceedings of the international conference on rehabilitation robotics, Chicago, IL

Bar-Haim S, Harries N, Belokopytov M, Frank A, Copeliovitch L, Kaplanski J, Lahat E (2006) Comparison of efficacy of Adeli suit and neurodevelopmental treatments in children with cerebral palsy. Dev Med Child Neurol 48:325–330

Baughman RH, Cui C, Zakhidov AA, Iqbal Z, Barisci JN, Spinks GM, Wallace GG, Mazzoldi A, Rossi DD, Rinzler AG, Casckinski O, Roth S., Kertesz M (1999) Carbon nanotube actuators. Science 284:340–344

Baughman RH, Zakhidov AA, de Heer WA (2002) Carbon nanotubes—the route toward applications. Science 297:787–792

Beenakker EAC, Oparina TI, Hartgring A, Teelken A, Arutjunyan AV, De Keyser J (2001) Cooling garment treatment in MS: clinical improvement and decrease in leukocyte nitric oxide (NO) production. Neurology 57:892–894

Bethke K, Carr CE, Pitts BM, Newman DJ (2004) Bio-suit development: viable options for mechanical counter pressure. SAE Transactions, SAE paper 2004–01–2294, 113(1):426–437

Bethke K, Newman DJ, Radovitzky R (2005) Creating a skin strain field map with application to advanced locomotion spacesuit design. In: Proceedings of the XXth congress of the international society of biomechanics (ISB), Cleveland, OH

Blaya J, Newman DJ, Herr H (2003) Comparison of a variable impedance control to a free and rigid ankle foot orthosis (AFO) in assisting drop foot gait. Proceedings of the International Society of Biomechanics (ISB) XIXth Congress, paper number 44, Dunedin

Blaya J, Herr H (2004) Adaptive control of a variable-impedance ankle-foot orthosis to assist drop-foot gait. IEEE Trans Neur Sys Reh 12(1):24–31

Boonstra MC, van der Slikke RMA, Keijsers NLW, van Lummel RC, de Waal Malefijt MC, Verdonschot N (2006) The accuracy of measuring the kinematics of rising from a chair with accelerometers and gyroscopes. J Biomech 39:354–358

Brodie M, Walmsley A, Page W (2008) Fusion motion capture: a prototype system using inertial measurement units and GPS for the biomechanical analysis of ski racing. Sports Technol 1:17–28

Broman G, Quintana M, Lindberg T, Jansson E, Kaijser L (2006) High intensity deep water training can improve aerobic power in elderly women. Eur J Appl Physiol 28:117–123

Brown RG, Hwang PYC (1997) Introduction to random signals and applied Kalman filtering, 3rd edn. Wiley, New York

Canina M, Newman DJ, Trotti GL (2006) Preliminary considerations for wearable sensors for astronauts in exploration scenarios. In: Proceedings of the 3rd IEEE-EMBS international summer school and symposium on medical devices and biosensors (ISSS-MDBS 2006), Boston, MA

Carpi F, Smela E (2009) Biomedical applications of electroactive polymer actuators. Wiley, New York

Carr CE, Hodges KV, Newman DJ (2003) Geologic traverse planning for planetary EVA. In: Proceedings of the international conference on environmental systems (ICES 2003), Vancouver, BC

Carr CE, Newman DJ (2005) When is running in a space suit more efficient than walking in a space suit? SAE T J Aerospace 114, 2005–01–2970, 114(1):403–408

Carr CE, Newman DJ (2007) Space suit bioenergetics: cost of transport during walking and running. Aviat Space Env Med 78:1093–1102

Clapp W (1983) Design and testing of an advanced spacesuit glove. SB Thesis, Massachusetts Institute of Technology, Cambridge, MA

Colombo G, Joerg M, Schreier R, Dietz V (2000) Treadmill training of paraplegic patients using a robotic orthosis. J Rehabil Res Dev 37(6):693–700

Davis S, Tsagarakis N, Canderle J, Caldwell DG (2003) Enhanced modelling and performance in braided pneumatic muscle actuators. Int J Robot Res 22:213–227

Dejnabadi H, Jolles BM, Casanova E, Fua P, Aminian K (2006) Estimation and visualization of sagittal kinematics of lower limbs orientation using body-fixed sensors. IEEE Trans Biomed Eng 53(7):1382–1393

De Rossi D, Carpi F, Scilingo EP (2005) Polymer based interfaces as bioinspired 'smart skins'. Adv Colloid Interface Sci 116:165–178

Favre J, Jolles BM, Siegrist O, Aminian K. (2006) Quaternion-based fusion of gyroscopes and accelerometers to improve 3D angle measurement. Electron Lett 42:612

Favre J, Jolles BM, Aissaoui R, Aminian K (2008) Ambulatory measurement of 3D knee joint angle. J Biomech 41:1029–1035

Ferris DP, Czerniecki JM, Hannaford B (2005) An ankle-foot orthosis powered by artificial pneumatic muscles. J Appl Biomech 21(2):189–197

Ferris DP, Gordon KE, Sawicki GS, Peethambaran A (2006) An improved powered ankle-foot orthosis using proportional myoelectric control. Gait Posture 23:425–428

Flouris AD, Cheung SS (2006) Design and control optimization of microclimate liquid cooling systems underneath protective clothing. Ann Biomed Eng 34(3):359–372

Gehlsen G, Grigsby S, Winant D (1984) Effects of an aquatic fitness program on the muscular strength and endurance of patients with multiple sclerosis. Phys Ther 64(5):653–657

Georgopoulos AP, Schwartz AB, Kettner RE (1986) Neuronal population coding of movement direction. Science 244:1416–1419

Giansanti D, Maccioni G, Macellari V (2005) The development and test of a device for the reconstruction of 3-D position and orientation by means of a kinematic sensor assembly with rate gyroscopes and accelerometers. IEEE Trans Biomed Eng 52(7):1271–1277

Grahne MS, Graziosi DA, Pauly RL (1995) Benefits of an EMU glove active heating system. SAE, international conference on environmental systems, document number 951549, San Diego, CA

Herr H, Kornbluh R (2004) New horizons for orthotic and prosthetic technology: artificial muscle for ambulation. In: Proceedings of the SPIE, vol 5385, smart structures and materials 2004: electroactive polymer actuators and devices (EAPAD), San Diego, CA

Hochberg LR, Serruya MD, Friehs GM, Mukand JA, Saleh M, Capalan AH, Branner A, Chen D, Penn RD, Donoghue JP (2006) Neuronal ensemble control of prosthetic devices by a human with tetraplegia. Nature 442:164–171

Hoffman SJ (2004) Advanced EVA capabilities: a study for NASA's revolutionary aerospace systems concept program. NASA Technical Publication 2004–212068, NASA Lyndon B. Johnson Space Center, Houston, TX

Iberall AS (1958) Development of a full pressure altitude suit, WADS Technical Report 58–236. Wright Air Development Center, Wright-Patterson Air Force Base, OH

Iberall AS (1964) The use of lines of non-extension to improve mobility in full-pressure suits, AMRL-TR-64–118. Aerospace Medical Research Lab, Wright-Patterson Air Force Base, OH

Iberall AS (1970) The experimental design of a mobile pressure suit. J Basic Eng 1970, 251–264

ILC Dover, Inc (1994) Space suit evolution from custom tailored to off-the-rack. http://www.hq.nasa.gov/office/pao/History/spacesuits.pdf

Jackson A, Moritz CT, Mavoori J, Lucas TH, Fetz EE (2006) The neurochip BCI: towards a neural prosthesis for upper limb function. IEEE Trans Neur Sys Reh 14(2):187–190

Jackson DK, Newman DJ (2000) Adaptive effects of space flight as revealed by short–term partial weight suspension. Aviat Space Environ Med 71(9, Suppl):A151–A160

Johnson AW, Newman DJ, Waldie JM, Hoffman JA (2009) An EVA mission planning tool based on metabolic cost optimization. International conference on environmental systems (ICES), Paper number 2009–01–2562, Savannah, GA

Johnson AW, Hoffman J, Newman DJ, Mazarico E, Zuber M (2010) An integrated EVA mission planner and support tool for future planetary exploration. In: Proceedings of the AIAA SPACE 2010 conference, Anaheim, CA

Jordan NC, Saleh JH, Newman DJ (2005a) The extravehicular mobility unit: case study in requirements evolution and the need for flexibility in the design of complex systems. In: Proceedings of the IEEE conference on requirements engineering, Paris

Jordan NC, Saleh JH, Newman DJ (2005b) The extravehicular mobility unit: A review of environment, requirements, and design changes in the US spacesuit. Acta Astronaut 59 (12):1135–1145

Julier SJ, Uhlmann JK (1997) A new extension of the Kalman filter to nonlinear systems. In: Signal processing, sensor fusion, and target recognition VI, vol 3068 of Proceedings of the SPIE, Orlando, FL

Korona F, Akin D (2002) Evaluation of a hybrid elastic EVA glove. In: Proceedings of the 32nd international conference on environmental systems (ICES-2002), San Antonio, TX

Kozlovskaya IB, Grigoriev AI, Stepantzov VI (1995) Countermeasure of the negative effects of weightlessness on physical systems in long-term space flights. Acta Astronaut 36(8–12):661–668

Krogman KC, Lowery JL, Zacharia NS, Rutledge GC, Hammond PT (2009) Spraying asymmetry into functional membranes layer-by-layer. Nat Mater 8:512–518

Ku YT, Montgomery LD, Lee HC, Luna B, Webbon BW (2000) Physiologic and functional responses of patients to body cooling. Am J Phys Med Rehabil 13:8994–9115

Lapinski M, Berkson E, Gill T, Reinold M, Paradiso JA (2009) A distributed wearable, wireless sensor system for evaluating professional baseball pitchers and batter. In: Proceedings of the IEEE international symposium on wearable computers (ISWC 2009), Linz

Li R, Di K, Wu B, Yilmaz A, Banks MS, Oman C, Bhasin K, Tang M. (2008) Enhancement of spatial orientation capability of astronauts on the lunar surface supported by integrated sensor network and information technology. NLSI lunar science conference, Moffett Field, CA

Li R, Wu B, He S, Shopljak B, Yilmaz A, Jiang J, Banks MS, Oman C, Bhasin KB, Warner JD, Knoblock EJ (2009) LASOIS: enhancing the spatial orientation capabilities of astronauts on the lunar surface. In: Proceedings of the 40th lunar and planetary science conference, The Woodlands, TX

Lorussi F, Rocchia W, Scilingo EP, Tognetti A, De Rossi D (2004) Wearable, redundant fabric-based sensor arrays for reconstruction of body segment posture. IEEE Sens J 4(6):807–818

Luinge HJ, Veltink PH (2005) Measuring orientation of human body segments using miniature gyroscopes and accelerometers. Med Biol Eng Comput 43(2):273–282

Luinge HJ, Veltink PH, Baten CTM (2007) Ambulatory measurement of arm orientation. J Biomech 40(1):78–85

Macias BR, Groppo ER, Bawa M, Cao HST, Lee B, Pedowitz RA, Hargens AR (2007) Lower body negative pressure treadmill exercise is more comfortable and produces similar physiological responses as weighted vest exercise. Int J Sports Med 28(6):501–505

Marquez JJ, Cummings ML, Roy N, Kunda M, Newman DJ (2005) Collaborative human-computer decision support for planetary surface traversal. In: Proceedings of the AIAA 5th aviation, technology, integration, and operations conference (ATIO), Arlington, VA

Miyazaki S (1997) Long-term unrestrained measurement of stride length and walking velocity utilizing a piezoelectric gyroscope. IEEE Trans Biomed Eng 44(8):753–759

Moreno JC, de Lima ER, Ruíz AF, Brunetti FJ, Pons JL (2006) Design and implementation of an inertial measurement unit for control of artificial limbs: application on leg orthoses. Sensors Actuator B 118(1–2):333–337

Moran DW, Schwartz AB (1999) Motor cortical representation of speed and direction during reaching. J Neurophysiol 82:2676–2692

NASA History Division (1959) NASA human spaceflight programs, mercury. http://www.hq.nasa.gov/office/pao/History/humansp.html

NASA (1971) Apollo operations handbook extravehicular mobility unit, volume 1, system description. Manned Spacecraft Center, Houston, TX

NASA (1998) Suited for spacewalking: a teacher's guide. Office of Human Resources and Education Division, Washington, DC

NASA (2009) Spinoff 2009. US Government Printing Office, Washington, DC

Newman DJ, Alexander HL, Webbon BW (1994) Energetics and mechanics for partial gravity locomotion. Aviat Space Env Med 65(9):815–823

Newman DJ, Barratt M (1997) Life support and performance issues for extravehicular activity (EVA). In: Churchill S (ed) Fundamentals of space life sciences, Krieger Publishing Company, Malabar, FL

Newman DJ, Bethke K, Carr C, Hoffman J, Trotti G (2004) Astronaut bio-suit system to enable planetary exploration. In: Proceedings of the international astronautical conference (IAC-2004), Vancouver, BC

Newman DJ (2005) Astronaut Bio-Suit System for Exploration Class Missions NIAC Phase II Final Report, NASA Institute for Advanced Concepts

Newman DJ, Canina M, Trotti GL (2007) Revolutionary design for astronaut exploration – beyond the bio-suit system. In: Proceedings of the Space technology and applications international forum—STAIF-2007, Albuquerque, NM

Nunneley SA (1970) Water cooled garments: a review. Space Life Sci 2:335–360

Oman CM (2007) Spatial Orientation and Navigation in Microgravity. In: Mast F, Janeke L (ed) Spatial Processing in Navigation, Imagery and Perception. Springer Verlag, New York

Pelrine R, Kornbluh R, Qibing P, Joseph J (2000) High-speed electrically actuated elastomers with strain greater than 100%. Science 287:836–839

Pollock NW, Fitzpatrick DT (2002) NASA flying after diving procedures. In: Flying after recreational diving workshop proceedings, Durham, NC

Radtka SA, Skinner SR, Johanson ME (2005) A comparison of gait with solid and hinged ankle-foot orthoses in children with spastic diplegic cerebral palsy. Gait Posture 21(3):303–310

Raheel M (ed) (1994) Protective clothing systems and materials. Marcel Dekker, New York

Raitt D (2008) Interdisciplinary knowledge transfer from aerospace to daily applications. In: Proceedings of the IFAI advanced textiles conference, Berlin

Rubin C (2001) Low mechanical signals strengthen long bones. Nature 412:603–604

Sabatini AM (2006) Quaternion-based extended Kalman filter for determining orientation by inertial and magnetic sensing. IEEE Trans Biomed Eng 53(7):1346–1356

Saleh JH, Hastings DE, Newman DJ (2004) Weaving time into system architecture: satellite cost per operational day and optimal design lifetime. Acta Astronaut 54(6):413–431

Schepers HM, Koopman HFJM, Veltink PH (2007) Ambulatory assessment of ankle and foot dynamics. IEEE Trans Biomed Eng 54(5):895–902

Schepers MH, Roetenberg D, Ventink PH (2010) Ambulatory human motion tracking by fusion and magnetic sensing with adaptive actuation. Med Biol Eng Comput 48(1):27–37

Schmidt P, Newman D, Hodgson E (2001) Modeling space suit mobility: applications to design and operations. In: Proceedings of the international conference on environmental systems (ICES 2001), Orlando, FL

Schneider PL, Crouter SE, Lukajic O, Basset DR (2003) Accuracy and reliability of 10 pedometers for measuring steps over a 400-m walk. Med Sci Sports Exerc 35(10):1779–84

Schwartz AB (2004) Cortical neural prosthetics. Annu Rev Neurosci 27:487–507

Schwid SR, Petrie MD, Murray R, Leitch J, Bowen J, Alquist A, Pelligrino R, Roberts A, Harper-Bennie J, Milan MD, Guisado R, Luna B, Montgomery L, Lamparter R, Ku YT, Lee H, Goldwater D, Cutter G, Webbon B (2003) A randomized controlled study of the acute and chronic effects of cooling therapy for MS. Neurology 60(12):1955–1960

Scilingo EP, Lorussi F, Mazzoldi A, De Rossi D. (2003) Strain-sensing fabrics for wearable kinaesthetic-like systems. IEEE Sens J 3(4):460–467

Senders CW, Tollefson TT, Curtiss S, Wong-Foy A, Prahlad H (2010) Force requirements for artificial muscle to create an eyelid blink with eyelid sling. Arch Facial Plast Surg 12(1):30–36

Shayler DJ (2001) Skylab: America's space station. Springer Praxis books in space exploration. Springer UK, London

Sim L, Bethke K, Jordan N, Dube C, Hoffman J, Brensinger C, Trotti G, Newman DJ (2005) Implementation and testing of a mechanical counterpressure bio-suit system. In: Proceedings of the international conference on environmental systems (ICES 2005), 2005–01–2968, Rome

Skoog, AI (1994) The EVA space suit development in Europe. Acta Astronaut 32(1):25–38

Tanaka K, Gotoh TM, Morita H, and Hargens AR (2003a). Skin blood flow with elastic compressive extravehicular activity space suit. Biol Sci Space 17(3):227

Tanaka K, Limberg R, Webb P, Reddig M, Jarvis C, Hargens A (2003b). Mechanical counterpressure on the arm counteracts adverse effects of hypobaric exposures. Aviat Space Env Med 74(8):827–832

Taylor DM, Tillery SIH, Schwartz AB (2002) Direct cortical control of 3D neuroprosthetic devices. Science 296:1829–1832

Tourbier D, Knudsen J, Hargens A, Tanaka K, Waldie J, Webb P, Jarvis C (2001) Physiological effects of a mechanical counter pressure glove. 31st International Conference on Environmental Systems, 2001-01-2165, Orlando, FL

Truccolo W, Friehs GM, Donoghue JP, Hochberg LR (2008) Primary motor cortex tuning to intended movement kinematics in humans with tetraplegia. J Neurosci 28(5):1163–1178

Turner AE (2006) The efficacy of Adeli suit treatment in children with cerebral palsy. Dev Med Child Neurol 48:324–324

Waldie J, Bus B, Tanaka K, Tourbier D, Webb P, Jarvis W, Hargens A (2002) Compression under a mechanical counterpressure space suit glove. J Gravit Physiol 9(2):93–97

Waldie JM, Newman DJ (2010) A gravity loading countermeasure skinsuit. Acta Astronaut doi:10.1016/j.actaastro.2010.07.022

Webb P (1968) The space activity suit: an elastic leotard for extravehicular activity. Aerospace Med 39:376–383

Weinberg MS, Wall C, Robertsson J, O'Neil E, Sienko K, Fields R (2006) Tilt determination in mems inertial vestibular prosthesis. J Biomech Eng 128(6):943–56

Wolfrum N, Newman DJ, Bethke K (2006) An automatic procedure to map the skin strain field with application to advanced locomotion space suit design. J Biomech 39(1):S393

Wu J, Zhou D, Too CO, Wallace GG (2005) Conducting polymer coated lycra. Synth Met 155 (3):698–701

Young D, Stirling L, Chamberlin S, Weinberg M (2009) Inertial sensing for estimating human kinematics. In: 15th international conference on perception and action (ICPA). Minneapolis, MN

Young D, D'Orey S, Opperman R, Hainley C, Newman DJ (2010) Estimation of lower limb joint angles during walking using extended Kalman filtering. In: Proceedings of the 6th world congress on biomechanics, Singapore

Part III
Environmental and Commercial Scenarios

Chapter 12
Scenarios for the Interaction Between Personal Health Systems and Chronic Patients

Maria Teresa Arredondo, Sergio Guillén, I. Peinado, and G. Fico

12.1 Introduction

Rising life expectancy and declining birthrate are globally resulting in an aging population. In particular, the western world will be particularly affected by the aging of its population during the next years. According to a report issued by Eurostat – the Statistical Office of the European Communities, in the European Union (EU) the share of population aged 65 years and over is projected to rise from 17.1% in 2008 to 30.0% in 2060, and those aged 80 years and over is projected to rise from 4.4% to 12.1% over the same period (Giannakouris 2008).

Elderly population is usually affected by chronic diseases, which usually imply different degrees of disability, a reduction in quality of life and increased costs for health care and long-term care. The prevalence of chronic diseases represents a significant burden on the public health care and social systems. Aiming to relieve the public health care and social systems of this burden, the European Union 27 (EU27) has promoted different initiatives aiming to look for new ways to face the socio-sanitary challenges that stems from this new social situation. The EU27 initiatives aim to create a framework for developing a fair provision of services, oriented to provide an adequate quality of service, focused on patients and affordable for Public Administrations and social and health companies involved in the care process of chronic patients.

Information and communication technologies (ICT) is the key for the success of this new paradigm. ICT has the potential to impact almost every aspect of the health sector, and will play a pivotal role in the creation and development of a sustainable framework. ICT will facilitate the development of new solutions, the optimization of processes, better response times, and better use of resources. Therefore, ICT will help improving efficiency, efficacy, and costs of the health care process, which will lead to an improvement of the quality of the services provided.

M.T. Arredondo (✉)
Technical University of Madrid, Madrid, Spain
e-mail: mta@lst.tfo.upm.es

A. Bonfiglio and D. De Rossi (eds.), *Wearable Monitoring Systems*,
DOI 10.1007/978-1-4419-7384-9_12, © Springer Science+Business Media, LLC 2011

The widespread advances in ICT, sensing technologies and network technologies, have led to the emergence of a new paradigm for health care that some refer to as e-health. The term e-health has been widely discussed during the last years, but few people have come up with a clear definition of this comparatively new term. A special interest group of the Healthcare Information and Management System Society (HIMSS) developed in 2003 a definition of e-health and that aims to encompass all aspects related to the application of ICT in the health care process, and which they expected to be adopted industrywide. The special interest group's definition of e-health is: *The application of Internet and other related technologies in the healthcare industry to improve the access, efficiency, effectiveness and quality of clinical and business processes utilized by healthcare organizations, practitioners, patients and consumers in an effort to improve the health status of patients* (Broderick and Smaltz 2003).

Within this new paradigm, new visions have emerged, focusing on different aspects of the health care process. One of these visions is called p-health, and encompasses all methodologies and systems for the personalization of health care. p-Health places the patient as the center of the health care process, empowering them to take a greater responsibility in the management of their own health condition, while taking into consideration all aspects and stages of the life of the patients. This new holistic, citizen-centered vision of social and medical care implies the design and development of solutions that can be grouped under the global definition of personal health systems (PHS).

The emergence of PHS and the subsequent shift of responsibility from health care professionals to citizens will pose new challenges to the designers and developers of such systems. In particular, the discretionary use of breakthrough technologies within the health care domain will pose a complex challenge for human–computer interaction (HCI) designers, as PHS should guarantee a high degree of usability and a satisfactory user experience to achieve the patients' compliance and adherence to the treatment and medical protocols.

In this chapter, some of these challenges are described and some scenarios describing the use of personal health solutions are proposed. In Sect. 2, the most relevant characteristics and needs of patient-centered care and PHS are described. In Sect. 3, the technologies that will make possible the implementation of PHS are described. Chapter 4 describes the challenges that HCI designers will have to face to provide a satisfactory user experience. And Chap. 5 proposes some scenarios describing the use of personal health solutions to manage chronic conditions that may become real in the near future.

12.2 The New Paradigm of Personalized Health: p-Health

The widespread availability of processing power and advanced sensing technologies and the rapid advances in the field of ICT have made possible the development of applications that help providing a more efficient and continuous health care for patients suffering from chronic conditions. These systems facilitate the continuous assessment of chronic patients by the continuous monitoring of vital signs, the improvement of communication between patients and health professionals, the provision of educational and motivation content that helps patients to take a greater responsibility in their own health care and the involvement of all actors involved in the care process. Nevertheless, this new distribution of responsibilities requires a more extensive and holistic vision of the health care process that places the patients at its center. This new health care paradigm is named patient-centered care, and the technological solutions and methodologies that make possible the realization of the patient-centered care vision can be grouped under the name of personal health (p-health) systems.

12.2.1 Patient-Centered Care: Toward a Holistic Vision of Care

In February 2004, the Picker Institute convened the Patient-Centered Care Vision Summit (Bezold et al. 2004). Parting from for four scenarios that described possible future situations, 27 key leaders shared their vision on the future of patient-centered care. The Picker institute detected seven principal dimensions of patient-health care:

1. Respect for patient's values, preferences, and expressed needs
2. Coordination and integration of care
3. Emotional support
4. Information, communication, and education
5. Involvement of family and friends
6. Physical comfort
7. Involvement of family and friends

The dimensions defined by the Patient-Centered Care Vision Summit provide a framework to develop a holistic vision of the health care process. According to these dimensions, patients will become the center of the health care process. This vision has been previously adopted by Health care solutions that aim to fulfill the goals of the patient-centered care paradigm should provide the following functionalities:

- *Knowledge*: Well-informed patients will be able to take a greater responsibility in their own health care. These systems and solution should provide access to educational and reference material, with a proper level of quality and credibility and adapted to the patient's needs and goals. The knowledge base of the system should also be available for health professionals and caregivers. Moreover, patients, health professionals, and caregivers should not be only information consumers, as these systems should permit all people involved in the health care process to share information about how to cope with chronic diseases, including practices and recommendations. Therefore, empathy and emotional support can also be addressed, improving the communication between all actors involved in the health care process.

- *Communication*: One of the most important contributions of ICT to the field of health care is the possibility to close the loop between patients and health professionals. Communication between patients and health professionals will facilitate the continuous assessment of the patient's health condition, the early detection of acute episodes, the promotion of healthy behaviors, or the possibility of consultations between clinical visits. Nevertheless, patient–physician communication is not the only communication that should be provided. Health professional to health professional communication is also fundamental to achieve a unified care strategy, where all specialists and medical professionals treating a patient can have a secure and immediate access to all the information regarding the patient's treatment and global health condition. These systems should also facilitate the communication between patients, caregivers, and health professionals, allowing them to share information as described in the previous point.
- *Personalization*: Patient-centered care system should provide personalization of the content and the presentation of the information, so the systems can satisfy the patient's goals and expectations while providing an optimized feedback for any kind of patient, depending on their previous knowledge, skills, cultural factors, and health situations, among others. Empathy and emotional well-being are as important as evidence-based medicine in a holistic approach.
- *Decision support*: Patients should be helped to understand the decisions that health care providers make daily. Patients should also be informed of the control-points that exist in their individual care situations, while receiving guidance and advice on their treatments, on the possible consequences of any changes on the medication.
- *Quality*: Quality of health care depends strongly on the quality of the information provided. Quality issues in health care – decreasing medical errors, improving facilities scheduling, and enhancing the match between patient needs and health care providers' strengths – can all be addressed to a large degree by increasing the quality of information exchange.

Any system that provides these functionalities should fulfill the goal of placing the patient at the center of the health care process. The combination of this new paradigm and the technological advances leads to the emergence of a new vision for the provision of health care: the personalized care, also called p-health.

12.2.2 p-Health

Personalized health, or individualized health or "p-health," is an incipient vision of health. The advances in ICT, the research in genetics, pharmacology, nanotechnologies, sensors, etc., allow the patients to become the center of the health care process. The promotion of a healthy lifestyle and the education on health are regarded as valuable experiences, and virtual hospitals are created, based on the physical decentralization of services and the distribution of work using the net.

This shift of paradigm emphasizes the pivotal role of new technologies, but also the integration of different disciplines such as bio and nanotechnology, genetics, ambient intelligence (AmI), and domotics, aiming to achieve a medicine strongly based on the promotion of a healthy lifestyle and the prevention of the illness, based on scientific evidence and centered on the patient and their environment.

p-Health is a tool that allows medical professionals, social professionals, and citizens to perform a complete management of the patient's life and health, expanding the support of the social and medical systems further form the control of acute episodes, but avoiding the workload and expense of resources. Within this framework, an individual can be characterized by means of five main dimensions that represent the main pillars of well-being, and therefore will be the basis for developing new strategies for social and medical care through p-health:

1. Reinforcing the global management of the illness and risk, supporting a more efficient management of the patient's chronic condition and treatment, allowing the patients to keep a better health status and improving their physical condition.
2. Supporting the patient to achieve a healthier lifestyle that allows them to improve their quality of life while maintaining their independence and the patients' ability to take decisions about their health status and condition.
3. Promoting psychological well-being, following a holistic approach and proposing solutions to the personalized and specific needs in every moment, in every different phase of the illness and every specific situation.
4. Promoting of autoefficacy, abilities and potential of the patients to self-manage their condition, and to take control and responsibility on their health status, and providing the tools and means to achieve this in an effective, informed and healthy way.
5. Contributing to the proactive and positive reconstruction of the patient's environment following a negative event or relapse, supporting a life reengineering process that supports the patient to achieve an effective social rehabilitation of the patient as an active member of society.

The application of ICT for the personalization of health care for patients, citizens, their environment, and health professionals not only allows a disruptive change in the management of health care processes in an effective and sustainable way, but also increases the portfolio of services available to the users and improves the user–system interaction. Moreover, these new models of health care make possible to incorporate new agents in the model of provision of services, funding a reimbursement, strengthening the offer at patients' disposal, and distributing the expenses among a larger number of actors and beneficiaries.

Moreover, p-health allows the provision of services to healthy population. These days, the care services provided to healthy population is poor and quite generic and vague. p-Health services will have a disruptive effect in the management of risk conditions among this population, anticipating possible worsening health conditions, improving their quality of life, and improving the knowledge of the different illness by the health care system. Moreover, p-health solutions do not only serve patients individually, but also provide services to their near environment, including family and friends, shifting the provision of health care from the hands of health professionals only to the overall society.

The development of p-health systems implies the need for new functionalities that include the continuous monitoring of vital signs, as well as a comprehensive characterization of the context of the patient. This context includes physical factors, such as location and environmental conditions, and personal and social factors, including health and mood status and social support net. Patients should also be able to access all information regarding his health status and related educational content, as well as to communicate with health professionals, caregivers or peers, anywhere and at any time. These needs can be fulfilled within the technological paradigms of ubiquitous computing and AmI.

12.3   The AmI Vision

During the past 15 years, technology's focus has gradually been shifting away from the computer to the user. Many authors have envisioned a future, where computers are invisible and integrated into people's daily lives. The gradual miniaturization of electronic components and the increase in their processing power and storage capabilities now permit the development of portable stations at affordable prices and intelligent sensors, which can even be embedded in textile clothes or implanted. The development of wireless network technologies – WiFi and Bluetooth – and advanced mobile communication – 2.5G and 3G networks – make possible for each user to send and receive large amounts of information almost anywhere and at any time. And the development of intelligent and natural user interfaces has made possible that people with no previous experience in using interactive systems can interact with complex systems in an easy and efficient way.

The result of these converging trends is AmI. AmI is a new paradigm in information technology, in which people are empowered through a digital environment that is aware or their presence and context, and is sensitive, adaptive, and responsive to their needs, habits, gestures, and emotions (Davide 2002). The AmI vision is part of a much wider emerging technological trend that goes by the name of ubiquitous (or pervasive) computing and that aims to take full advantage of the relentless miniaturization and the convergence of consumer electronics, (wireless) networking and mobile communication.

The concept of AmI was introduced by the European's Commission's IST Advisory Group (ISTAG), a group whose mission is to advise the European Commission on the overall strategy to be followed in carrying out the IST thematic priority under the Framework Programmes for Research and Technological Development. According to ISTAG, AmI refers to a vision of the future of the information society stemming from the convergence of ubiquitous computing, ubiquitous communication, and intelligent user-friendly interfaces.

According to ISTAG vision statement, humans will, in an AmI environment, be surrounded by intelligent interfaces supported by computing and networking technology that is embedded in everyday objects, such as furniture, clothes, vehicles, roads and smart materials. AmI implies a seamless environment of computing, advanced networking technology, and specific interfaces. This environment should be aware of the specific characteristics of human presence and personalities; adapt to the needs of users; be capable of responding intelligently to spoken of gestured indications of desire; and even result in systems that are capable of engaging in intelligent dialog.

To fulfill the AmI vision goals, three major functionalities need to be developed: ubiquitous access, context awareness, and intelligent user interfaces. Ubiquitous access refers to the creation of extensive networks that guarantee seamless and trusted communication between a host of devices and support for ubiquitous access to communication, information, and services. The concepts of context awareness and intelligent user interfaces are described in the next sections.

12.3.1   Context Awareness

Context awareness refers to the idea that computers can both sense and react to information captured from their environment. To achieve its premises, context awareness and hence AmI environments should comprise a plethora of embedded sensors that would be used to capture information about the patient's identity and context.

One of the main problems that have slowed down the development of context aware applications has been the lack of a common definition and characterization of the term *context*. In the early work related to context awareness, context was related to the close physical environment around the system, and more specifically to *location* and *surroundings*. In the work that first used the concept "context awareness," Schilit and Theimer (1994) defined the term *context* as location, identities of nearby people and objects, and changes to those objects. In a similar definition, Brown et al. (Brown 1997) defined context as location, identities of the people around the use, the time of day, season, temperature, etc. Ryan et al. defined context as the user's location, environment, identity, and time.

The choice of context information used in applications is very often driven by the context acquisition mechanisms available. So, as technology evolved, it became possible to expand the definition of context, allowing a more comprehensive definition of the physical context and including information on human factors. Dey and Abowd (2000) proposed a definition of context that aimed to encompass the definitions given by previous authors. Later, Dey defined context as *any information that can be used to characterize the situation of entities (i.e. whether a person, place or object) that are considered relevant to the interaction between a user and an application, including the user and the application themselves. Context is typically the location, identity and state of people, groups and computational and physical objects* (Dey 2000). Therefore, the context is divided into four main categories: identity, location, time, and activity that are similar to the Who, Where, When, and What of an entity. Identity (*Who*) describes the user in its – almost – static properties: identity, physical and psychological conditions, social network, social roles, etc. Location (*Where*) describes the physical environment of the user, the user's coordinates, the surrounding assets (human or machines), environmental variables such as the weather, light condition, etc. Time (*When*) could be a simple timestamp or also include extra information such as the season, if it is holiday, etc. Finally, Activity (What) describes what the user is actually doing and assumptions about what the user is expected to do, whether explicitly or implicitly.

The role of sensors networks in a context-aware environment is to capture context signals that provide information aimed to characterize the four dimensions of context: the who (sensors able to track and identify entities), where (location sensors, environmental sensors), what (sensors able to recognize activities, interactions, spatiotemporal relations, and verbal and nonverbal communication recognition). As it is almost impossible to make a comprehensive list of all available AmI sensors, four main categories can be defined:

1. *Audiovisual sensors*: Traditionally, research has focused mainly on audiovisual observations as these modalities provide almost all signals that humans make use of in interpersonal communication. Most prominent in the visual domain are face detection and identification, person and object tracking, facial expression recognition, body posture recognition, attention direction sensing, and hand gesture recognition. For example, the Kira Cell Television manufactured by Toshiba includes a technology called 3D Motion Gesture Control, which allows users to control the TV through gestures. Audio has been used mainly to detect speech and identify and localize speakers, speech recognition and estimation of auditory features relevant to communication.
2. *Passive infrared sensors (PIRs)*: PIRs (sometimes called pyroelectric detectors) detect the infrared energy emitted by objects that generate heat (living organisms or machines), allowing to monitor changes in the environment. PIRs are used for motion detection.
3. *Radiofrequency identification (RFID)*: During the last years, the deployment of RFID tags (RFID) has become not just commonplace but ubiquitous. Passive RFID tags are small, flexible, and do not need external power to operate. They can be placed on everyday objects, woven into fabric and even implanted in persons and animals. RFID tags can be used to identify objects and persons that pass in the proximity. Tagging the users of an AmI system with RFID is also very useful for collecting ground data for face, gesture, body posture, and speech recognition applications.
4. *Intelligent sensors*: Thanks to the advances in solid state electronics it is now possible to develop miniaturized sensors and to integrate a variety of sensors

into daily use devices, such as furniture and home appliances, and embedded into textile clothes. The sensors can be broadly classified into three main groups:

(a) *Physiological sensors*: Ambulatory blood pressure, continuous glucose monitoring, core body temperature, blood oxygen, ECG, EEG
(b) *Biokinetic sensors*: Acceleration and angular rate of rotation
(c) *Ambient sensors*: Environmental phenomena, etc.

12.3.2 Intelligent User Interfaces

As in AmI systems the environment becomes the interface, all traditional inter-action techniques have become obsolete for designing the interaction within AmI systems. In most traditional systems, HCI was performed explicitly in a Graphical User Interface (GUI), through the direct manipulation of interaction widgets using input methods, such as the keyboard of mouse, and the output was provided through visual or audio feedback. For many years, the focus of user interaction techniques had been put on improving the usability of GUIs which, in most cases, followed the Windows, Icons, Menus and Pointing Devices (WIMP) paradigm. So far, the user interfaces of mobile devices have reproduced most of the characteristics and interaction modalities of the GUIs of traditional computers, still using the WIMP paradigm and traditional input methods (keyboard, stylus).

Nevertheless, in AmI environments there may not even be a GUI to interact with, and the user is likely to be interacting with a set of devices that are connected loosely across a network. Moreover, users may have to interact directly with physical objects, such as furniture and even clothes. Therefore, within this context the term HCI can be considered obsolete, and the term human–context interaction would be more accurate. As it is possible that within this context, the patient will not have any display to interact, new interaction modalities have to be provided. This new generation of user interfaces should support multiple modalities of interaction, as well as natural interaction. These interfaces should therefore be multimodal and natural.

According to W3C, multimodal interfaces allow users to dynamically select the most appropriate mode of interaction for their current needs, including any disabil-ities, while enabling developers to provide an effective user interface for whichever modes the user selects. Multimodal interfaces combine various input modes beyond the traditional keyboard and mouse input/output, such as speech, pen, touch, manual gestures, gaze, and head and body movements. The most common such interface combines a visual modality (e.g., a display, keyboard, and mouse) with a voice modality (speech recognition for input, speech synthesis, and recorded audio mes-sages for output). However, other modalities such as pen-based input or haptic input/output may be used. The advantage of multiple input modalities improves the accessibility of the user interfaces. Not only for users with disabilities, but for users "situationally impaired," such as a user trying to talk on a mobile phone in a noisy environment, or driving in a car and unable to dial a number or take a call.

Natural user interfaces are user interfaces that are effectively invisible to the users, and that interact with the users in a similar way to the ways human interact with other humans. Multitouch interaction allows users to move objects through a display in the same way they move objects in the real world. The development of high definition cameras, accelerometers, and microphones make possible the recognition of gestures and movements of the users. For instance, the Wii developed by Nintendo allows the reproduction of the user's movements through the accelerometers installed in its WiiMote (Schlömer, 2008). Microsoft has also developed Project Natal for its Xbox 360 game console. This console will include a bar that includes a camera and a microphone that register every movement and utterance the user makes. The system is capable of recognizing the distance between the user and itself, and can interpret and analyze each movement of the user and their orientation in 3D. The system comprises a RGB camera (able to perform face recognition), a depth sensor, a multidirectional microphone and recognition software. These examples just illustrate that natural interfaces have already become a reality, and their use for health care purposes open a whole world a possibilities.

Nevertheless, it will take awhile until discretionary users get used to interact with natural and multimodal user interfaces. Classic user interaction design techniques have become obsolete within this new paradigm, and new considerations and problems arise with these new interaction modalities, such as the need to avoid the "big brother effect" or "ubicomp dystopia" associated with context-aware and AmI environments.

12.4 Challenges of User Interaction Within the Patient-Centered Care Paradigm

The way people interact with interactive device is vital for the success of such devices. This aspect becomes critical when dealing with personal health devices used for managing chronic condition, as a poor user experience can lead to the lack of adherence and compliance to the treatment and medical protocols, even putting the patient's life at stake.

Interaction techniques are limited by technology available. So far, telemedicine and e-health applications have been based on pre-existing technologies, such as the WWW, Web cameras and communication capabilities, with a little participation of the patient. Nevertheless, the widespread advances in mobile technologies and sensing technologies have shifted the paradigm from ambulatory and reactive care to proactive and preventive care, providing patients with the possibility to participate in their own health care. But a bigger participation also involves a bigger responsibility, and therefore PHS should provide adaptation capabilities to adapt to the patients' preferences and goals, prevention and minimization errors, and a satisfactory user experience that motivates patients to comply with using the system.

Traditional user interaction design techniques and methodologies are becoming obsolete when designing for these new environments described in the previous section. New concepts and methodologies have emerged aiming to capture and solve the interaction needs arisen from the new paradigms of ubiquitous computing and AmI. Among others, the concept of "implicit interaction" appeared, as opposed to the traditional explicit experience that required the willingness of users to start the interaction and expecting a response that satisfies a goal or need. In the next section, a definition of HCI will be provided. Then, traditional HCI methodologies will be described and then, the paradigm of implicit HCI will be introduced.

12.4.1 What is HCI

In a broad sense, HCI can be defined as the discipline that studies the exchange of communication between people and computers. HCI methodologies aim at improving the effectiveness of this exchange of information: minimize errors, increase satisfaction, decrease frustration and, in a nutshell, to make all tasks that involve people and computers more easy and effective. To sum up, the fundamental objective of HCI research is to make systems more usable, more useful, and to provide users with experiences that fit their specific background knowledge and goals.

HCI is a multidisciplinary field that covers many areas (Hewett et al. 1996). These areas include Applied Psychology, Ergonomics, Graphical and Industrial Design, Etniography, Artificial Intelligence and/or Software Design, among others.

HCI is a young discipline. Since the advent of the first computers, it has been necessary to design a human–computer communication system. Nevertheless, it took some time for HCI to become a relevant field of research. The first specific studies related to HCI referred to human–computer symbiosis (Licklider 1960). This author anticipated that the main problem of HCI was not to create computers that just provide answers, but computers that are able to anticipate and participate in the formulation of the questions. The first journals, conferences, and professional associations related to HCI appeared in the late 1970s and 1980s.

In the first 10–15 years of its history, HCI focused mainly on the possibilities and design criteria for GUIs using the Windows, Icons, Menus and Pointing devices (WIMP) paradigm developed in the first personal computers, and aiming at facilitating the tasks of professionals using a workstation. As GUI problems were better understood and the widespread advances in ICT allowed different forms of communication between humans and interactive systems beyond the keyboard, mouse and screen, the primary HCI concerns stated to shift its focus from usability and ease of use to more complex concepts, such as task completion, shared understanding, and prediction and response to user's actions. The new essential challenges are improving the way people use computers to work, think, communicate, learn, critique, explain, argue, debate, observe, decide, calculate, simulate, and design. (Fischer 2001).

As mobile computing becomes the prevalent computer usage paradigm, the range of challenges facing HCI practitioner increases. So far, the communication between human and computers usually happened in an explicit way. Keeping in mind current and upcoming technologies described before, as well as the discretionary use of these technologies, will start the shift from explicit HCI.

12.4.2   Traditional HCI Models

One of the first HCI models proposed in the literature was the basic model for HCI proposed by Schomaker (Schomaker et al. 1995). This model proposes a taxonomy that defines the most relevant components of an interaction model. This model assumes that there are at least two separate agents involved, one human and one computer that are physically remote but able to exchange information through different information channels. On the human side, two basic processes are involved: perception and control. Regarding the perceptive process, two different information channels can be identified: human input channels (HICs) and computer output modalities (COMs). Within the control process, two different channels can be identified: human output channels (HOCs) and computer input modalities (CIMs). Within the intermediate cognitive level lies a cognitive or computational component. This component handles the incoming input information and prepares the output. The Schomaker model is shown in the next figure.

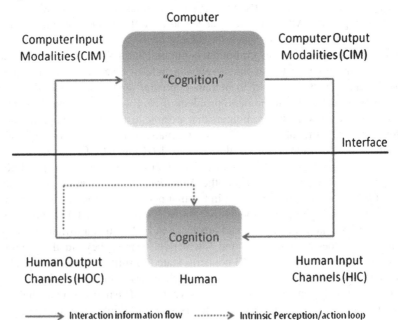

The Schomaker model is useful for describing the main processes that take place during the interaction between a man and a machine from a traditional point of view, but does not take into account some aspects that are fundamental in the new scenarios that arise from the advances in mobile technologies and artificial intelligence. For example, it does not take into account environmental factors that may influence the interaction, nor the physical characteristics of the user. Some of these factors are considered in the model presented by the Human Factors Research and Engineering Group from the Federal Aviation Administration.

This model is adapted from David Meister's "Human Factors: Theory and Practice (1971)" and illustrates a typical information flow between the "human" and "machine" components of a system.

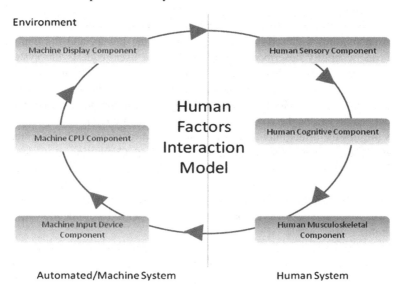

Unlike the Shoemaker model, this model takes into account the different factors that affect each of its components, allowing, for example, a more detailed characterization and modeling of both the "human" and the "machine" and the environment, and supporting different modalities of interaction in different environments. Nevertheless, this model is not completely valid to describe the HCI within the new paradigm of AmI, as it assumes implicitly that every interaction between the human and the system implies an explicit intention of the user (explicit interaction), which may not always happen in Ambient Intelligent environments. Thus, the concept of implicit interaction has recently emerged as opposite to explicit interaction. Implicit interaction will be described in the next section.

12.4.3 Implicit Interaction

When two people are communicating face to face, some of the communication is verbal and some nonverbal. Although it is not still very clear, many studies have tried to calculate the relative importance of verbal and nonverbal communication within the communication process. One of the most cited studies, developed by Argyle in 1970, stated that nonverbal cues had 4.3 times the effect of verbal cues (Argyle et al. 1970). According to Argyle, the most important effect was that body posture communicated superior status very effectively. Although the results of this study have been refuted and discussed by several subsequent studies, it is worthy to remark the importance of nonverbal communication, not only between two human interlocutors, but also between humans and interactive systems too.

As described in the previous section, traditional HCI is based on the explicit exchange of information between a human and an interactive system. This explicit interaction requires a closed-loop dialog between the human and the computer, which can be briefly described as follows: (1) the user requests the system to perform a certain action, then (2) the system processes the command, and finally (3) the system generates a response that aims to satisfy the patient's goals as adequately as possible. Moreover, the exchange of information could be initiated by either the human or the computer. Nevertheless, this communication modality always implies the direct intention of the user to communicate with the system.

Within the new paradigms of AmI and Ubiquitous Computing, it is clear that new HCI models are needed. These models should consider two different types of interaction: one where the user interacts consciously and explicitly with a software application (the traditional, explicit interaction), and a new kind of interaction where the user interacts unconsciously or implicitly with the application (implicit interaction).

Albert Schmidt defined the implicit user interaction as *an action, performed by the user that is not primarily aimed to interact with a computerized system but which such a system understands as input* (Schmidt 2000). Later in 2005, Schmidt proposed his abstract model that aimed to support the creation of systems that use iHCI. The model is shown in the next figure:

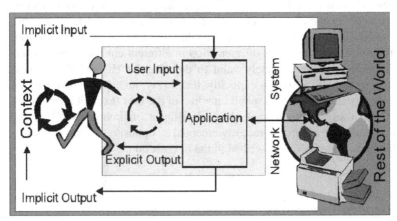

The basic idea of implicit interaction implies that the system can perceive the user interaction with the physical environment and also the context, where the interaction takes place. The basic premise is that iHCI allows a transparent interaction with computer systems, turning the environment into the interface. Any interaction may have an element of both the explicit and the implicit modalities. For example, consider the case where a user encounters some difficulty with a software application. In some cases, but certainly not at all, the application monitors the user activities. On noticing, a pause that extends beyond a predefined threshold, the application may proactively intervene and ask the users whether they need any assistance. Thus the users, by the length of time they take contemplating their next course of action, implicitly signal to the application that they might be experiencing some difficulty.

Therefore, it could be stated that Implicit HCI is based on two fundamental concepts: perception and interpretation. In all but the simplest cases, implicit interaction is multimodal. It may require the parallel capture of distinct modalities, such as audio and gesture. For example in the case of the audio modality, semantic meaning and emotional characteristics may be extracted in effort to develop a deeper meaning of the interaction.

All applications that implement iHCI should be able to:

1. Perceive use, the environment (context) and the circumstances
2. Develop mechanisms to understand and interpret what the sensors see and hear.
3. Generate an appropriate response depending on the information interpreted by the system.

12.5 Scenarios for the Application of AmI to p-Health

This section proposes three scenarios that describe a day in the life of people with chronic patients using a personal health application. The chronic diseases that will be addressed are heart failure, diabetes, and bipolar disorder. Different projects funded by the European Commission are currently developing applications for the management of these chronic conditions. The main objective of these projects is to develop personal health applications that aim to empower the patients to take greater responsibility in the management of their health condition and to close the loop between them and health professionals. These projects have also in common that they put focus on patients since the first stages of the development process, involving patients in the decisions following the principles of user-centered design.

12.5.1 AmI for Patients with Heart Failure

Chronic heart failure (CHF) is a chronic condition that prevents the heart from pumping enough blood to meet the demands of the human body. Patients who suffer from CHF experience a dramatic decrease in their quality of life, as CHF is associated with a terrible burden of symptoms that include breathlessness, general tiredness – even at rest – and swelling of the feet and ankles. CHF patients have also to stick to a complex medication regime. A lack of compliance to the medication regime can lead to a decompensation and a subsequent hospitalization or even death. Furthermore, CHF affects mainly people over 65 years old and is commonly associated with several comorbidities, such as diabetes, sleep disorders, and chronic renal failure. Unlike most cardiovascular diseases, CHF is becoming more common, and nearly one million new cases are diagnosed every year worldwide, making it the most rapidly growing cardiovascular disorder.

12.5.1.1   Current Research in the EU

The European Commission has funded different projects that address the challenge of the prevalence of cardiovascular diseases in the EU. Within the 6th Framework Program, the EU funded the project MyHeart (IST-2002-507816) (MyHeart 2002), whose objective was to fight cardiovascular diseases through preventive lifestyle and early diagnosis. One of the main innovations within the MyHeart project was the use of wearable sensors that made possible the continuous monitoring of vital signs, such as the ECG. The MyHeart project aimed to provide solutions for cardiovascular risks and problems in all the stages in the life of patients, and therefore the consortium developed four different products focusing on different stages of cardiovascular status: healthy population (Activity Coach), patients at risk (TakeCare), patients under rehabilitation (NeuroRehab), and CHF patients (Heart Failure Management).

The HFM system comprises a set of sensors that included a weigh scale, a blood pressure cuff, a bed monitoring equipment, and a wearable sensor to measure their ECG, and a PDA that provided instructions and clinical questionnaires for feedback. The patients were compelled to follow a routine adapted to the patients' lifestyle and habits. The HFM PDA was developed following the principles of user-centered design, and several patients were interviewed during the different phases of the development process to refine the define process. The usability and usability tests performed during the different development phases showed that patients considered the system as useful and easy to use (Villalba 2008). During the final validation phase, a pilot involving 120 patients has been performed in different hospitals of the EU, with acceptable results in terms of usability and user acceptability.

Nevertheless, although the monitoring functionality of the HFM system was widely appreciated, the interviews with the patients showed that there were needs that still needed to be addressed and were not supported by the HFM system. The results of these confrontations with patients served as the starting point of the proyect HeartCycle (FP7-216695) (HeartCycle 2007).

HeartCycle aims to effectively close the loop between patients and health professionals. Hell professionals will receive relevant monitoring data so they can prescribe personalized therapies and lifestyle recommendations. Patients will be empowered to get involved in the daily management of their disease through the provision of education and the use of motivational strategies, aiming to guarantee the patients' compliance with their treatment and medical protocols. The user platform in development will have context-aware capabilities, adapting its response and feedback to the patient depending on usage patterns and environmental information (implicit) and response to clinical and motivational questionnaires (explicit).

12.5.1.2 Future Scenarios for the Care of Chronic Heart Patients

Peter is 73 years old. Almost 1 year ago, he was diagnosed with heart failure after he suffered a decompensation and had to be hospitalized. After the diagnosis, the cardiologist proposed Peter to join a novel heart failure management program that would allow hospital staff to remotely assess his health status. Peter said yes immediately, and then his cardiologist asked him to fill some questionnaires. The questionnaires comprised questions on his knowledge about heart failure and heart functioning and some questions about his personal routines. After Peter filled the questionnaires, his cardiologist configured a personal mobile device, and gave it to Peter together with a briefcase that includes a set of sensors. Then, Peter went home and the next day he started using the system.

It has been almost a year since that. Now, it is 9:00 a.m. and Peter's lights start to slowly turn on. Beside his bed, his personal mobile device is also flashing. The night monitoring system has noticed that Peter did not sleep well last night, so the personal mobile device asks him a couple of questions to find out why Peter did not sleep well. The system also noticed that Peter had to wake up to go to the bathroom a couple of times, so it makes a couple of questions about the diuretics Peter had to take yesterday. After finishing his questionnaires, Peter goes to the bathroom. Once in the bathroom, Peter steps on his bathroom rug, which automatically detects his weight. He stands in front of the mirror, where he can see his last weight measurement and which also displays a text reminding him of the medicines he has to take before he has breakfast. The images displayed in the mirror points him to the cabinet where he stores his medicines, and reminds him to reduce his dose of diuretics this morning. Peter confirms that he will reduce his dose by nodding in front of the mirror. Then, Peter opens his medicine cabinet, where a small screen displays the list of medicines he has to take this morning, and confirms he has taken them all. After finishing breakfast, Peter puts on the T-shirt he got with the system and his personal mobile device and goes out to run some errands. Once he is running his errands, Peter starts feeling a little tired. The system, which has detected Peter's tiredness, then beeps and suggests Peter some places nearby where he can sit and rest for some minutes. Peter goes to a park nearby and, while he is sitting in a bench, he searches through the educational material stored in his personal mobile device to find out some reasons for his tiredness. After reading some tips and advice, he fills a brief questionnaire describing his symptoms. Later that night, Peter receives a message from his cardiologist. The staff from the hospital has been revising Peter's data for the last weeks and has noticed nothing strange or wrong, but in case he is feeling worried, the cardiologist fixes an appointment for the day after.

12.5.2 AmI for Patients with Diabetes

Diabetes mellitus is a chronic metabolic disturbance that has no cure. It is characterized by increased blood glucose concentrations and decreased insulin secretion and/or action. All these clinical manifestations of diabetes are linked to a poor glycemic control, and the risk of developing chronic complications can be greatly reduced by restoring normal or near-normal blood glucose levels. However, the intensive use of glucose-lowering medications (especially insulin) aiming to optimal metabolic control leads to a significant increase in episodes of hypoglycemia (that may lead to cognitive dysfunction, de-coordination, memory loss, seizures, and sudden loss of consciousness, and even coma). Diabetes can affect almost every organ system of the body. These include effects on the heart, brain, kidneys, eyes, stomach, bowel, bladder, sexual organs, peripheral nerves, and other organs.

Diabetes has become a worldwide epidemic, with increasing costs for health and welfare system and for the community. It affects approximately a 5% of population in developed countries (150 million people currently and 300 in 2025). High percentage of sick people remains still without diagnosis. The International Diabetes Federation (IDF) estimates that the equivalent of an additional 23 million years of life are lost each year to the disability and to the reduced quality of life caused by the preventable complications of diabetes. In Europe, IDF estimates (IDF Diabetes Atlas 2007) that the yearly cost of diabetes is around 67 million international dollars.

There are two major types of diabetes: Type I diabetes mellitus (T1DM-ID) and type 2 diabetes mellitus (T2DM). T1DM is caused by an absolute deficiency of insulin secretion. The most distinctive aspects of T1DM are as follows: (1) the patients lack pancreatic secretion of insulin and depend on exogenous insulin injection; (2) the main risk factors are autoimmune, genetic, and environmental; (3) it is usually diagnosed in children and young adults; and (4) it represents between 5% and 10% of the total diabetic population. T2DM is a much more prevalent category (90–95%). The most distinctive aspects of T2DM are: (1) patients have an endogen secretion of insulin, but insufficient and (2) it usually appears in overweighed people over 40 years old, though it has been detected a "new" tendency to appear in overweighed children and young people.

The complexity of diabetes pathophysiology and the relatively high number of environmental factors influencing daily metabolic profile do not allow to clearly define a "single" diabetic patient and, in turn, a "single" scope. As a consequence of the diversity in the diabetes conditions among different types of patients and the different pathologies associated with the disease, the difficulty to control all the variables playing a role in the glycemic levels of patients is high and requires a high level of commitment from patients to follow adequately the prescriptions from the doctors, as well as a high level of understanding of the different factors susceptible of affecting their glycemic levels.

12.5.2.1 Current Research in the EU

METABO (2007) is an ICT research project partially financed by the European Commission under the seventh framework program. METABO project tries to define a situation/problem-oriented vision of the disease, to focus its features and try to address most of the issues the diabetic patients have to deal with while treating the disease.

Diabetes treatment requires a control over a compound of clinical and nonclinical variables in the patients. For this, laboratory analysis and clinical measurements are done periodically, as well as some information is asked to the patient on their daily activities and food intake regime and capillary glycemic levels (these will be asked in more or less detail depending on the patients diagnoses). Based on all this information, the treatment for any patient will be tuned by the doctor to adapt it to each patient's condition but it will always be based on four aspects: medication, diet, physical activity, and lifestyle changes.

12.5.2.2 Future Scenarios for the Care of Chronic Diabetes Patients

Ana is a 59-year-old engineer, independent, living alone. She goes to the office every day and works in her computer and in the lab for the experiments she carries out. She has been a type 2 diabetic for 6 years. She is a happy person and has an intense social life, which includes traveling in the weekends to visit friends (homeland or abroad), going to concerts, theaters enjoying a drink or a coffee every other afternoon. Some time ago, Ana used to keep a diary where she noted her meals and measurements.

During her daily activities, she controls the monitoring data through a handheld device and receives instructions for corrections in lifestyle and/or treatment. Through this closed short-loop, Ana makes significant changes in lifestyle and treatment and the average glycemic control improves. In addition, the system alerts her when there is a high possibility of developing a hypoglycemic or hyperglycemic episode in the next few minutes/hours suggesting possible short-term corrections. After using the METABO system, she learnt that exercise affects her during a couple of days and she manages her insulin doses accordingly. Sometimes, it is hard because changing one aspect of her treatment, such as food, insulin doses, activity, or schedules, will affect her for a couple of weeks more or less until she adapts to those changes, but METABO was crucial for her to understand that the only key to succeed and be healthy was in her hands and was her self-management.

The METABO system recorded several episodes of initial hypoglycemia and several warnings were sent to the patient during the last period of recording. In a few occasions, the hypoglycemic episode was only perceived by the system as the patient was without symptoms (unaware hypoglycemia). Ana George needs to travel long distances by car almost every day because of her work. Since she wants to change her car, she decides to buy a METABO-compliant car to have a constant hypoglycemia monitoring while driving. After registration of his ID,

Ana starts using the car every day. The car alerts her of several possible hypoglycemic episodes in the following weeks. Anytime she is alerted, Ana pulls the car over and measures her blood glucose concentration with a glucometer, then, if blood glucose is into a hypoglycemic range, she takes some sugar, feeds the information back to the system and continues her journey. After a few weeks, the system improves significantly its capacity to identify with accuracy when Ana is actually in hypoglycemia (short patient-system loop).

Meanwhile, the medical team is informed of frequency and pattern of hypoglycemic episodes, and, at the next visit, George, the physician and the dietitian agree on eating a small snack in case of long driving in the afternoon. George's hypoglycemic episodes decrease significantly in the following months and the result is achieved without loss of the optimal degree of glycemic control obtained using the wearable METABO platform (long patient-system-caregiver loop).

When she needs to travel long distances by car she uses a METABO-compliant car to have a constant hypoglycemia monitoring while driving.

12.5.3   AmI for Patients with Mental Disorders (Bipolar Disorder)

Bipolar disorder (also called manic-depressive disorder) is a psychiatric illness that falls under the category of mood disorders. Patients suffering from bipolar disorder experience dramatic mood swings, going through repetitive phases of depression alternated with episodes of euphoria (manic phases). Although bipolar disorders have been usually identified with the classic manic-depressive illness, research suggests that there are more conditions that fall under the bipolar spectrum. These condition disorders may include bipolar I (the classic manic-depressive illness), bipolar II, cyclothymia, and other related conditions such as the schizoaffective disorder and borderline personality disorders, among others. All these conditions range significatively in symptoms and severity. International studies about lifetime prevalence have stated that approximately 8% of the population may have a bipolar spectrum disorder over the course of their lifetime. What is more, international studies indicate lifetime prevalence rates for all mood disorders, including depression, range from 7% to 19% (Brondolo 2007). Moreover, bipolar disorder comes usually with more symptoms that just mood swings, and may include problems with information processing, problems with sleep quality, or difficulties with mood regulation and stability. Therefore, patients suffering from bipolar spectrum disorders experience serious impairments to lead a normal life.

Bipolar spectrum disorders are neither easy to diagnose nor easy to treat. There are still divergences on the symptoms and their severity that allow us to classify the patients in one of the previously cited groups. Moreover, these patients are usually not compliant to the medical treatment and protocols, especially when they are entering the manic phase. The unpredictable and episodic nature of bipolar spectrum disorders makes it also difficult to adjust the treatments to prevent the onset of a depressive or manic episode. Therefore, within the field of mood disorders it is needed to develop solutions and systems that provide continuous monitoring, patient participation, and medical prediction.

12.5.3.1 Current Research in the EU

The PSYCHE project (PSYCHE 2010) aims to develop a personalized health system for the long- and short-term acquisition of data from selected class of patients affected by mood disorders. This project comprises a set of sensors that will monitor biochemical markers, voice analysis, and a behavioral index correlated with the patient status. All the data gathered by these sensors will be further processed in a remote server and sent to the physicians, who will be able to verify the diagnosis and help in prognosis of the illness. Finally, communication and feedback to the patient will be given through a closed-loop approach that aims to empower the patients to take responsibility in the self-management of their disease.

12.5.3.2 Future Scenarios for the Care of Bipolar Disorders Patients

Julie is now 26 years old. She was diagnosed with bipolar II when she was 16 years old. This last week, Julie has been fatigued most of the time, she has felt low and had a great deal of difficulty concentrating on her work. The night monitoring system she had recently installed has detected some disturbances in her circadian rhythms. The voice detection system installed in her mobile phone has also detected that her voice tone has become more linear during the last week, and the environmental sensors installed at her home have also detected that she has spent more time than usual lying on her sofa in front of the TV. This morning, as Julie is watching TV, a warning appears in one of the corners of the TV. Then she agrees to enter a private channel where she reads some suggestions on some concentration exercises she can perform now. She does not feel like doing the exercise now, so she declines. Then, the system proposes her some activities she could do that afternoon. As she defined herself as a music lover when she filled in her profile at her first visit to her psychiatrist, the system recommends her she could go to a concert of a band she loves in a venue nearby. She then confirms her interest, and immediately the system shows a list of people who are also using the system and who may be interested in attending the same concert, according to their profiles. She notices that her acquaintance Arnaud is now online, so she asks him whether he would like to go with her to the concert. They set up a rendezvous time and then she decides to turn the TV off. Before shutting off, the system reminds her to fill her diary, as she has not filled it for the last couple of days. That afternoon, Julie takes her personal mobile device and heads for the concert 1 h early. When she is near the venue, the system alerts Julie that her friend Elodie is in a cafeteria nearby, so she decides to stop by and say hello.

12.6 Conclusions

The objective of the three scenarios described is to show the advantages and possibilities associated with the continuous monitoring of vital signs and the use of ambient intelligent environments for the treatment of chronic diseases. PHS can take the most of the rapid advances in sensing technologies and user interaction modalities, as they allow the continuous monitoring, the provision of feedback, and the improvement of the communication between patients, relative, and health professionals in a noninvasive, easy, and effective way. As stated in the previous sections, there is still a huge work in progress for developing PHS, both from Public Administrations, such as the European Community, and from private companies. Nevertheless, there are still many issues that should be addressed to assure the patient's acceptance of this system, especially in terms of invisibility of the sensors and privacy.

References

Argyle M, Salter V, Nicholson H, Williams M, Burgess P (1970) The communication of inferior and superior attitudes by verbal and non-verbal signals. Br J Soc Clin Psychol 9:222–231

Bezold CP, Peck J, Rowley W, Rhea M (2004) The Institute for Alternative Futures on behalf of The Picker Institute. Patient-centered care 2015: Scenarios, vision, goals & next steps. Camden, ME: The Picker Institute; July 2004 Available on: 174.120.202.186/~pickerin/wp-content/uploads/.../PCC-2015.pdf Visited October, 12, 2010

Broderick MS, Smaltz D (2003) HIMSS e-health white paper: e-health defined. Healthcare Information and Management Systems Society, Chicago, IL

Brondolo, Elizabeth and Amador, Xavier (2007). Break the Bipolar Cycle: A Day-by-Day Guide to Living with Bipolar Disorder. McGraw-Hill. ISBN 978-0-07-148153-3

Brown PB (1997) Context-aware applications: from the laboratory to the marketplace. IEEE Pers Comm 4:58–64

Davide FL (2002) Communication through virtual technologies. In: Gregori EA (ed.) Advances lectures on networking. Springer, pp 124–154

Dey AK, Abowd G (2000) Towards a better understandint of context and context-awareness. Workshop on the what, who, where, when and how of context-awareness, affiliated with the CHI 2000 conference on human factors in computer systems. ACM, New York

Dey AA (2000) Towards a better understanding of context and context-awareness. Proceedings of the workshop on the what, who, where, when and how of context-awareness, affiliated with the CHI 2000 conference on human factors in computer systems. ACM, New York

European Commission. Information Society Technologies Program. HeartCycle project. Compliance and effectiveness in HF and CHD closed-loop management. ICT-216695. http://heartcycle.med.auth.gr/. Visited October 12, 2010

European Commission. Information Society Technologies Program. METABO project. Chronic diseases related to metabolic disorders. ICT-26270. www.metabo-eu.org. Visited October 12, 2010

European Commission. Information Society Technologies Program. PSYCHE project. Personalised monitoring SYstems for Care in mental HEalth. ICT-247777. http://www.psyche-project.org/. Visited October 12, 2010

Fischer G (2001) User modeling in human-computer interaction. User Model User-Adapt Interact 11:65–86

Giannakouris, K. (2008), Ageing Characterises the Demographic Perspectives of the European Societies. Statistics in Focus 72, pp. 1–12.

Hewett,T.T.Baecker,R.,Card,S.,Carey,T.,Gasen,J.,Mantei,M.,Perlman,G.,Strong,G.&Verplank (1996) Chapter 2: human-computer interaction. ACM SIGCHI curricula for human-computer interaction. Association for Computing Machinery, New York

International Diabetes Federation. IDF Diabetes Atlas (2007), 3th edn. Brussels, Belgium: International Diabetes Federation

Licklider J (1960) Man-computer symbiosis. IRE Trans Hum Fact Electron 1:4–11

Schilit BT, Theimer MM (1994) Disseminating active map information to mobile hosts. IEEE Netw 8:22–32

Schlömer TPB (2008) Gesture recognition with a Wii controller. In: Proceedings of the second international conference on tangible and embedded interaction (TEI 08), Bonn, pp 11–14

Schmidt A (2000) Implicit human computer interaction through context. In: London S (ed) Personal and ubiquitous computing, vol. 4 (2&3), pp 191–199

Schomaker L, Münch S, Hartung K (1995) A taxonomy of multimodal interaction in the human information processing system. Technical Report, Nijmegen Institue of Cognition and Information, The Netherlands, Miami

Villalba Mora, Elena (2008) Holistic interaction model for peoble living with a chronic disease. Thesis (Doctorate), E.T.S.I. Telecomunicación (UPM)

Chapter 13
The Commercialization of Smart Fabrics: Intelligent Textiles

George Kotrotsios and Jean Luprano

What is the relationship between hearing aids, corrective lenses, and smart textiles[1]? At first glance, none – but one thread, often invisible– is revealing in terms of the future of smart textiles. Hearing aids address a niche market, albeit a large niche – those persons (mostly elderly) suffering from what we still call today "impairment" – a hearing impairment. But is it not the same for corrective lenses? Can a young person or adolescent be considered an "impaired person"? Today, both corrective lenses and hearing aids improve human body functioning: corrective lenses are mainstream products that correct basic functions; and hearing aids are on the way to becoming mainstream products.

Extrapolating from this situation to smart textiles, one can see a similar, global trend developing. Today, smart textiles address niche markets, with relatively small market penetration to begin with, but tomorrow they will cover mainstream markets and help to improve the quality of life of continuously growing – and ageing – populations.

The underlying technology of smart textiles is multifaceted and integrating. It is best perceived as a platform that aggregates technological breakthroughs coming from fields as diverse as textile engineering, nanotechnology, microsystems, polymers and displays, communication engineering (including ad hoc protocols), multisensor data fusion, and many others. Each one of the aforementioned technologies brings unique features, while continuously added features reveal new applications and new benefits for the end-user, and therefore new business opportunities.

13.1 Analysis of the Markets: Today and Tomorrow

This analysis is an attempt to categorize the markets and applications for smart textiles today and to predict their evolution tomorrow. A secondary function of this

---

[1]We are going to use the terminology Smart Textiles to refer to Smart Fabrics and Intelligent Textiles; within this chapter, the understanding and the meaning of the terminology will be discussed.

G. Kotrotsios (✉)
CSEM SA, Jaquet Droz 1, 2002 Neuchâtel, Switzerland
e-mail: georges.kotrotsios@csem.ch

A. Bonfiglio and D. De Rossi (eds.), *Wearable Monitoring Systems*,
DOI 10.1007/978-1-4419-7384-9_13, © Springer Science+Business Media, LLC 2011

text is to give some background and clues to understand what the initial niche markets might be, and the advantages and drawbacks of each market segment. The principal objective is to understand which factors will accelerate (or decelerate) the transition toward the mainstream markets.

In this analysis, we address each one of the Smart Fabric, Interactive Textile (SFIT[2]) market sub-segments, as well as their nascent structures. An important objective of this article is to understand the value-generation processes and the interaction of the interests of the different parties. This understanding is the basis for identifying commonalities and levers that can contribute to the acceleration of the aforementioned transition. A secondary objective is to understand the potential of the "smartness" in terms of opportunities that never previously existed, and to try to foresee the directions in which such opportunities could evolve.

Such an endeavor is continuously evolving, and will continue to do so, because this integrating technology enables brand new services. The generation and evolution of such services depend on the overall technological and business climate. Let us take the example of telemedicine: will smart textiles be used for telemedicine services? Probably, but acceptance depends on non-technological factors, such as the international regulatory environment, liability of harmonization policies, and cross-country agreement on social security reimbursement.

13.1.1 What is a Smart Textile, as Seen from the Technology Perspective?

For simplicity, so far in this text we have used the terminology "smart textiles". The community which developed this domain usually applies the terminology Smart Fabrics and Interactive Textiles (SFIT). What is it? This question seems innocent enough. However, it is not so simple. We will explore this question further below.

The term "smart fabrics" relates to the behavior of the fibre, the yarn and the fabric itself: it has to do with the first three links of the value chain. Adding "smartness" at this level means modifying the reactivity of the material and even making some of this material part of a "programmed" machine, a microprocessor or a network of microprocessors. In contrast, the term "interactive textiles" usually denotes the capability of a system to integrate sensors, processors, and possibly also actuators, and to behave in a programmed way.

For instance, we can define as a "smart fabric" a device that changes color (using the properties of a nano-coating) in the presence of methane; this can be very interesting for security clothing in mining environments for example. We can qualify as an "interactive textile" a wearable device that concentrates information from multiple sensors, processes the signal using a microprocessor (or several microprocessors), and informs, for example, the emergency services of the health status of the wearer.

---

[2] www.csem.ch/sfit/html/background.html

Why is this distinction important? Because it underlines the increasing complexity of the already complex value chain, with the addition of new players that add "smartness" or "intelligence". The question is how these new players in the value chain can shape credible business models and which of them will endure and dominate. We will still use the terminology "smart textiles", understanding them as the overall domain of SFIT.

13.1.2   What is a Smart Textile Seen from the User's Perspective?

The definition of a smart textile, usually in terms of its function, varies considerably from source to source. Usually, the definition is determined by the function of the materials (e.g., carbon-nanotube-based materials), embedded intelligence, sensing capabilities, etc.

An important factor concerns what the function offers in terms of, e.g., sensing, thinking, etc., and consequently what added value is brought to the user.

The role of smart textiles as a technological support for the "augmented person" of the future is becoming today apparent as common denominator. In a very similar way that corrective lenses improve optical acuity, smart textiles can be seen as tools to improve functions such as:

- Perception of the environment and contextual awareness
- Monitoring of human health status
- Generation of energy (e.g., energy harvesting)
- Adding cognitive capabilities
- Interacting and interfacing
- Increasing human functioning

Functions performed by "wearables" were not anticipated two decades ago. It began with the pulse-monitoring chest belt (e.g., the one manufactured by Polar), which performs the online monitoring of human physiological parameters, and creates new roles and new market niches. The expectation is that the new functions of the smart textiles – this new shell of the human being – is going, by virtue of its functionalities, to contribute to the creation of new market segments for products and services. Therefore, we will attempt to map the market segments and related opportunities as they appear, while anticipating new – previously unimagined – market segments.

The global functions expected from a smart textile are outlined below and illustrated in Fig. 13.1.

13.1.2.1   Sensing

Sensing can be observed through multiple prisms:

- Sensing of the person
- Sensing of the environment

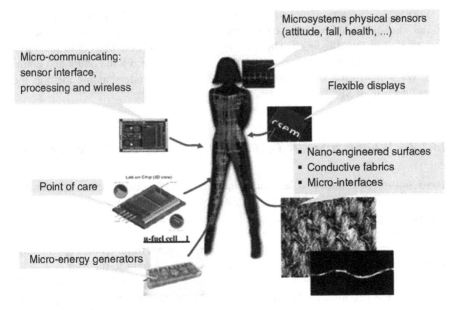

Fig. 13.1 Outline of main functions of wearable systems

- Sensing of the location of the person
- All or perhaps a subset of these

Sensing can also be categorized by types of measurements, and more particularly for the human body in monitoring of physiological parameters (body or skin temperature, posture, and gesture) and monitoring of biochemical parameters (e.g., perspiration).

13.1.2.2 Energy Harvesting

Energy harvesting can arise from the interior of the textile or from the exterior world. The latter mainly concerns solar, thermal, or mechanical energy, while the former concerns the mechanical and thermal energy produced by the human body.

13.1.2.3 Acting: Actuating

Two types of actuating mechanisms are present: physiological (acting on the human body, e.g., through piezoelectric actuators) or chemical, which can be either surface or bulk property changes, as triggered by external stimuli or even chemical delivery (e.g., controlled drug delivery).

13.1.2.4 Intelligence

Intelligence can be based on the use of conventional microprocessor(s) or in the longer term use of the textile itself as part of the microprocessor. These considerations imply the natural separation between distributed versus concentrated intelligence, whose implications are analyzed below.

13.1.2.5 Interface: Including Displaying

Concerning terminology interfaces, a number of parameters are implied. A first aspect is the interface which informs the user. It can be a display or an acoustic stimulus (e.g., voice) or a tactile stimulus (e.g., vibration) or another type of stimulus of our nervous system. Another consideration with regard to interfaces pertains to sensor-to-human, as well as sensor-to-environment, interfaces, which due to the stochastic behavior of humans require particular care. Finally, one might consider interfaces in terms of telecommunication links to remote locations (e.g., for medical telemonitoring of first responder in emergency situations).

As an ultimate vision, we can consider the continuous increase of performance of each of the aforementioned functions, up to a reconfigurable human augmentation type, according to the environment, function, moment, etc. Such reconfigurability could be performed either by dynamically modifying the textile, for instance by dynamically changing the parameters to be monitored: during one part of the day the measured parameters could be focused on medical diagnostics needs, while during another part of the day they could be focused on professional needs.

Coming back to the present, or even to the recent past, we see the ancestors of such futuristic systems in our everyday life: a pulse-monitoring chest belt and a pedometer are existing tools that allow us to monitor the human being and contribute to improving his own capabilities through conscious feedback (e.g., improving training conditions for athletes). The path toward increasingly complex smart textiles is nothing more than the increase in density of sensors or actuators and their ubiquitous and seamless integration on the immediate "human shell" (i.e., the smart textile) to augment the human potential.

Taking this view of the smart textile as an ultimate shell, the main roles of any shell in terms of protection, helping, healing, entertaining, or enjoying should be integrated and mapped into specific market segments. The functions of such a shell are continuously evolving, thus creating new markets and new opportunities for both products and services. In this logical first step, a higher diversity of the applications for smart textiles is going to emerge, as new opportunities will serve innovative applications currently not served or even existing today.

13.2   Common Backbone of Applications

In parallel, we have begun to observe a continuous convergence of application needs toward a common backbone: These form the basis of a future market consolidation and transition of smart textiles from a high-end product to a commodity. These areas of convergence are described below.

13.2.1   SFIT Configuration

13.2.1.1   Elementary Functions Without Embedded Intelligence (e.g., Reactive Color Change)

We understand by this category functions that operate without the use of microprocessor intelligence. To better describe this category, we can use the example of micelles, that open or close depending on the chemical environment, releasing pharmaceuticals into the body.

Another good example consists of nanostructured patches, embroidered on textiles, that, for example, can change periodicity (and therefore color of diffracted light) when absorbing human liquids (e.g., detection of wound healing).

13.2.1.2   Intelligence Embedded in the Textile

We understand by this configuration a complex function achieved by the process of sensing and interfacing of data devices that are using some kind of digital intelligence.

The most straightforward of digital intelligence systems are tiny microprocessors that can be equipped with communication features. Such devices can only be interfaced to the smart textiles either as unique devices that control the whole textile or as multiple devices that exchange information from multiple points on the same textile.

The newest configurations addressing intelligence are no longer embedded on conventional microprocessors, but rather on the textile itself. Integration can occur through the use of fibers that have semiconducting properties, and can therefore behave as elementary nodes of a large microprocessor.

Another aspect is the use of rapidly developing polymer electronics. Flexible polymers can be part of the smart textile itself as displays, but also as microprocessors.

13.2.1.3   Distributed Versus Localized

In terms of application approaches, one can see two large categories: distributed or localized (or obviously a combination of both). This is the architecture of the microprocessor(s) and the embedded intelligence itself. However, the essential

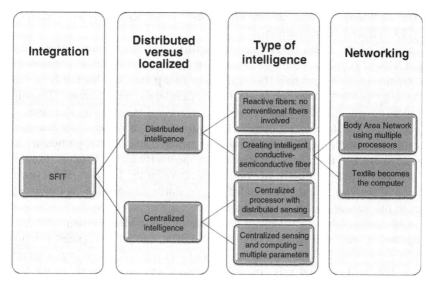

Fig. 13.2   Type of smart fabrics and intelligent textiles according to the type of sensing/actuating and processing

point is that microprocessor architecture should follow the needs of the application for sensing and actuating.

Distributed sensing means that almost each part of the textile is in itself a sensor. The example of the motion sensor Capri (Carpi and De Rossi 2005) is an excellent example. Natural Sensor redundancy is an important asset of such configurations, since with such high numbers of sensors, failure or sensor misplacement can be covered by signals coming from neighboring sensors.

The advantage of having localized "smart sensors" lies in the simultaneous measurement at the same location of several parameters; a combination of multiple measurands can lead to extremely useful conclusions. The European Space Agency's LTMS program (Krauss 2009) for the monitoring of astronauts in future manned missions has adopted precisely this approach.

We expect in the future a combination of these architectures, ultimately on the same textile, therefore offering the better of the two to the user.

These functions are outlined in Fig. 13.2 above.

13.3   Present Situation and Competitors in Terms of R&D and Commercialization

The market today is fragmented – this is often the nature of nascent markets. This fragmentation results in the presence of several small companies; in some cases, these small organizations emerge as spin-offs from much larger organizations.

The traditional value chains developed over centuries of textile development (Fig. 13.3) have been revolutionized:

In terms of intelligent fibers, one can identify a number of corporations. In the UK, Eleksen[3] is producing pressure-sensitive fibers and is therefore positioned on the left side of the value chain (Fig. 13.3), as a new player. Further left is the basic technology provider Paratech,[4] which has developed the Quantum Tunneling Component, which in everyday terms means electrical conductivity increases.

Eleksen is not alone in this domain. Companies such as Auxetix[5] are attempting to capture value through ground-breaking developments (in this particular case, auxetic fibers, i.e., fibers with negative expansion coefficient). These two examples do not fully represent the list of companies active in the left part. However, the citation of their positioning is intended only to illustrate the value capture mechanism in this part of the new value chain.

In the center of the value chain are companies that provide subsystems that can be integrated in the textile. Such subsystems can offer different integrated functions. Energy generation is the first, very important function, and solar energy capture is among the most efficient ways of capturing such energy. Konakra[6] in the US is positioned in this segment. Other companies such as Switzerland's Flexcell[7] are also making valuable contributions. Thermoelectric energy generation (i.e., the temperature difference between the human body and the exterior world) has been investigated for years with significant technical success by Infineon.

Another area of interest is the communication of information in a visual way: this is enabled by polymer displays. Numerous companies are active in this arena, aiming to penetrate the market of flexible displays. One of them applies such displays to wearable systems. Philips, with its Lumalive[8] textile integrating displays, has made a very interesting demonstration. This particular development is positioned in the middle of the value chain; that is, the integration of high-end components. However, one can easily imagine that such a large corporation can vertically integrate complete solutions.

Going to the right-hand part of the value chain in Fig. 13.3, we identify a number of subsystem integrators in or on textiles: typically, Ohmatex,[9] based in Denmark, Clothing+[10] of Finland and industrial R&D and smaller-scale production facility Smartex.[11] All these companies are targeting the integration of fibers and yarns in the fabric with a clear trend towards full systems.

---

[3] www.eleksen.com

[4] www.peratech.com

[5] www.auxetix.com/science.htm

[6] www.konakra.com

[7] www.flexcell.com

[8] www.lumalive.com

[9] www.ohmatex.dk

[10] www.clothingplus.fi

[11] www.smartex.it

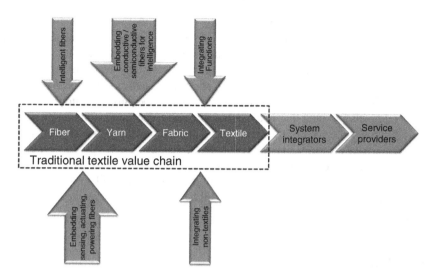

Fig. 13.3 Outline of value chain of textile and changes introduced by the advent of smart fabrics and intelligent textiles

Early phases of consolidation are starting to appear, including Textronics,[12] a spin-off of Invista, and targeting the exploitation of technologies from that company. Consolidation of the company within Adidas is a considerable step towards mass marketing.

Vivometrics,[13] one of the market's early movers, has made enormous efforts to create impetus for smart textiles in the market place. Vivometrics was among the first companies having addressed both products (Lifeshirt device) and services (Vivosoft), in particular services for data consolidation and reporting to business customers.

Apart from Vivometrics, one of the pioneering companies in this field is Sensatex,[14] which at the beginning of the decade produced the Smartshirt systems. Both Vivometrics and Sensatex were positioned closer to the final user – Vivometrics by providing elaborated signal analysis was also positioned as a service provider.

The UK-based company SmartLife[15] is also moving in the same direction, focusing on a dry sensor technology that improves reliability of the sensing system of physiological signals.

WearTech,[16] a company created in Valencia, Spain, is tackling the market of wellness. WearTech emerged from companies and research centers having participated in the European Commission's integrated project MyHeart.[17] This company is positioned closer to service provision.

---

[12] www.physicventures.com/textronics

[13] www.vivometrics.com

[14] www.sensatex.com

[15] www.smartlifetech.com

[16] www.weartech.es

[17] www.extra.research.philips.com/euprojects/myheart/

Sensecore[18] from Switzerland is initially targeting the markets of professional sport, in a vertically integrated manner, focusing its strategy on smart electrodes (as opposed to distributed sensing). A competing technology, albeit not in full-shirt configuration but rather in belt configuration, is commercially available from the US company Zephyr.[19] To complete the chain, companies such as the Swiss Athlosoft[20] use devices from, e.g., Zephyr and capture their own value through customer service.

It is interesting to note that in a number of cases some actors completely new to the traditional value chain are appearing. Furthermore, these actors compete in different parts of the value chain, producing a large diversity of product types.

The introduction of some major corporations indicates the first stages of market consolidation: however, taking into account the aforementioned structural market variability, it is expected that the market will be maintained fragmented for quite a long period, thus hindering fast growth.

13.4   Market Segmentation

Market segmentation in this rapidly changing environment is not simply an exercise in style; it is rather the foundation of understanding how the field is going to evolve and the key forces behind it.

Among different ways of segmenting the market, we propose to accept the splitting in:

- Medical
- Military
- Sport

  - Elite athletes and
  - High level amateurs

- Wellness
- Health
- Professional and protective
- Consumer and Fashion

13.4.1   Medical

Medical is a large market segment that can be subdivided into multiple ways. Since the objective here is to understand the value-creation mechanism, a logical way of segmenting the market is by using the criterion as the sources of revenue. One could be regular

---

[18] www.sense-core.com

[19] www.nzherald.co.nz

[20] www.athlosoft.com

medicine, involving professional devices, and the other could be self-healthcare devices, which can be purchased in pharmacies (such as blood-glucose meters).

In the case of SFIT, by "regular medicine" is meant the use of such smart textiles in disease prevention, monitoring, and rehabilitation processes, though while being under treatment by a medical doctor. Efforts to introduce such systems have been widely deployed in the past. In such cases, the objective was to use discrete measuring systems, such as ECG holders or belts. One of the most interesting cases was that of Philips Telemedical Services, deployed at the beginning of 2000 and targeting Denmark, Germany, Switzerland, and Italy. The model was initially based on the payment of the person and/or his family of a relatively small daily amount.

By self-healthcare is meant the use of SFIT by the individual, without medical prescription or follow-up, to monitor and improve her own health. In this case, there is certainly a good overlap with what we refer to below as the "wellness" market segment; however, the overlap is not complete and here we will attempt to address the medical aspect.

Medical CE and FDA approval will certainly be a delaying factor in terms of market penetration in this segment. This argument becomes even more pertinent if one considers that the majority of the smart textiles that are on the market today, or close to the market, suffer from poor interfaces between the textile and the wearer, due to the natural movement of the body. Reliability of the interface between the textile electrodes and the body requires general improvement.

The challenge is, however, certainly enormous. Control of the increasing costs, while maintaining comfort for the patient and ensuring security, will be a key market penetration factor. Rapidly changing demographics will moreover create excellent opportunities.

An interesting case that has not been previously considered in the literature is the integration of the elderly in society. This is not considered as a mainstream "medical" issue. However, lack of integration and increased isolation leads to risk of rapidly increasing psychological diseases of the elderly. Then we can see the average elderly individual as a person at risk, not only because of somatic diseases (e.g., cardiovascular, diabetic, or neurodegenerative diseases) but also for psychological conditions that can appear alone or in conjunction with neurodegenerative diseases. In this case, smart textiles can and will play an important role, not only in the monitoring of the physiological parameters and/or the physiological impact of psychological diseases (e.g., perspiration, tremor, lack of sleep, etc.) but also as tools to communicate and interact with society.

Infant monitoring, beyond avoiding sudden death, but also in optimizing monitoring of the vital functions, can open considerable new markets. In this field, the business models will in the near future follow two paths. The first one is of regular medicine, where the physician will, within an appropriate legal and financial framework, monitor the child. The second path is the one of the consumer market, where the family purchases such systems to monitor the infant's situation and to prevent sudden death. Respiration, skin and body temperature, as well as perspiration, are among the most important parameters to monitor.

The requirements for success are multiple, including:

- CE and/or FDA certification
- Introduction of an appropriate business model

Another extremely interesting development is the increasing appearance of diseases characterized by the degeneration of main functions. The need for artificial organs such as artificial pancreases, artificial livers, etc., will steeply increase over the following decades, in particular due to the ageing of the population and the shift of disease types from acute to chronic.

Obviously, a lot more than just smart textiles will be required for these treatments; however, the smart textiles, if developed to an acceptable level of reliability and quality, will be an important complement to implantable devices.

One of the additional exploitation paths in the market is the use by the pharmaceutical industry of wearable devices, for instance in clinical tests. This is just one example of its application in pharmaceuticals.

13.4.2 Wellness

Wellness is considered here as the market segment of individual consumers who are willing to use SFIT as tools to contribute to their own everyday well-being. This is a consumer market segment, where previous successes of predecessors of smart textiles have proven their feasibility. The example of the Polar belt is the most interesting. Its evolution and the cooperation with Textronics and Adidas (Textronics[21]) seem a very natural step forward where the probability of success seems high, due to the potential of each individual market.

Yet another case worthwhile to underline is the Nike pedometer system, associated with the Apple iPod; the most interesting aspect of this is that an internet community, through a dedicated website, succeeding in strengthening the market appeal of the concept.

In both cases, it appears that the business model can be feasible and successful. However, the parameters monitored (ECG – pulse in the case of Adidas, and distance and running rate in the case of Nike) can certainly be enriched and this will certainly accelerate market penetration.

In such configurations, and as opposed to the elite sport market segment, the ubiquitous communication feature is not a must. It may become that such ubiquitous measurement becomes a market need in the future, but for the time being the person caring about her wellness is more inclined to display and store data and possibly share data after exercise, but not necessary in real-time. Real-time display (or acoustic communication, through, for example, artificial voice communication) is an interesting tool.

---

[21] www.physicventures.com/textronics

Several start-ups are currently active here, one of which is the aforementioned very interesting case of WearTech, in Valencia,[22] which is a spin-off of the European Commission funded project MyHeart.[23]

An interesting feature of this business-to-consumer segment is the uniformity expected in terms of performance monitoring. The appeal of the online community is going to generate requirements for focusing on a defined set of monitoring parameters that all users will share. Today, such parameters are pulse (e.g., Polar, Suunto, Decathlon belts) or distance running (Nike pedometer). It is quite likely that the list will lengthen, but it is also likely that a common denominator of parameters, though limited in number, will remain.

For the time being, this segment does not require medical quality data. However, it is likely that the user level of requirements will increase during the coming years. Engineers will need to identify ways of providing this quality.

13.4.3 Military

The Military is one of the segments considered to be among the most demanding. It has been considered, and is still considered, to be a potential killer application. The reason is twofold:

1. In terms of protecting personnel, smart textiles can add value from multiple points of view: monitoring of physiological situations (before, during and after combat, optimization of training) is one of the most obvious applications. Obviously, this is valid for infantry, but also for pilots of aircraft and vehicles.
2. In terms of business potential, it is a large segment, but it is also a segment that can produce, uniform, large-scale orders.

The main weakness of this segment is that the penetration is time-demanding; this demands large commercial structures that have the resources to face the stringent needs and the long-time requirement; this can be incompatible with the structure of this nascent market, which is based on small companies.

13.4.4 Professional/Protective

These are large business-to-business markets, with a high degree of variability and enormous potential for evolution. Professional and protective market segments might seem at first glance unattractive – this is not actually the case. High replacement rates, in particular in specific markets (e.g., industry) can create interesting opportunities. In this case, smart textiles are expected to increase the productivity, performance, and

---

[22] www.weartech.es

[23] www.extra.research.philips.com/euprojects/myheart/

security of professionals. The difference in motivation for a purchase by a customer, between productivity and usefulness, is very important.

The nature of this segment is concerned with a high degree of quality and reliability in terms of measuring and interacting; no error can be tolerated when human life is at stake. However, taking into account the imperfect interfaces between the textiles and the human body, the factor of quality and reliability is expected to delay somewhat the introduction of smart textiles in this segment.

As far as security is concerned, one might consider a number of examples, for example mining or protective clothing for industry (e.g., metallurgic industry, electricity industry). In all of these examples, security is very important. Sensing devices, which for instance rapidly detect methane, can save hundreds of lives every year in mining. However, security is expected to come at minimal cost, since it does not bring additional revenues (i.e., no additional productivity).

In other market segments, where the smart textiles are expected to increase productivity, increased cost is more easily justified by purely economic rationale.

High renewal rates of professional clothing, together with potential increased productivity, could be ideal factors for fast market penetration. As far as we know, no such application has been identified up to now. In some ways, the aforementioned "elite athlete" segment can be considered as a segment with such characteristics, but with limited market volume.

It is expected that disruptive technology will quickly lead to applications, where the combined characteristics of increased productivity, high renewal rate, and business-to-business configuration will contribute to rapid market breakthroughs.

13.4.5 Sport

Within sport, we define here two subsegments: the professional market subsegment and the high-end amateur sport sub-segment.

Both subsegments of this larger sport market segment represent a natural extension of the wellness market (or *vice versa*, wellness can be considered as the natural extension of the sport market segment). Despite this natural continuity, the market segment structures are completely different: elite athletes have their own medical teams to follow them up; individual in wellness centers are monitoring themselves; the first segment is business-to-business the second business-to-consumer.

For professional sports, whether this has to do with individual sports or team sports, monitoring is today performed by professional medical teams that continuously monitor and optimize every single aspect of performance. This market is ready to upgrade the current tools toward more seamless devices, integrated in the textile, on condition that they provide quality medical data.

Each sport has its specific needs in terms of training and competition. Due to this specialization, this is a market that requires a high degree of flexibility and customization that is not adapted to mass production.

A deep medical understanding of sport medicine is required within those companies that commercialize smart textiles, to intelligently adapt to the needs of the users. Specific technical features are particularly important. Real-time communication and processing is of major importance, for training and in several cases, depending on the rules, for the activity itself.

Due to the limited number of top-level athletes, this market is by nature restricted. No single sport can sustain adequate business levels of companies working in this sector. Even the possibility of building a company focusing on elite sport teams or individuals for several sports is questionable, though not impossible. This possibility exists due to the increasing use of scientific/medical optimization through monitoring of the athletes' performance. Obviously, this subsegment is considered to obey the rules of a business-to-business relation, since customers are not the individuals but the teams, albeit with a limited number of individuals within each team.

For the subsegment of high-end amateur athletes, the situation is completely different. Here, the individuals who are potential users are usually extremely well-informed people, who manage their own performance.

The market itself is very different from the one comprising elite athletes. Here, the equipment is financed by the individuals themselves; the degree of price-sensitivity increases, however, due to the passion of such individuals for sport, and price therefore remains less important than performance and quality criteria.

Obviously, here we are moving toward a business-to-consumer market, and product standardization is important in reducing production costs. Again, the differentiation of needs of different sports requires a high degree of flexibility and customization. This is, however, in contradiction with mass production. Embedding flexibility and customization within the individual smart textile is the main challenge of this market.

13.4.6 Consumer and Fashion Segments

Consumer and fashion is the ultimate market segment. It will eventually flourish when manufacturers have gone through the learning curve, and when costs are reasonable, enabling competitive, and adequate consumer and fashion pricing.

Smart textiles, at first glance, are expected to play only a marginal role in high-end fashion; instead, they are expected to be more and more an integral part of mid-range garments. Fun and image are expected to be the key marketing drivers.

In terms of market type, business-to-business markets and business-to-consumer have fundamentally different characteristics with regard to market approach and penetration. An initial categorization is outlined in Fig. 13.4 below.

An outline of the relative importance of the factors presented here is presented in Table 13.1, below.

Market penetration, in particular for high technology markets such as the smart fabrics and intelligent textiles segments, starts from high added-value and high-margin market segments. Over time, these markets are moving toward high-volume, lower-margin products. Among the indicators of such transitions is the

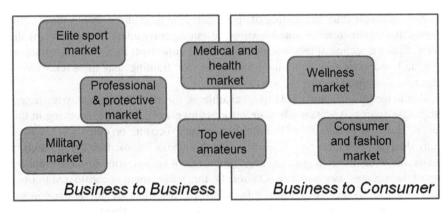

Fig. 13.4 Market segments and their character in terms of approach

Table 13.1 Main characteristics of the market segments

Market segment	Main characteristics
Sport (elite and high-end amateurs)	Needs applications with high visibility and high innovation level
	Small market volume
	Increasing demand from sport teams for optimization of performance
	Use of external tools, such as cameras, for monitoring of important (but not all) parameters
Wellness	Clearly proven – indicated for smart textile wellness
	Only a small part of potential functionalities
	Several competitors can establish themselves using existing tools
Professional and protective	High renewal rate
	Security is a "must" that does not justify cost increase
	Cases where smart textiles increase productivity and not only security are not yet clearly identified
	High cost inhibits market entrance
Health and medical	SFIT might be a factor in limiting sharp increases in health costs
	Multiple markets
	Requires long homologation periods
	Ageing society may in medium-term require tools to medically monitor and integrate individuals
	Newly conceived services based on new features
	e-Health, which uses such tools, is expected to get increased importance over the next years
	Legal, reimbursement and regulatory situation not resolved
Military	Big homogeneous market
	Market difficult to access
	Increased demand in western countries for soldier security
	Financial limitations
	Competing technologies
Consumer and fashion	Image of fun and modern
	Unpredictability of consumer markets
	Customer goods can only be low cost and very attractive
	Fragmented market

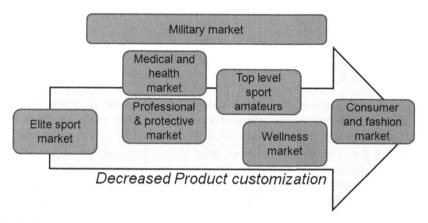

Fig. 13.5   Degree of customization as a function of market segment. This viewgraph is only a high level picture that due to the "smartness" of textiles can be rapidly modified

customization degree in each market segment. However, due to the "smartness" of the textile the customization can be higher at every level, as we attempt to illustrate in Fig. 13.5.

13.5   Market Volumes

To define a market volume, one needs to initially define the market perimeter, taking into account the numerous players and their revenue capture.

Several recent and less recent analyses can be cited. IntertechPira (2009) made a prediction of a $642 million market for 2008, with an expected growth of 28%. Venture Development Corporation (VTC), British Chamber of Commerce (BCC[24]), or Reportlinker[25] announce double figure growth as well as worldwide market volume forecasts of the order of the one billion US dollars. Last but not least, one other forecast of Frost and Sullivan (2009) mentions a growth rate of 76% for the SFIT market (SFIT[26]).

Market volume forecast can vary a lot; the main message from market and analysts is that a strong growth is expected, in a market which is going to be in the order of billion US dollars worldwide.

---

[24] info.hktdc.com/imn/06120501/clothing219.htm

[25] www.reportlinker.com/p096832/Global-Markets-for-Smart-Fabrics-and-Interactive-Textiles-2008-edition.html

[26] www.csem.ch/sfit/html/background.html

13.6   Conclusions

If one can draw a conclusion, it is that the world market is significant, but still relatively small in terms of overall capitalization, and taking into account market fragmentation, it means that survival is difficult.

A second conclusion is that forecasts seem to lack coherence. This is explained by the definition of the market. Including services or not, and addressing systems or sub-components, can alter the perception of market volume and growth rate considerably.

In our understanding, the market for smart textiles (or more correctly expressed SFIT) is a significant albeit relatively small segment. We anticipate a rapid growth over the years to follow. We expect to see significant growth rates that will in the future establish SFIT as a main market, once a strong penetration is achieved in specific market segments.

As any product with high technology content penetrates first in specific niches, the additional presence of an adequate business model, including services, accelerates market penetration. Features that seem important are: (1) the potential of high renewal rates; (2) clear increases in productivity and/or performance; (3), business-to-business environment, and finally (4) existing or upcoming services.

The market is fragmented, and this situation will be maintained until a killer application leads to fast market growth. Some signs of consolidation are starting to appear.

Traditional value chains are getting more complex. The revenue-generation mechanism changes as new players enter this nascent arena. The continuous increase in technological features will continuously modify these mechanisms and is expected to create business opportunities requiring a high degree of flexibility and customization from industry.

References

Carpi F, De Rossi D (2005) Electroactive polymer-based devices for e-textiles in biomedicine. IEEE Trans Inf Technol Biomed 9(3):295–318

Frost and Sullivan (2009) Technical insights' high-tech materials alert. Frost and Sullivan

IntertechPira (2009) www.innovationintextiles.com/articles/Smart015.php. Accessed on 6 oct 2010

Krauss J (2009) LTMS: smart and wearable life sign monitoring. International Congress on Wearable Technologies, Münich

Index